Landolt-Börnstein / New Series

Landolt-Börnstein
Numerical Data and Functional Relationships
in Science and Technology

New Series
Editor in Chief: W. Martienssen

Units and Fundamental Constants in Physics and Chemistry

Elementary Particles, Nuclei and Atoms (Group I)
(Formerly: Nuclear and Particle Physics)

Molecules and Radicals (Group II)
(Formerly: Atomic and Molecular Physics)

Condensed Matter (Group III)
(Formerly: Solid State Physics)

Physical Chemistry (Group IV)
(Formerly: Macroscopic Properties of Matter)

Geophysics (Group V)

Astronomy and Astrophysics (Group VI)

Biophysics (Group VII)

Advanced Materials and Technologies (Group VIII)

Some of the group names have been changed according to a better description of their contents.

Landolt-Börnstein
Numerical Data and Functional Relationships in Science and Technology
New Series / Editor in Chief: W. Martienssen

Group VIII: Advanced Materials and Technologies
Volume 1

Laser Physics and Applications

Subvolume A: Laser Fundamentals

Part 1

Editors: H. Weber, G. Herziger, R. Poprawe

Authors:
H.J. Eichler, B. Eppich, J. Fischer, R. Güther, G.G. Gurzadyan,
A. Hermerschmidt, A. Laubereau, V.A. Lopota, O. Mehl, C.R. Vidal,
H. Weber, B. Wende

ISSN 1619-4802 (Advanced Materials and Technologies)

ISBN-10 3-540-44379-7 Springer Berlin Heidelberg New York
ISBN-13 978-3-540-44379-7 Springer Berlin Heidelberg New York

Library of Congress Cataloging in Publication Data:
Landolt-Börnstein: Numerical Data and Functional Relationships in Science and Technology, New Series.
Editor in Chief: W. Martienssen.
Group VIII, Volume 1: Laser Physics and Applications. Subvolume A: Laser Fundamentals. Part 1.
Edited by H. Weber, G. Herziger, R. Poprawe.
Springer-Verlag, Berlin, Heidelberg, New York 2004.
Includes bibliographies.
1. Physics - Tables. 2. Chemistry - Tables. 3. Engineering - Tables.
I. Börnstein, Richard (1852-1913). II. Landolt, Hans (1831-1910).
QC 61.23 502'.12   62-53136

This work is subject to copyright. All rights are reserved, whether the whole or part of the material is concerned, specifically the rights of translation, reprinting, reuse of illustrations, recitation, broadcasting, reproduction on microfilm or in other ways, and storage in data banks. Duplication of this publication or parts thereof is permitted only under the provisions of the German Copyright Law of September 9, 1965, in its current version, and permission for use must always be obtained from Springer-Verlag. Violations are liable for prosecution act under German Copyright Law.

Springer is a part of Springer Science+Business Media.
springeronline.com
© Springer-Verlag Berlin Heidelberg 2005
Printed in Germany

The use of general descriptive names, registered names, trademarks, etc. in this publication does not imply, even in the absence of a specific statement, that such names are exempt from the relevant protective laws and regulations and therefore free for general use.

Product Liability: The data and other information in this handbook have been carefully extracted and evaluated by experts from the original literature. Furthermore, they have been checked for correctness by authors and the editorial staff before printing. Nevertheless, the publisher can give no guarantee for the correctness of the data and information provided. In any individual case of application, the respective user must check the correctness by consulting other relevant sources of information.

Cover layout: Erich Kirchner, Heidelberg
Typesetting: Authors and Redaktion Landolt-Börnstein, Darmstadt
Printing and Binding: AZ Druck, Kempten (Allgäu)

SPIN: 1050 7868   63/3020 - 5 4 3 2 1 0 – Printed on acid-free paper

Editors

Weber, Horst
Technische Universität Berlin, Optisches Institut, Berlin, Germany

Herziger, Gerd
Rheinisch-Westfälische Technische Hochschule, Aachen, Germany

Poprawe, Reinhart
Fraunhofer-Institut für Lasertechnik (ILT), Aachen, Germany

Authors

Eichler, Hans Joachim
Technische Universität Berlin, Optisches Institut, Berlin, Germany

Eppich, Bernd
Technische Universität Berlin, Optisches Institut, Berlin, Germany

Fischer, Joachim
Physikalisch-Technische Bundesanstalt, Abteilung Temperatur und Synchrotronstrahlung, Berlin, Germany

Güther, Reiner
Ferdinand-Braun-Institut für Höchstfrequenztechnik, Berlin, Germany

Gurzadyan, Gagik
Technische Universität München, Institut für Physikalische und Theoretische Chemie, Garching, Germany

Hermerschmidt, Andreas
Technische Universität Berlin, Optisches Institut, Berlin, Germany

Laubereau, Alfred
Technische Universität München, Physik Department E11, München, Germany

Lopota, Vitalyi A., member of Russian Academy of Sciences
Central R & D Institute of Robotics and Technical Cybernetics, Saint Petersburg, Russian Federation

Mehl, Oliver
Technische Universität Berlin, Optisches Institut, Berlin, Germany

Vidal, Carl Rudolf
Max-Planck Institut für Extraterrestrische Physik, Garching, Germany

Weber, Horst
Technische Universität Berlin, Optisches Institut, Berlin, Germany

Wende, Burkhard
Physikalisch-Technische Bundesanstalt, Abteilung Temperatur und Synchrotronstrahlung, Berlin, Germany

Landolt-Börnstein

Editorial Office
Gagernstraße 8
D-64283 Darmstadt, Germany
fax: +49 (6151) 171760
e-mail: lb@springer-sbm.com

Internet
http://www.landolt-boernstein.com

Preface

The three volumes VIII/1A, B, C document the state of the art of "Laser Physics and Applications". Scientific trends and related technological aspects are considered by compiling results and conclusions from phenomenology, observation and experience. Reliable data, physical fundamentals and detailed references are presented.

In the recent decades the laser source matured to a universal tool common to scientific research as well as to industrial use. Today a technical goal is the generation of optical power towards shorter wavelengths, shorter pulses and higher power for application in science and industry. Tailoring the optical energy in wavelength, space and time is a requirement for the investigation of laser-induced processes, i.e. excitation, non-linear amplification, storage of optical energy, etc. According to the actual trends in laser research and development, Vol. VIII/1 is split into three parts: Vol. VIII/1A with its two subvolumes 1A1 and 1A2 covers laser fundamentals, Vol. VIII/1B deals with laser systems and Vol. VIII/1C gives an overview on laser applications.

In Vol. VIII/1A1 the following topics are treated in detail:

Part 1: Fundamentals of light-matter interaction

This part compiles the basic elements of classical electromagnetic wave theory, non-relativistic quantum mechanics of the two-level system and its interaction with the non-quantized radiation field. The relevant relations with their approximations and range of validity are discussed. It starts with Maxwell's equations, wave equation and SVE-approximations, presents the Schrödinger equations, the field/atom interaction including the Einstein coefficients and cross-sections. The main parameters characterizing the two-level system with typical numbers are given in several tables. Finally, the coherent interaction is briefly discussed. This semiclassical approach is sufficient for most applications in laser technology. The fully quantized theory is offered in Vol. VIII/1A2, Chap. 5.

Part 2: Radiometry

In the first section the definitions of the radiometric quantities and their measurement are summarized. In the second part the main elements of laser beam characterization are compiled with a detailed discussion of the theoretical background. The experimental determination of the essential quantities according to the ISO-normalizations is given.

Part 3: Linear optics

The design of optical resonators and beam handling requires a broad knowledge in optics. In this part the fundamentals of beam propagation, Gaussian beams, diffraction, refraction, lens design and crystal optics are presented. The extensive references give access to detailed information.

Part 4: Nonlinear optics

Nonlinear effects are widely used in laser technology to generate new wavelengths or to improve beam quality.In four sections the essential nonlinear optical effects are discussed: frequency conversion in crystals, frequency conversion in gases and liquids, stimulated scattering and phase conjugation. In extensive tables the coefficients of the nonlinear processes are compiled.

August 2005   The Editors

Contents

Part 1 Fundamentals of light-matter interaction

1.1	**Fundamentals of the semiclassical laser theory** V.A. LOPOTA, H. WEBER	3
1.1.1	The laser oscillator	3
1.1.2	The electromagnetic field	5
1.1.2.1	Maxwell's equations	5
1.1.2.2	Homogeneous, isotropic, linear dielectrics	6
1.1.2.2.1	The plane wave	7
1.1.2.2.2	The spherical wave	8
1.1.2.2.3	The slowly varying envelope (SVE) approximation	9
1.1.2.2.4	The SVE-approximation for diffraction	9
1.1.2.3	Propagation in doped media	10
1.1.3	Interaction with two-level systems	11
1.1.3.1	The two-level system	11
1.1.3.2	The dipole approximation	12
1.1.3.2.1	Inversion density and polarization	12
1.1.3.2.2	The interaction with a monochromatic field	14
1.1.3.3	The Maxwell–Bloch equations	15
1.1.3.3.1	Decay time T_1 of the upper level (energy relaxation)	15
1.1.3.3.1.1	Spontaneous emission	15
1.1.3.3.1.2	Interaction with the host material	15
1.1.3.3.1.3	Pumping process	16
1.1.3.3.2	Decay time T_2 of the polarization (entropy relaxation)	16
1.1.4	Steady-state solutions	17
1.1.4.1	Inversion density and polarization	17
1.1.4.2	Small-signal solutions	19
1.1.4.3	Strong-signal solutions	19
1.1.5	Adiabatic equations	20
1.1.5.1	Rate equations	20
1.1.5.2	Thermodynamic considerations	21
1.1.5.3	Pumping schemes and complete rate equations	22
1.1.5.3.1	The three-level system	23
1.1.5.3.2	The four-level system	24
1.1.5.4	Adiabatic pulse amplification	25
1.1.5.5	Rate equations for steady-state laser oscillators	26
1.1.6	Line shape and line broadening	26
1.1.6.1	Normalized shape functions	27
1.1.6.1.1	Lorentzian line shape	27
1.1.6.1.2	Gaussian line shape	27

1.1.6.1.3	Normalization of line shapes	27
1.1.6.2	Mechanisms of line broadening	28
1.1.6.2.1	Spontaneous emission	28
1.1.6.2.2	Doppler broadening	28
1.1.6.2.3	Collision or pressure broadening	28
1.1.6.2.4	Saturation broadening	29
1.1.6.3	Types of broadening	29
1.1.6.3.1	Homogeneous broadening	29
1.1.6.3.2	Inhomogeneous broadening	30
1.1.6.4	Time constants	31
1.1.7	Coherent interaction	31
1.1.7.1	The Feynman representation of interaction	32
1.1.7.2	Constant local electric field	33
1.1.7.3	Propagation of resonant coherent pulses	34
1.1.7.3.1	Steady-state propagation of $n\pi$-pulses	35
1.1.7.3.1.1	2π-pulse in a loss-free medium	35
1.1.7.3.1.2	π-pulse in an amplifying medium	36
1.1.7.3.2	Superradiance	37
1.1.8	Notations	37
	References for 1.1	40

Part 2 Radiometry

2.1 Definition and measurement of radiometric quantities
B. WENDE, J. FISCHER ... 45

2.1.1	Introduction	45
2.1.2	Definition of radiometric quantities	45
2.1.3	Radiometric standards	47
2.1.3.1	Primary standards	47
2.1.3.2	Secondary standards	48
2.1.4	Outlook – State of the art and trends	50
	References for 2.1	51

2.2 Beam characterization
B. EPPICH ... 53

2.2.1	Introduction	53
2.2.2	The Wigner distribution	53
2.2.3	The second-order moments of the Wigner distribution	55
2.2.4	The second-order moments and related physical properties	56
2.2.4.1	Near field	56
2.2.4.2	Far field	58
2.2.4.3	Phase paraboloid and twist	59
2.2.4.4	Invariants	60
2.2.4.5	Propagation of beam widths and beam propagation ratios	60
2.2.5	Beam classification	61
2.2.5.1	Stigmatic beams	62
2.2.5.2	Simple astigmatic beams	63

2.2.5.3	General astigmatic beams	64
2.2.5.4	Pseudo-symmetric beams	64
2.2.5.5	Intrinsic astigmatism and beam conversion	65
2.2.6	Measurement procedures	66
2.2.7	Beam positional stability	67
2.2.7.1	Absolute fluctuations	67
2.2.7.2	Relative fluctuations	69
2.2.7.3	Effective long-term beam widths	69
	References for 2.2	70

Part 3 Linear optics

3.1 Linear optics
R. GÜTHER 73

3.1.1	Wave equations	73
3.1.2	Polarization	75
3.1.3	Solutions of the wave equation in free space	78
3.1.3.1	Wave equation	78
3.1.3.1.1	Monochromatic plane wave	78
3.1.3.1.2	Cylindrical vector wave	78
3.1.3.1.3	Spherical vector wave	78
3.1.3.2	Helmholtz equation	79
3.1.3.2.1	Plane wave	79
3.1.3.2.2	Cylindrical wave	79
3.1.3.2.3	Spherical wave	79
3.1.3.2.4	Diffraction-free beams	79
3.1.3.2.4.1	Diffraction-free Bessel beams	79
3.1.3.2.4.2	Real Bessel beams	80
3.1.3.2.4.3	Vectorial Bessel beams	80
3.1.3.3	Solutions of the slowly varying envelope equation	80
3.1.3.3.1	Gauss-Hermite beams (rectangular symmetry)	81
3.1.3.3.2	Gauss-Laguerre beams (circular symmetry)	83
3.1.3.3.3	Cross-sectional shapes of the Gaussian modes	83
3.1.4	Diffraction	84
3.1.4.1	Vector theory of diffraction	85
3.1.4.2	Scalar diffraction theory	85
3.1.4.3	Time-dependent diffraction theory	89
3.1.4.4	Fraunhofer diffraction patterns	89
3.1.4.4.1	Rectangular aperture with dimensions $2a \times 2b$	89
3.1.4.4.2	Circular aperture with radius a	90
3.1.4.4.2.1	Applications	92
3.1.4.4.3	Gratings	92
3.1.4.5	Fresnel's diffraction figures	93
3.1.4.5.1	Fresnel's diffraction on a slit	93
3.1.4.5.2	Fresnel's diffraction through lens systems (paraxial diffraction)	94
3.1.4.6	Fourier optics and diffractive optics	94
3.1.5	Optical materials	95
3.1.5.1	Dielectric media	96
3.1.5.2	Optical glasses	97

3.1.5.3	Dispersion characteristics for short-pulse propagation	97
3.1.5.4	Optics of metals and semiconductors	98
3.1.5.5	Fresnel's formulae	98
3.1.5.6	Special cases of refraction	101
3.1.5.6.1	Two dielectric isotropic homogeneous media ($\hat{n}$ and $\hat{n}'$ are real)	101
3.1.5.6.2	Variation of the angle of incidence	101
3.1.5.6.2.1	External reflection ($n < n'$)	101
3.1.5.6.2.2	Internal reflection ($n > n'$)	101
3.1.5.6.3	Reflection at media with complex refractive index (Case $\hat{n} = 1$ and $\hat{n}' = n' + ik'$)	103
3.1.5.7	Crystal optics	104
3.1.5.7.1	Classification	104
3.1.5.7.2	Birefringence (example: uniaxial crystals)	106
3.1.5.8	Photonic crystals	107
3.1.5.9	Negative-refractive-index materials	108
3.1.5.10	References to data of linear optics	108
3.1.6	Geometrical optics	108
3.1.6.1	Gaussian imaging (paraxial range)	108
3.1.6.1.1	Single spherical interface	109
3.1.6.1.2	Imaging with a thick lens	110
3.1.6.2	Gaussian matrix (ABCD-matrix, ray-transfer matrix) formalism for paraxial optics	111
3.1.6.2.1	Simple interfaces and optical elements with rotational symmetry	112
3.1.6.2.2	Non-symmetrical optical systems	112
3.1.6.2.3	Properties of a system	112
3.1.6.2.4	General parabolic systems without rotational symmetry	112
3.1.6.2.5	General astigmatic system	116
3.1.6.2.6	Symplectic optical system	116
3.1.6.2.7	Misalignments	116
3.1.6.3	Lens aberrations	117
3.1.7	Beam propagation in optical systems	120
3.1.7.1	Beam classification	120
3.1.7.2	Gaussian beam: complex q-parameter and its ABCD-transformation	120
3.1.7.2.1	Stigmatic and simple astigmatic beams	120
3.1.7.2.1.1	Fundamental Mode	120
3.1.7.2.1.2	Higher-order Hermite-Gaussian beams in simple astigmatic beams	123
3.1.7.2.2	General astigmatic beam	123
3.1.7.3	Waist transformation	124
3.1.7.3.1	General system (fundamental mode)	124
3.1.7.3.2	Thin lens (fundamental mode)	124
3.1.7.4	Collins integral	126
3.1.7.4.1	Two-dimensional propagation	126
3.1.7.4.2	Three-dimensional propagation	127
3.1.7.5	Gaussian beams in optical systems with stops, aberrations, and waveguide coupling	127
3.1.7.5.1	Field distributions in the waist region of Gaussian beams including stops and wave aberrations by optical system	127
3.1.7.5.2	Mode matching for beam coupling into waveguides	128
3.1.7.5.3	Free-space coupling of Gaussian modes	128
3.1.7.5.4	Laser fiber coupling	129
	References for 3.1	131

Part 4 Nonlinear optics

4.1 Frequency conversion in crystals
G.G. GURZADYAN 141

- 4.1.1 Introduction 141
- 4.1.1.1 Symbols and abbreviations 141
- 4.1.1.1.1 Symbols 141
- 4.1.1.1.2 Abbreviations 142
- 4.1.1.1.3 Crystals 142
- 4.1.1.2 Historical layout 143
- 4.1.2 Fundamentals 144
- 4.1.2.1 Three-wave interactions 144
- 4.1.2.2 Uniaxial crystals 145
- 4.1.2.3 Biaxial crystals 145
- 4.1.2.4 Effective nonlinearity 147
- 4.1.2.5 Frequency conversion efficiency 151
- 4.1.2.5.1 General approach 151
- 4.1.2.5.2 Plane-wave fixed-field approximation 152
- 4.1.2.5.3 SHG in "nonlinear regime" (fundamental wave depletion) .. 154
- 4.1.3 Selection of data 154
- 4.1.4 Harmonic generation (second, third, fourth, fifth, and sixth) .. 156
- 4.1.5 Sum frequency generation 167
- 4.1.6 Difference frequency generation 172
- 4.1.7 Optical parametric oscillation 176
- 4.1.8 Picosecond continuum generation 186
- References for 4.1 187

4.2 Frequency conversion in gases and liquids
C.R. VIDAL 205

- 4.2.1 Fundamentals of nonlinear optics in gases and liquids 205
- 4.2.1.1 Linear and nonlinear susceptibilities 205
- 4.2.1.2 Third-order nonlinear susceptibilities 206
- 4.2.1.3 Fundamental equations of nonlinear optics 207
- 4.2.1.4 Small-signal limit 207
- 4.2.1.5 Phase-matching condition 208
- 4.2.2 Frequency conversion in gases 209
- 4.2.2.1 Metal-vapor inert gas mixtures 209
- 4.2.2.2 Mixtures of different metal vapors 209
- 4.2.2.3 Mixtures of gaseous media 209
- References for 4.2 212

4.3 Stimulated scattering
A. LAUBEREAU 217

- 4.3.1 Introduction 217
- 4.3.1.1 Spontaneous scattering processes 217
- 4.3.1.2 Relationship between stimulated Stokes scattering and spontaneous scattering .. 219

4.3.2	General properties of stimulated scattering	219
4.3.2.1	Exponential gain by stimulated Stokes scattering	219
4.3.2.2	Experimental observation	220
4.3.2.2.1	Generator setup	220
4.3.2.2.2	Oscillator setup	220
4.3.2.2.3	Stimulated amplification setup	221
4.3.2.3	Four-wave interactions	221
4.3.2.3.1	Third-order nonlinear susceptibility	221
4.3.2.3.2	Stokes–anti-Stokes coupling	222
4.3.2.3.3	Higher-order Stokes and anti-Stokes emission	222
4.3.2.4	Transient stimulated scattering	222
4.3.3	Individual scattering processes	223
4.3.3.1	Stimulated Raman scattering (SRS)	223
4.3.3.2	Stimulated Brillouin scattering (SBS) and stimulated thermal Brillouin scattering (STBS)	227
4.3.3.3	Stimulated Rayleigh scattering processes, SRLS, STRS, and SRWS	228
	References for 4.3	232
4.4	**Phase conjugation** H.J. EICHLER, A. HERMERSCHMIDT, O. MEHL	235
4.4.1	Introduction	235
4.4.2	Basic mathematical description	236
4.4.3	Phase conjugation by degenerate four-wave mixing	236
4.4.4	Self-pumped phase conjugation	237
4.4.5	Applications of SBS phase conjugation	240
4.4.6	Photorefraction	242
	References for 4.4	245
Index		247

Part 1

Fundamentals of light-matter interaction

1.1 Fundamentals of the semiclassical laser theory

V.A. LOPOTA, H. WEBER

A rigorous description of light–matter interaction requires a fully quantized system of field equations, which is the content of quantum optics [70Hak, 95Wal, 97Scu, 95Man, 01Vog]. This theory is well developed and the results are confirmed perfectly by many experiments (see Chap. 5.1). But most problems of laser design and laser technology can be solved in a satisfactory way by applying the semiclassical theory. This means a non-relativistic quantum-mechanical approach for the electronic system and a non-quantized, classical electromagnetic field.

Non-relativistic means that the velocity of the interacting electrons is small compared with the velocity of light. This holds for the outer shell electrons of the atoms and molecules, which are relevant in laser physics. It is not true for the free-electron laser and for the interaction of strong fields with plasmas, which demand a relativistic treatment.

A non-quantized electromagnetic field implies that the photon is neglected. In laser technology the photon flux in most applications is extremely high and the granulation of light beams is of no importance. It is of significance for metrology, where the lower limit of detectability is partly given by photon statistics. There are some other effects, which are not covered by the semiclassical theory:

– Planck's law, related to photon statistics,
– squeezed states,
– entangled photons,
– zero-point energy effects,
– spontaneous emission,

and some spectral line shifts (Lamb-shift [47Lam]), of minor importance for laser technology, although of great experimental interest for the confirmation of the fundamental theory. The spontaneous emission of excited atoms/molecules is responsible for the lower limit of laser line width [74Sar, 95Man] and for the on-set of laser oscillation. Therefore, spontaneous emission has to be included in the semiclassical theory by a phenomenological term as shown in Fig. 1.1.1.

It is the intention of this chapter to compile the relevant relations of laser dynamics, their application in laser design and to discuss the limitations and approximations. The mathematical derivations can be taken from the references.

1.1.1 The laser oscillator

The laser oscillator is based on the principle of the feed-back amplifier, a principle invented by A. Meissner 1913 and patented 1919 [19Mei]. All coherent electromagnetic waves are generated by such self-sustained oscillators, from radio frequencies to microwaves and finally lasers. Basov and Prokhorov published 1954 a theoretical paper on masers [54Bas], Schawlow and Townes in 1958 [58Sch] a theoretical paper discussing the possibility of masers in the visible range of the spectrum, and Maiman realized 1960 the first laser [60Mai].

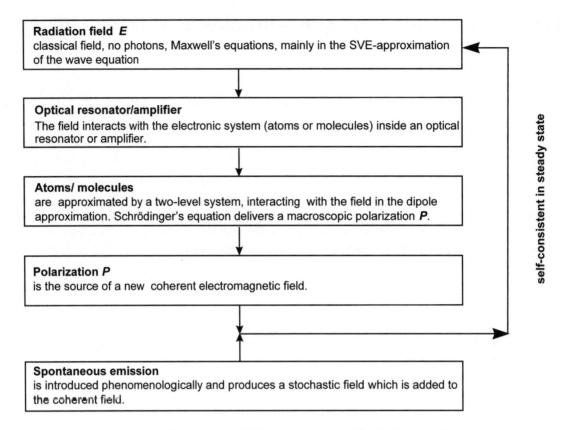

Fig. 1.1.1. The semiclassical laser theory (SVE-approximation: Slowly Varying Envelope approximation, see Sect. 1.1.2.2.3).

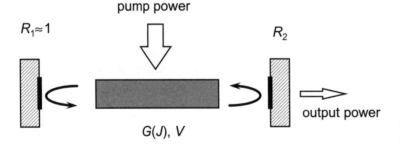

Fig. 1.1.2. Schematic set-up of a laser oscillator.

The principle set-up of a laser oscillator is plotted in Fig. 1.1.2. Light is amplified by induced emission in an active medium (gas discharge, doped crystals or liquids, pn-transitions). The active medium is characterized by an intensity- and frequency-dependent gain factor $G(J)$ (with J: intensity). The beam bounces forth and back between the two mirrors of an optical resonator. On-set of laser oscillation requires a gain factor exceeding the total losses per round trip:

$$G_0 RV > 1 \quad \text{(threshold condition)} \tag{1.1.1}$$

with

G_0: small-signal gain factor for the intensities,
$R = \sqrt{R_1 R_2}$: average reflection factor of the mirrors,
V: internal loss factor of the resonator.

With increasing intensity J the gain decreases due to saturation of the amplifier

$G(J) \leq G_0$.

In steady state the gain has to compensate the losses:

$G(J)RV = 1$ (steady-state condition) . (1.1.2)

If the relation $G(J)$ is known, depending on the specific amplifier, (1.1.2) gives the internal intensity of the laser system in steady state.

The wavelength of the field is determined by the resonance condition. After one round trip the phase shift $\Delta\varphi$ of the field must be

$\Delta\varphi = 2\pi p$, $p = 1, 2, 3, \ldots$ (resonance condition) , (1.1.3)

otherwise the field would be reduced by destructive interference. The resonator is mainly responsible for the mode structure of the output field and can be described by a non-quantized field. Details are given in Chap. 8.1. For the interaction field–amplifier a plane wave is assumed and diffraction is neglected.

1.1.2 The electromagnetic field

Light is a special case of propagating electromagnetic waves, as was predicted by Maxwell 1856 [54Max] and confirmed experimentally by Hertz [88Her]. The electromagnetic field is characterized by the electric/magnetic vector fields $\boldsymbol{E}$, $\boldsymbol{H}$. In this section the propagation of quasi-monochromatic waves with frequency ω and wavelength λ is investigated. The wavelength range from the infrared ($\lambda \approx$ some 10 µm) to the UV ($\lambda \approx 0.1$ µm) is normally called light.

1.1.2.1 Maxwell's equations

The electromagnetic field is used in the classical representation, neglecting the quantization. The materials equations, based on quantum mechanics, are introduced phenomenologically. The final result is a wave equation, describing the propagation of electromagnetic waves.

The classical electromagnetic field is completely described by Maxwell's equations:

$$\operatorname{curl} \boldsymbol{E} = -\frac{\partial \boldsymbol{B}}{\partial t} , \qquad (1.1.4)$$

$$\operatorname{curl} \boldsymbol{H} = \frac{\partial \boldsymbol{D}}{\partial t} + \boldsymbol{j} , \qquad (1.1.5)$$

$$\operatorname{div} \boldsymbol{D} = \rho , \qquad (1.1.6)$$

$$\operatorname{div} \boldsymbol{B} = 0 \qquad (1.1.7)$$

with

$\boldsymbol{E}$: electric field (SI-unit: V/m),
$\boldsymbol{H}$: magnetic field (SI-unit: A/m),
$\boldsymbol{D}$: electric displacement (SI-unit: As/m^2),
$\boldsymbol{B}$: magnetic induction (SI-unit: Vs/m^2),
$\boldsymbol{j}$: current density (SI-unit: A/m^2),
ρ: density of electric charges (SI-unit: As/m^3).

For all quantities the complex notation is used [99Bor], the real quantities are $Q_{\text{real}} = \frac{1}{2}(Q+Q^*)$. The relations between D, E and B, H are given by the material equations. Under the action of an external electric/magnetic field atomic or molecular electric/magnetic dipoles are generated in matter. The dipole moment per unit volume is called the electric or magnetic polarization $P(E,H)$ or $J(E,H)$, respectively. The resulting material quantities are the electric displacement D and the magnetic induction B given as:

$$D = \varepsilon_0 E + P(E,H) = \varepsilon_0 \varepsilon(E,H) \cdot E, \tag{1.1.8}$$

$$B = \mu_0 H + J(E,H) = \mu_0 \mu(E,H) \cdot H \tag{1.1.9}$$

with

$P = \varepsilon_0 \chi_e(E,H) E$: electric polarization (SI-unit: As/m^2),

$J = \mu_0 \chi_m(E,H) H$: magnetic polarization (SI-unit: Vs/m^2),

$\chi_e(E,H)$, $\chi_m(E,H)$: electric/magnetic susceptibility, in general a tensor and a function of the fields,

$\varepsilon = 1 + \chi_e$, $\mu = 1 + \chi_m$: permittivity/permeability number, in general tensors, 1 : unit tensor,

$\varepsilon_0 = 8.8542 \times 10^{-12}$ As/Vm: electric constant,

$\mu_0 = 4\pi \times 10^{-7}$ Vs/Am: magnetic constant.

The current inside a medium is caused by the electric field and Ohm's law holds

$$j = \sigma_e E \tag{1.1.10}$$

with

σ_e: electric conductivity, in general a tensor and function of the field, (SI-unit: A/Vm).

Electric and magnetic polarization depend in general on both generating fields, E and H. In many cases this relation is linear, but quite often a very complicated relation occurs, as in non-linear optics, ferro-magnetism or ferro-electricity. The material equations can only be evaluated by quantum mechanics. In the following non-conducting ($\sigma_e = 0$), charge-free ($\rho = 0$) and non-magnetic ($\chi_m = 0$, $\mu = 1$) media are assumed, which holds for dielectrics. The magnetic field can be eliminated and a wave equation results from Maxwell's equations:

$$\text{grad div } E - \Delta E + \frac{1}{c_0^2} \frac{\partial^2}{\partial t^2}\left(E + \frac{1}{\varepsilon_0} P\right) = 0, \tag{1.1.11}$$

$$\text{div } D = 0 \tag{1.1.12}$$

with

$$c_0 = \frac{1}{\sqrt{\varepsilon_0 \mu_0}} = 2.99792458 \times 10^8 \text{ m/s : vacuum velocity of light}.$$

Equation (1.1.11) is the fundamental equation, describing the propagation of optical fields. It includes diffraction as well as amplification of light and non-linear effects. It has now to be adapted and simplified for the different applications in optics and laser technology.

1.1.2.2 Homogeneous, isotropic, linear dielectrics

The propagation of light in homogeneous media as gases, liquids, glasses or cubic crystals is investigated. These materials are assumed to be homogeneous (permittivity ε does not depend on the spatial coordinates), isotropic (ε does not depend on the polarization of light), and linear (ε does not depend on the intensity of the field). The last assumption holds for low-intensity fields only.

The permittivity ε is a scalar and (1.1.11)/(1.1.12) reduces to the *standard wave equation*:

$$\Delta \boldsymbol{E} - \frac{\varepsilon}{c_0^2} \frac{\partial^2 \boldsymbol{E}}{\partial t^2} = 0 \,, \tag{1.1.13}$$

$$\operatorname{div} \boldsymbol{E} = 0 \,. \tag{1.1.14}$$

Simple solutions are the plane and the spherical waves.

1.1.2.2.1 The plane wave

The infinite, monochromatic wave with a plane phase front and constant amplitude reads:

$$\boldsymbol{E} = \boldsymbol{E}_0 \exp[\mathrm{i}(\omega t - n\boldsymbol{k}_0\boldsymbol{r})] \,, \tag{1.1.15}$$

$$\boldsymbol{H} = \boldsymbol{H}_0 \exp[\mathrm{i}(\omega t - n\boldsymbol{k}_0\boldsymbol{r})] \,; \tag{1.1.16}$$

$$\boldsymbol{H}_0 = \frac{[\boldsymbol{k}_0 \times \boldsymbol{E}_0]}{k_0 Z} \,.$$

It is a transversely polarized field with $\boldsymbol{E} \perp \boldsymbol{H} \perp \boldsymbol{k}_0$, as plotted in Fig. 1.1.3.

$n = \sqrt{\varepsilon} = \sqrt{1 + \chi_\mathrm{e}}$: the refractive index of the medium, in general complex, (1.1.17)

$k_0 = 2\pi/\lambda_0$: wave number in vacuum,
$\boldsymbol{k}_0$: wave vector, direction of propagation,
λ_0: wavelength in vacuum,

$Z = \sqrt{\dfrac{\mu\mu_0}{\varepsilon\varepsilon_0}}$: impedance, $Z_0 = \sqrt{\dfrac{\mu_0}{\varepsilon_0}} = 376.7 \ \Omega$: vacuum impedance.

The Poynting vector or energy flux is a real quantity with

$$\boldsymbol{S} = [\boldsymbol{E}_\mathrm{real} \times \boldsymbol{H}_\mathrm{real}] \quad (\text{SI-unit: W/m}^2).$$

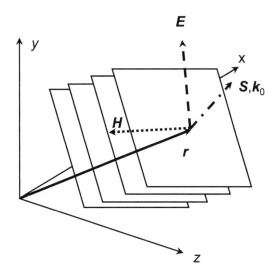

Fig. 1.1.3. The plane wave in a homogeneous, isotropic medium.

Table 1.1.1. Values of refractive index n_r and absorption coefficient α at wavelength λ_0 [85Pal, 82Gra, 78Dri].

Material	λ_0 [μm]	n_r	α [m^{-1}]
Fused quartz	0.54	1.46	very small
Sapphire	0.50	1.765/1.764	very small
Water	0.54	1.332	0.8
Water	1	1.328	80
Copper	0.54	0.7	11.6×10^6
Gold	0.54	0.3	11.1×10^6
Iron	0.54	2.4	16.4×10^6

The intensity is the time average over one period $T = 2\pi/\omega$ and results in:

$$J = \langle \boldsymbol{S} \rangle_T = \frac{1}{4} \left(\frac{1}{Z} + \frac{1}{Z^*} \right) \langle \boldsymbol{E}_0 \boldsymbol{E}_0^* \rangle \,. \tag{1.1.18}$$

For dielectrics without losses ($\mu = 1$, $n = n_r$ is real), (1.1.18) reduces to

$$J = \frac{1}{2} c_0 n_r \varepsilon_0 |E_0|^2 \tag{1.1.19}$$

with both quantities, $\boldsymbol{E}_0$ and J, inside the medium. For vacuum applies

$$J_{\text{W/m}^2} = 1.33 \times 10^{-3} |E_{0,\text{V/m}}|^2 \,, \quad |E_{0,\text{V/m}}| = 27.4 \sqrt{J_{\text{W/m}^2}} \,.$$

For a homogeneous dielectric, low-absorbing medium the complex refractive index is given by [99Bor, p. 739]:

$$\hat{n} = n_r - i \frac{\alpha}{2 k_0} \,, \quad \alpha \ll k_0 \tag{1.1.20}$$

with

n_r: real part of the refractive index,

α: absorption coefficient, in general the non-resonant broad-band absorption.

For a field propagating in z-direction (1.1.15)/(1.1.20) deliver an exponentially damped amplitude:

$$\boldsymbol{E}(z,t) = \boldsymbol{E}_0 \exp \left[i(\omega t - n_r k_0 z) - \frac{\alpha z}{2} \right] \,. \tag{1.1.21}$$

Some numbers of n_r, α are compiled in Table 1.1.1.

1.1.2.2.2 The spherical wave

One solution of the wave equation (1.1.13) in spherical coordinates is the quasi-spherical wave, generated by an oscillating dipole (Hertz's dipole), see Fig. 1.1.4. The far field reads [99Jac]:

$$\boldsymbol{E}(r, \vartheta, t) = \frac{\lambda_0 \boldsymbol{E}_\vartheta}{r} \exp \left[i(\omega t - \hat{n} \boldsymbol{k}_0 \boldsymbol{r}) \right] \sin \vartheta \,, \quad |\boldsymbol{E}_\vartheta| = \frac{|\boldsymbol{\mu}| \, 4\pi^2 k_0^3}{\varepsilon_0} \,, \quad r \gg \lambda_0$$

with μ the dipole moment and ϑ the angle between the dipole axis and beam propagation $\boldsymbol{k}_0$. In the paraxial approach ($\vartheta \cong \pi/2$, $\theta \ll 1$) the well-known spherical wave, useful for applying Huygens' principle, results:

$$\boldsymbol{E}(z,t) \cong \frac{\lambda_0}{r} \boldsymbol{E}_0 \exp \left[i(\omega t - \hat{n} \boldsymbol{k}_0 \boldsymbol{r}) \right] \,, \quad \theta \ll 1 \,, \tag{1.1.22}$$

where $\boldsymbol{E}$ is approximately parallel to the dipole axis.

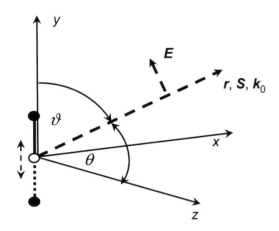

Fig. 1.1.4. A quasi-spherical wave, emitted by an oscillating dipole.

1.1.2.2.3 The slowly varying envelope (SVE) approximation

In the Slowly Varying Envelope approximation (1.1.11) is solved approximately with the ansatz of a quasi-monochromatic, quasi-plane wave

$$\boldsymbol{E} = \boldsymbol{E}_0(x,y,z,t)\exp[\mathrm{i}(\omega t - n_\mathrm{r} k_0 z)]\,, \quad \boldsymbol{P} = \boldsymbol{P}_0(x,y,z,t)\exp[\mathrm{i}(\omega t - n_\mathrm{r} k_0 z)]\,. \quad (1.1.23)$$

The wave propagates mainly in z-direction and the amplitude is slowly varying with x, y, z, t, which means:

- slowly varying in time (quasi-monochromatic): $\partial|\boldsymbol{E}_0|/\partial t \ll \omega|\boldsymbol{E}_0|$, or spectral bandwidth $\Delta\omega \ll \omega$,
- slowly varying in space (quasi-plane wave): $\partial|\boldsymbol{E}_0|/\partial z \ll k_0|\boldsymbol{E}_0|$, which means low divergence of the beam $\Delta\theta \ll 1$ (paraxial approach), and a smooth transverse profile,
- slowly varying polarization $\partial|\boldsymbol{P}_0|/\partial t \ll \omega|\boldsymbol{P}_0|$,
- slowly varying electric susceptibility $\partial|\chi_\mathrm{e}|/\partial t \ll \omega|\chi_\mathrm{e}|$ and $|\mathrm{grad}\,\chi_\mathrm{e}| \ll k_0|\chi_\mathrm{e}|$.

Then second order terms can be neglected and the SVE-approximations are obtained [84She, p. 47], [66War, 86Sie].

1.1.2.2.4 The SVE-approximation for diffraction

Steady-state propagation in vacuum means $\partial|\boldsymbol{E}_0|/\partial t = 0$ and $\boldsymbol{P} = 0$. Equation (1.1.11) delivers with the ansatz (1.1.23) and neglecting $\partial^2 \boldsymbol{E}_0/\partial t^2$ the SVE-approximation used in diffraction theory, also called the *Schrödinger equation of optics*:

$$\left(\Delta_\mathrm{tr} - 2\mathrm{i}k_0\frac{\partial}{\partial z}\right)\boldsymbol{E}_0 = 0\,, \quad \mathrm{div}\,\boldsymbol{E} = 0\,. \quad (1.1.24\mathrm{a})$$

Δ_tr is the transverse delta-operator, which in rectangular coordinates reads

$$\Delta_\mathrm{tr} = \frac{\partial^2}{\partial x^2} + \frac{\partial^2}{\partial y^2}\,.$$

The field in (1.1.24a) is a vector field, and the Δ-operator in cylinder coordinates is rather complicated, because the unit-vectors are no longer constant [99Jac], especially for non-uniform polarization in circular birefringent media [82Fer, 93Wit]. In most cases (except birefringence) the scalar version of the *SVE-approximation* is sufficient. It reads in rectangular/cylindrical coordinates

$$\left(\frac{\partial^2}{\partial x^2} + \frac{\partial^2}{\partial y^2} - 2ik_0\frac{\partial}{\partial z}\right)E_0 = 0\,, \tag{1.1.24b}$$

$$\left(\frac{\partial^2}{\partial r^2} + \frac{1}{r}\frac{\partial}{\partial r} + \frac{1}{r^2}\frac{\partial^2}{\partial \varphi^2} - 2ik_0\frac{\partial}{\partial z}\right)E_0 = 0\,. \tag{1.1.24c}$$

This is the fundamental equation in paraxial diffraction optics. It gives the Fresnel-integral and the eigenmodes of free propagation (Gauss-Hermite/Gauss-Laguerre polynomials, see Chaps. 3.1 and 8.1). Equations (1.1.24a)/(1.1.24b)/(1.1.24c) hold for a homogeneous medium, but can be extended to quadratic index media [86Sie].

1.1.2.3 Propagation in doped media

The active medium of a laser amplifier consists of a host material, doped with the active atoms (molecules). Host and doping interact differently with the laser radiation.

A plane wave without transverse structure interacts with active atoms or molecules and induces a polarization $\boldsymbol{P}_A$. In most cases the active atoms are embedded in a host medium (glass, crystal, liquid, gas), which is also polarized by the field, generating an additional polarization $\boldsymbol{P}_H$. The total polarization is:

$$\boldsymbol{P} = \boldsymbol{P}_A + \boldsymbol{P}_H - (\boldsymbol{P}_{A0} + \boldsymbol{P}_{H0})\exp[i(\omega t - n_r k_0 z)]\,. \tag{1.1.25}$$

The response of the host medium is in most cases very fast ($10^{-12} \ldots 10^{-14}$ s), no transient behavior occurs and nonlinear effects are assumed to be small. Then the host polarization is proportional to the applied field:

$$\boldsymbol{P}_H = \varepsilon_0 \chi_H \boldsymbol{E}\,.$$

χ_H is the complex susceptibility of the host material and is related to the refractive index n_r and the loss coefficient α according to (1.1.17)/(1.1.20) [99Ber]:

$$\chi_H = (n_r^2 - 1) - i\frac{n_r \alpha}{k_0}\,, \quad \alpha \ll k_0\,. \tag{1.1.26}$$

The imaginary part of χ_H is called extinction coefficient. Some values of refractive indices n_r and absorption coefficients α are given in Table 1.1.1. For the polarization of the active atoms one has

$$\boldsymbol{P}_A = \varepsilon_0 \chi_A(\boldsymbol{E}_0)\boldsymbol{E}\,, \tag{1.1.27}$$

where χ_A depends on the field and has to be evaluated quantum-mechanically. Neglecting first and second order derivations of $\boldsymbol{P}_{A0}$ and second order derivations of $\boldsymbol{E}_0$, the SVE-approximation for the interaction is obtained, assuming a plane wave without transverse structure:

$$\left(\frac{\partial}{\partial z} + \frac{1}{c}\frac{\partial}{\partial t} + \frac{\alpha}{2}\right)E_0 = -i\frac{k_0}{2\varepsilon_0 n_r}\boldsymbol{P}_{A0}(\boldsymbol{E}_0)\,, \quad \text{div}\,\boldsymbol{E} = 0 \tag{1.1.28}$$

(SVE-approximation for the amplitude of a plane wave in an active medium)

with $c = c_0/n_r$ the phase velocity of the wave in the host medium. The above equation describes the amplification/attenuation of cw-fields and pulsed radiation by an active medium. It provides also the widely used rate-equation approach, as will be shown in Sect. 1.1.5.1. It fails for fields with amplitudes varying very rapidly in time or space (fs-pulses). If the intensity J (1.1.19) and the susceptibility of the active medium (1.1.27) are introduced, (1.1.28) reduces to:

$$\left(\frac{\partial}{\partial z} + \frac{1}{c}\frac{\partial}{\partial t}\right) J + \left(\alpha - \frac{k_0}{n_\mathrm{r}} \operatorname{Im} \chi_\mathrm{A}\right) J = 0 \ . \tag{1.1.29}$$

The active atoms enhance or reduce the losses of the medium, depending on the sign of the imaginary part $\operatorname{Im} \chi_\mathrm{A}$ of the susceptibility, which is a function of the intensity. In steady state and for constant χ_A, which holds for low intensities, (1.1.29) can be integrated and delivers for the intensity

$$J(z) = J(0) \exp\left[-\alpha + \frac{k_0}{n_\mathrm{r}} \operatorname{Im}(\chi_\mathrm{A})\right] z \ .$$

The amplifying factor is called the small-signal gain factor G_0 of the medium and the exponent the small-signal gain coefficient g_0:

$$G_0 = \exp\left[\frac{k_0}{n_\mathrm{r}} \operatorname{Im}(\chi_\mathrm{A}) z\right] = \exp[g_0 z] \ , \quad g_0 = \frac{k_0}{n_\mathrm{r}} \operatorname{Im}(\chi_\mathrm{A}) \ . \tag{1.1.30}$$

Some typical values of g_0 are compiled in Table 1.1.4.

1.1.3 Interaction with two-level systems

Most quantum systems as atoms or molecules have an infinite number of energy levels. To demonstrate the essential features of light–matter interaction, a simplified model with only two levels is presented.

1.1.3.1 The two-level system

The relevant parameters are the energy difference ΔE of the two levels, the inversion Δn, the dipole moment $\boldsymbol{\mu}$, and the polarization $\boldsymbol{P}_\mathrm{A}$.

The two-level system can be part of an atom, ion, molecule, or something more complicated. A monochromatic electric field $\boldsymbol{E}$ of frequency ω in the SVE-approximation according to (1.1.23) acts via the Coulomb force on the bound electrons of the active medium. In linear systems (parabolic potential) the negative electrons will oscillate sinusoidally, whereas the heavy positive nucleus remains more or less at rest. An oscillating dipole is induced with a dipole moment $\boldsymbol{\mu}(t)$, which is given by

$$\boldsymbol{\mu} = -e\boldsymbol{x} \tag{1.1.31}$$

with

e: electron charge,
$\boldsymbol{x}$: displacement of the electron.

The dipole moment per volume is the macroscopic polarization $\boldsymbol{P}_\mathrm{A}$ of the active medium. As all single dipoles are aligned by the electric field, the resulting polarization reads:

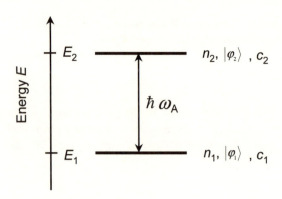

Fig. 1.1.5. The two-level system.

$$\boldsymbol{P}_A = n_0 \boldsymbol{\mu} \tag{1.1.32}$$

with

n_0: dipole density (m^{-3}),
$\boldsymbol{\mu}$: expectation value of the dipole moment (Asm).

In this section the induced dipole moment will be evaluated quantum-mechanically, which requires some simplifications. It is not the intention to discuss in detail the mathematics, but only to summarize briefly the main results of interest for laser technology and to emphasize the approximations and the range of validity. A consistent presentation of the interaction light–matter, starting from first principles, is given in many textbooks [61Mes, 68Sch, 77Coh, 95Man].

From the infinite number of energy levels of an electronic system only two, E_1 and E_2, are taken into account for the interaction [75All, 89Yar, 69Are], see Fig. 1.1.5. This is a reasonable approach if the field is nearly resonant with the transition from E_1 to E_2. In this case the other levels of the system will not or only very weakly interact with the field.

It applies

$$|\omega_A - \omega| \ll \Delta\omega_A$$

with

ω_A: resonance frequency of the transition,
$\Delta\omega_A$: bandwidth of the transition,
ω: frequency of the radiation field,
$\hbar = 1.0546 \times 10^{-34}$ Ws^2: Planck's constant.

1.1.3.2 The dipole approximation

The oscillating electric field $\boldsymbol{E}$ deforms the electron cloud of the two-level system and generates a complicated, oscillating charge distribution. A first order approximation is an oscillating dipole. The interaction of this dipole with a monochromatic wave is evaluated quantum-mechanically.

1.1.3.2.1 Inversion density and polarization

The interaction of an electromagnetic field with a two-level system was first investigated by Bloch [46Blo] and extensively discussed by Allen and Eberly [75All]. It is characterized by its dipole moment and the population densities in the two levels:

n_1, n_2 : density of states (atoms, molecules) in the lower/upper level,
$\Delta n = n_2 - n_1$: inversion density,
$n_0 = n_1 + n_2$: total density, const.

The following assumptions are made:

- Non-relativistic interaction. The velocity of the electrons is small compared with the velocity of light. This does not hold for inner-shell electrons, hot plasmas and free-electron lasers.
- The wavelength of the light is large compared with the diameter of the atoms/molecules. It means that in the domain of the atomic wave function the electromagnetic field is locally constant. Bohr's radius with $r_B = 5.3 \times 10^{-5}$ μm is a typical atomic dimension. The wavelength in the visible range of the spectrum is about 0.5 μm, thus this condition is fulfilled in the visible and UV-part of the spectrum. It is called the dipole approximation [97Scu].
- The permanent dipole moments of the two-level system $\boldsymbol{\mu}_{11} = \boldsymbol{\mu}_{22}$ are zero. Even if larger molecules have a permanent dipole moment, their response to the high-frequency field is small. Only for very strong fields are the permanent dipole moments of importance (see Part 4 on nonlinear optics). A dipole moment exists only for the transition from level 1 to 2 and vice versa. Non-degenerated levels are assumed with $\boldsymbol{\mu} = \boldsymbol{\mu}_{12} = \boldsymbol{\mu}_{21}$.

The two-level system is completely described by its state vector $|\varphi\rangle$, which in general is time-dependent:

$$|\varphi\rangle = c_1(t)|\varphi_1\rangle \exp\left(-\mathrm{i}\frac{E_1 t}{\hbar}\right) + c_2(t)|\varphi_2\rangle \exp\left(-\mathrm{i}\frac{E_2 t}{\hbar}\right), \tag{1.1.33}$$

with $|\varphi_1\rangle$, $|\varphi_2\rangle$ the eigenfunctions and E_1, E_2 the energy eigenstates. The eigenfunctions are normalized, orthogonal and depend on the position vector $\boldsymbol{r}$:

$$\int \varphi_i^* \varphi_j \mathrm{d}r = \langle \varphi_1 \varphi_2 \rangle = \delta_{ij}. \tag{1.1.34}$$

The state vector has to fulfill the time-dependent Schrödinger equation:

$$\mathrm{i}\hbar \frac{\partial |\varphi\rangle}{\partial t} = (H_0 + H_{\mathrm{int}})|\varphi\rangle, \tag{1.1.35}$$

with H_0 the Hamilton operator of the undisturbed system ($H_{\mathrm{int}} = 0$) and H_{int} the interaction energy. For the undisturbed system holds [89Yar]:

$$H_0 |\varphi_i\rangle = E_i |\varphi_i\rangle, \quad i = 1, 2, \tag{1.1.36}$$

which follows directly from (1.1.35) by replacing $|\varphi\rangle$ by $|\varphi_i\rangle \exp(-\mathrm{i}E_i t/\hbar)$. The parameters of interest, the inversion density $\Delta n = n_2 - n_1$ and the macroscopic polarization

$$\boldsymbol{P}_A = n_0 \boldsymbol{\mu} \tag{1.1.37}$$

are determined by the coefficients c_1, c_2. The probability of the system to be in the lower/upper state is given by $|c_1|^2$, $|c_2|^2$, respectively, which requires:

$$|c_1|^2 + |c_2|^2 = 1. \tag{1.1.38}$$

The number of atoms in the lower/upper level is then given by:

$$n_1 = n_0 |c_1|^2, \quad n_2 = n_0 |c_2|^2, \quad n_1 + n_2 = n_0$$

and hence the inversion density :

$$\Delta n = n_0 \left(|c_2|^2 - |c_1|^2 \right) . \tag{1.1.39}$$

The expectation value of the dipole moment $\langle \boldsymbol{\mu} \rangle = -e \langle \varphi \boldsymbol{r} \varphi \rangle$ is obtained from (1.1.33). Using the afore mentioned assumptions:

$$\langle \boldsymbol{\mu}_{11} \rangle = -e \langle \varphi_1 \boldsymbol{r} \varphi_1 \rangle = 0 , \quad \langle \boldsymbol{\mu}_{22} \rangle = -e \langle \varphi_2 \boldsymbol{r} \varphi_2 \rangle = 0$$

one obtains finally for the polarization from (1.1.33), (1.1.34), (1.1.38)

$$\boldsymbol{P}_A = n_0 \left\{ \langle \boldsymbol{\mu}_{12} \rangle c_1^* c_2 \exp(-\mathrm{i}\omega_A t) + \langle \boldsymbol{\mu}_{21} \rangle c_1 c_2^* \exp(+\mathrm{i}\omega_A t) \right\} \tag{1.1.40}$$

with $\langle \boldsymbol{\mu}_{12} \rangle$, $\langle \boldsymbol{\mu}_{21} \rangle$ the dipole moment of the transition $E_1 \leftrightarrow E_2$ and vice versa. For non-degenerated transitions one has $\langle \boldsymbol{\mu}_{12} \rangle = \langle \boldsymbol{\mu}_{21} \rangle = \boldsymbol{\mu}_A$. In the following only $\boldsymbol{\mu}_A$ will be used, which is a characteristic parameter of the specific transition:

$$\boldsymbol{\mu}_A = -e \langle \varphi_1 \boldsymbol{r} \varphi_2 \rangle . \tag{1.1.41}$$

1.1.3.2.2 The interaction with a monochromatic field

The interaction Hamiltonian for a non-quantized real field $\boldsymbol{E}_{\mathrm{real}}$ corresponds to the classical energy of an electric dipole in an electric field. It reads [97Scu]:

$$H_{\mathrm{int}} = \boldsymbol{\mu}_A \boldsymbol{E}_{\mathrm{real}} = \boldsymbol{\mu}_A \frac{(\boldsymbol{E} + \boldsymbol{E}^*)}{2} . \tag{1.1.42}$$

Substitution of (1.1.42) into (1.1.35), using the orthogonality (1.1.34) and (1.1.41) provides two differential equations for the coefficients c_1, c_2 of the state vector:

$$\begin{aligned}
\frac{\mathrm{d}c_1}{\mathrm{d}t} &= \frac{\mathrm{i}}{\hbar} c_2 \exp(-\mathrm{i}\omega_A t) \boldsymbol{\mu}_A \frac{(\boldsymbol{E} + \boldsymbol{E}^*)}{2} , \\
\frac{\mathrm{d}c_2}{\mathrm{d}t} &= \frac{\mathrm{i}}{\hbar} c_1 \exp(+\mathrm{i}\omega_A t) \boldsymbol{\mu}_A \frac{(\boldsymbol{E} + \boldsymbol{E}^*)}{2} .
\end{aligned} \tag{1.1.43}$$

The time dependence of inversion density and polarization is obtained from (1.1.39), (1.1.40) by differentiating and applying (1.1.43). After some simple mathematics the following two equations for the macroscopic parameters of the two-level-system result are obtained:

$$\frac{\partial \Delta n}{\partial t} = \frac{\mathrm{i}}{\hbar} \left\{ (\boldsymbol{E} + \boldsymbol{E}^*)(\boldsymbol{P}_A - \boldsymbol{P}_A^*) \right\} , \tag{1.1.44a}$$

$$\frac{\partial \boldsymbol{P}_A}{\partial t} = \mathrm{i} \left\{ \omega_A \boldsymbol{P}_A + \frac{\boldsymbol{\mu}_A}{\hbar} \langle \boldsymbol{\mu}_A (\boldsymbol{E} + \boldsymbol{E}^*) \rangle \Delta n \right\} . \tag{1.1.44b}$$

For $\boldsymbol{E}$ and $\boldsymbol{P}_A$ the SVE-approximations of (1.1.23), (1.1.25) are used. Then in (1.1.44a), (1.1.44b) terms with the frequency 2ω appear, which are neglected. This approach is called the rotating-wave approximation [97Scu, 72Cou]. The above equations simplify to

$$\frac{\partial \Delta n}{\partial t} = \frac{\mathrm{i}}{2\hbar} \left\{ \boldsymbol{E}_0^* \boldsymbol{P}_{A0} - \boldsymbol{E}_0 \boldsymbol{P}_{A0}^* \right\} , \tag{1.1.45a}$$

$$\frac{\partial \boldsymbol{P}_{A0}}{\partial t} = -\mathrm{i}\delta \boldsymbol{P}_{A0} + \frac{\mathrm{i}\boldsymbol{\mu}_A}{\hbar} \langle \boldsymbol{\mu}_A \boldsymbol{E}_0 \rangle \Delta n , \quad \delta = \omega - \omega_A \tag{1.1.45b}$$

(rotating-wave approximation)
with

$\boldsymbol{\mu}_A$: electric dipole moment of the transition,
ω: frequency of the interacting field,
ω_A: resonance frequency of the two-level system,
$\hbar = 1.0546 \times 10^{-34}$ Ws^2: Planck's constant.

Some typical values of dipole moments are given in Table 1.1.2.

Table 1.1.2. Typical values of dipole moments [01Men].

Transition			$|\boldsymbol{\mu}_A|$ [As m]
Bohr's radius × electron charge			10^{-29}
Hydrogen	1s – 2p	$\lambda_0 = 121$ nm	0.8×10^{-29}
	4f – 5g	$\lambda_0 = 4053$ nm	8.3×10^{-29}
Chromium ions in ruby	$4A_2(3/2)$ – E levels	$\lambda_0 = 694$ nm	10^{-29}

1.1.3.3 The Maxwell–Bloch equations

The idealized rotating-wave approximation is adapted to the real situation and combined with the SVE wave equation. Incoherent perturbations by the environment are taken into account.

So far the interaction of the two-level system with the electromagnetic field is purely coherent, no perturbations by external influences on the system are considered. Stochastic processes will modify the interaction considerably. Here only a very basic description is presented. A detailed analysis of these statistical processes is given in [70Hak, 97Scu].

1.1.3.3.1 Decay time T_1 of the upper level (energy relaxation)

Three incoherent processes reduce or increase the upper-level population and have to be considered in (1.1.45a), (1.1.45b):

– spontaneous emission,
– interaction with the host material (collisions, lattice vibrations),
– increase of the population by pumping (light, electron collisions, or other processes).

1.1.3.3.1.1 Spontaneous emission

The two-level system is coupled to the modes of the optical resonator or to the free-space modes. Spontaneous emission into these modes reduces the upper-level population. Moreover, by each spontaneous emission process the phase relation between the field and the two-level eigenfunction is destroyed. If the dimensions of the resonator are large compared with the wavelength, the decay is given by $\partial n_2/\partial t = -n_2/T_{\text{sp}}$, with $A_{21} = 1/T_{\text{sp}}$, the Einstein coefficient of spontaneous emission. If the resonator dimensions are comparable with the wavelength, spontaneous emission is strongly influenced by the resonator geometry, it can be enhanced or reduced (see Chap. 8.1).

1.1.3.3.1.2 Interaction with the host material

This interaction reduces the population density. Energy is transferred to the host material and converted into heat. A simple approach for this decay is again an exponential ansatz $\partial n_2/\partial t = -n_2/T_{\text{H}}$. This decay time together with the spontaneous decay time delivers a resulting decay T_1 of the upper-level population, also called energy relaxation time or longitudinal relaxation time.

1.1.3.3.1.3 Pumping process

The dynamics of upper-level excitation depend on the special pumping scheme and are discussed in Sect. 1.1.5.3 and in Vol. VIII/1B, "Solid-state laser systems". In any case the pump produces in steady state and without a coherent field ($\boldsymbol{E}_0 = 0$) an inversion density Δn_0.

These three processes are included into (1.1.45a) by the term:

$$\frac{\partial \Delta n}{\partial t} = -\frac{\Delta n - \Delta n_0}{T_1} \tag{1.1.46}$$

with

T_1: the resulting time constant.

1.1.3.3.2 Decay time T_2 of the polarization (entropy relaxation)

An external field $\boldsymbol{E}$ induces dipoles, which generate the macroscopic polarization $\boldsymbol{P}_\mathrm{A}$. If the external field is switched off, the polarization will disappear for several reasons:

The energy of the two-level system decays with T_1, which means that the polarization disappears at least with the same time constant.

Due to incoherent interaction with the host material (collisions), the single dipoles are disoriented in their direction or dephased. The resulting polarization becomes zero, although the single dipole still exists. This process can be much faster than T_1 (see Table 1.1.6) and is characterized by a time constant T_2. This decay strongly depends on the interaction process. The simplest approach is:

$$\frac{\partial \boldsymbol{P}_{\mathrm{A}0}}{\partial t} = -\frac{\boldsymbol{P}_{\mathrm{A}0}}{T_2}, \tag{1.1.47}$$

and (1.1.45b) has to be completed by (1.1.47). T_2 is called the transverse relaxation time, the entropy time constant or the dephasing time. Finally, the two-level equations together with the SVE-approximation, (1.1.28), of the wave equation read:

$$\frac{\partial \Delta n}{\partial t} = -\frac{\mathrm{i}}{2\hbar} \left(\boldsymbol{E}_0^* \boldsymbol{P}_{\mathrm{A}0} - \boldsymbol{E}_0 \boldsymbol{P}_{\mathrm{A}0}^* \right) - \frac{\Delta n - \Delta n_0}{T_1}, \tag{1.1.48a}$$

$$\frac{\partial \boldsymbol{P}_{\mathrm{A}0}}{\partial t} = -\left(\mathrm{i}\delta + \frac{1}{T_2} \right) \boldsymbol{P}_{\mathrm{A}0} + \mathrm{i}\frac{\boldsymbol{\mu}_\mathrm{A} \langle \boldsymbol{\mu}_\mathrm{A} \boldsymbol{E}_0 \rangle}{\hbar} \Delta n, \quad \delta = \omega - \omega_\mathrm{A}, \tag{1.1.48b}$$

$$\left(\frac{\partial}{\partial z} + \frac{1}{c}\frac{\partial}{\partial t} + \frac{\alpha}{2} \right) \boldsymbol{E}_0 = -\mathrm{i}\frac{k_0}{2\varepsilon_0 n_\mathrm{r}} \boldsymbol{P}_{\mathrm{A}0} \tag{1.1.48c}$$

(Maxwell–Bloch equations).

They describe the propagation of radiation in two-level systems and are called Maxwell–Bloch equations. Equation (1.1.48c) holds, if the transition frequency ω_A for all two-level atoms is the same (homogeneous system). In inhomogeneous systems (see Sect. 1.1.6.3, Fig. 1.1.13) different groups of atoms exist with center frequencies ω_A of each group and a center frequency ω_R of the ensemble. Therefore (1.1.48c) has to be replaced by [81Ver]:

$$\left(\frac{\partial}{\partial z} + \frac{1}{c}\frac{\partial}{\partial t} \right) \boldsymbol{E}_0 = -\mathrm{i}\frac{k_0}{2\varepsilon_0 n_\mathrm{r}} \int h(\omega_\mathrm{A}, \omega_\mathrm{R}) \boldsymbol{P}_{\mathrm{A}0}(\boldsymbol{E}_0, \omega_\mathrm{A}) \mathrm{d}\omega_\mathrm{A}. \tag{1.1.48d}$$

$h(\omega, \omega_\mathrm{A})$ is the spectral density of atoms with the transition frequency ω_A according to (1.1.92)/(1.1.93). For the solution of these equations, three different regimes are distinguished:

Steady-state equations
The temporal variations of the radiation field are slow compared with T_1.
$$\frac{\partial \Delta n}{\partial t} = 0 \qquad \frac{\partial \boldsymbol{P}_{A0}}{\partial t} = 0$$

Adiabatic equations
no transient effects of the atom, $T_2 \ll T_1$.
$$\frac{\partial \Delta n}{\partial t} \neq 0 \qquad \frac{\partial \boldsymbol{P}_{A0}}{\partial t} = 0$$

Coherent equations
The width τ of the interacting pulses is short compared with T_1, T_2; (1.1.45a), (1.1.45b) can be applied.
$$\frac{\partial \Delta n}{\partial t} \neq 0 \qquad \frac{\partial \boldsymbol{P}_{A0}}{\partial t} \neq 0$$

1.1.4 Steady-state solutions

In steady state inversion density Δn_0, polarization $\boldsymbol{P}_{A0}$, and intensity J of the field are constant in time, but may depend on the spatial coordinates.

1.1.4.1 Inversion density and polarization

The stationary solutions of (1.1.48a), (1.1.48b) are obtained immediately:

$$\Delta n = \frac{\Delta n_0}{1 + (J/J_s) f(\omega)} \quad \text{(\textit{inversion density}, homogeneously broadened)}, \tag{1.1.49}$$

$$\chi_A = \frac{n_r \sigma}{k_0} \left[\frac{\omega - \omega_A}{\Delta \omega_A / 2} + i \right] \Delta n \quad (\textit{susceptibility}), \tag{1.1.50}$$

$$\boldsymbol{P}_{A0} = \varepsilon_0 \chi_A \boldsymbol{E}_0 \quad (\textit{polarization}) \tag{1.1.51}$$

with

$$J = \frac{1}{2} \varepsilon_0 c_0 n_r |\boldsymbol{E}_0|^2 \quad (\textit{intensity of the field}), \tag{1.1.52}$$

$$J_s = \frac{\hbar \omega_A}{2 \sigma_0 T_1} \quad (\textit{saturation intensity of the two-level transition}), \tag{1.1.53}$$

$$\sigma = \sigma_0 f(\omega, \omega_A) \quad (\textit{frequency-dependent cross section of the transition}), \tag{1.1.54}$$

$$\sigma_0 = \frac{|\boldsymbol{\mu}_A|^2 \omega_A T_2}{\varepsilon_0 c_0 n_r \hbar} \quad (\textit{cross section in resonance}), \tag{1.1.55}$$

$$f_L(\omega, \omega_A) = \frac{(\Delta \omega_A/2)^2}{(\omega_A - \omega)^2 + (\Delta \omega_A/2)^2} \quad (\textit{spectral line shape, Lorentzian}), \tag{1.1.56}$$

$$\Delta \omega_A = 2/T_2 \quad (\textit{line width of the transition}), \tag{1.1.57}$$

$$g_\mathrm{h}(\omega,\omega_\mathrm{A}) = \Delta n \sigma = \frac{\Delta n_0 \sigma_0 f(\omega,\omega_\mathrm{A})}{1+(J/J_\mathrm{s})f(\omega,\omega_\mathrm{A})} \quad \text{(gain coefficient, homogeneously broadened)}, \tag{1.1.58a}$$

$$g_\mathrm{inh}(\omega,\omega_\mathrm{R}) = \frac{\Delta n_0 \sigma_0}{\sqrt{1+J/J_\mathrm{s}}} h(\omega,\omega_\mathrm{R})\frac{\Delta\omega_\mathrm{A}}{2} \quad \text{(gain coefficient, inhomogeneously broadened, see Sect. 1.1.6.3).} \tag{1.1.58b}$$

In Table 1.1.3 some numbers of relevant laser transitions are compiled, in Table 1.1.4 some typical values of the small-signal gain coefficient in resonance are given. The susceptibility strongly depends on the frequency as shown in Fig. 1.1.6. According to (1.1.26) the real part of χ_A produces an additional refractive index, and the imaginary part absorption or amplification:

$$\operatorname{Re}\chi_\mathrm{A} = n_\mathrm{r}^2 - 1 = \frac{n_\mathrm{r}\sigma}{k_0}\left(\frac{\omega-\omega_\mathrm{A}}{\Delta\omega_\mathrm{A}/2}\right)\Delta n, \tag{1.1.59a}$$

$$\operatorname{Im}\chi_\mathrm{A} = -n_\mathrm{r}\alpha k_0 = \frac{n_\mathrm{r}\sigma}{k_0}\Delta n. \tag{1.1.59b}$$

The steady-state propagation of the electric field is obtained from (1.1.48c):

$$\frac{\mathrm{d}\boldsymbol{E}_0}{\mathrm{d}z} = \left[-\frac{\alpha}{2} + \frac{\sigma\Delta n}{2} + \mathrm{i}\sigma\Delta n\frac{\omega-\omega_\mathrm{A}}{\Delta\omega_\mathrm{A}}\right]\boldsymbol{E}_0, \tag{1.1.60}$$

where Δn is a function of the field or the intensity.

Table 1.1.3. Examples of resonance wavelength λ_0, resonance cross section σ_0, upper-level lifetime T_1 and saturation intensity J_s. The simple relation (1.1.53) for the saturation intensity holds for two-level systems only and is not applicable in general [01Men].

	λ_0 [µm]	σ_0 [m^2]	T_1 [s]	J_s [W/m^2]
Amplifiers				
CO$_2$-gas (1300 Pa)	10.6	10^{-20}	10^{-5}	2×10^5
Neodymium-ion in glass	1.06	4×10^{-24}	3×10^{-4}	$8\ldots 12\times 10^7$
Neodymium-ion in YAG	1.06	5×10^{-23}	2×10^{-4}	2×10^7
Chromium-ion in Al$_2$O$_3$ (ruby, $T=300$ K)	0.69	2×10^{-24}	3×10^{-3}	2.4×10^7
Neon (25 Pa)	0.63	3×10^{-17}	10^{-8}	5.3×10^5
Rhodamine 6G in ethanol	0.57	4×10^{-20}	5×10^{-9}	10^9
Absorbers				
SF$_6$	10.6	8×10^{-22}	4×10^{-4}	2.5×10^5
KODAK dye 9860	1.06	4×10^{-20}	$\sim 10^{-11}$	5.6×10^{11}
KODAK dye 9740	1.06	6×10^{-20}	$\sim 10^{-11}$	4×10^{11}
Cryptocyanine-dye in methanol	0.7	5×10^{-20}	5×10^{-10}	2×10^{10}

Table 1.1.4. Typical values of the small-signal gain coefficient $g_0 = \Delta n_0 \sigma_0$ in resonance. The exact values depend on pumping, doping, and other parameters of operation [01Men].

System	λ_0 [nm]	g_0 [m^{-1}]
He/Ne laser	632.8	0.1
Nd-doped glass	1060	5
Nd-doped YAG	1060	50
GaAs-diode	880	4×10^3

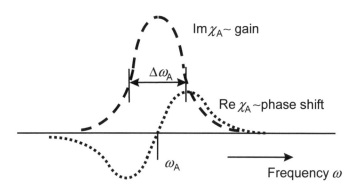

Fig. 1.1.6. Real and imaginary part of the susceptibility vs. frequency.

1.1.4.2 Small-signal solutions

The solutions for low intensities are discussed. Low means that the intensity J is small compared with the characteristic parameter J_s of the system (see Table 1.1.3).

At low intensities $J \ll J_s$, the inversion density is not affected by the intensity,

$$\Delta n = \Delta n_0 \;,$$

and (1.1.60) can be integrated. Together with (1.1.23), the complete field is obtained:

$$\boldsymbol{E}(z) = \boldsymbol{E}_0(0) \exp[\mathrm{i}(\omega t - n_\mathrm{t} k_0 z) - \frac{1}{2}(\alpha - \Delta n_0 \sigma) z] \tag{1.1.61}$$

with a total refractive index n_t

$$n_\mathrm{t} = n_\mathrm{r} \left(1 + \frac{\sigma \Delta n_0}{n_\mathrm{r} k_0} \frac{\omega - \omega_\mathrm{A}}{\Delta \omega_\mathrm{A}} \right) \;. \tag{1.1.62}$$

The active atoms of the two-level system cause an additional phase shift or refractive index and an additional absorption or amplification, depending on the sign of Δn_0. The small-signal gain factor according to (1.1.30)/(1.1.50) is:

$$G_0 = \exp[\sigma(\omega) \Delta n_0 z] \;. \tag{1.1.63}$$

Amplification, $G_0 > 1$, requires inversion $\Delta n_0 > 0$. The complex amplitude transmission factor A is defined as the ratio of the monochromatic field amplitudes and can be written:

$$A = \frac{E_0(z)}{E_0(0)} = \exp\left[\mathrm{i} \frac{\sigma_0 \Delta n_0}{2} \frac{\Delta \omega_\mathrm{A}/2}{(\omega - \omega_\mathrm{A}) + \mathrm{i} \Delta \omega_\mathrm{A}/2} \right] z \;. \tag{1.1.64}$$

It depends on the frequency of the field, which means dispersion. Time-dependent fields and especially short pulses are distorted by the amplifying system, pulse broadening and chirping occur.

1.1.4.3 Strong-signal solutions

The steady-state solutions are discussed for intensities which saturate the inversion, see Fig. 1.1.7.

The inversion now depends on the intensity. For the propagation of the intensity, (1.1.48c) gives in steady state

$$\frac{\mathrm{d}J}{\mathrm{d}z} = (g(J) - \alpha) \, J \,, \tag{1.1.65}$$

where $g(J)$ is the saturated gain coefficient of (1.1.58a), (1.1.58b). For a homogeneously broadened transition and without losses ($\alpha = 0$) this equation can be can be integrated and provides a transcendental relation for the gain factor G:

$$\frac{G_0}{G} = \exp\left[\frac{J(0)}{J_\mathrm{s}} f(\omega) \, (G-1)\right] \tag{1.1.66}$$

with G_0 the small-signal gain factor of (1.1.62) and G the ratio of output/input intensities

$$G = J(z)/J(0) \,.$$

For inhomogeneously broadened transitions a more complicated relation is obtained [81Ver].

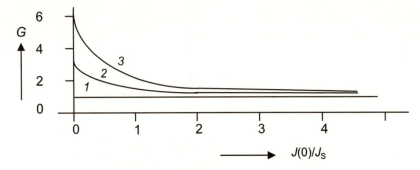

Fig. 1.1.7. Saturation of the gain factor G for a homogeneously and inhomogeneously broadened transition. *1*: $G_0 = 1$, *2*: $G_0 = 4$, *3*: $G_0 = 6$.

1.1.5 Adiabatic equations

If the polarization is in equilibrium with the applied field, without transient oscillations of the electronic system, the interaction is called adiabatic.

1.1.5.1 Rate equations

The field is replaced by the intensity, most spectral effects are neglected and the rate equations are obtained. They represent an energy balance.

T_2 is the time constant, which characterizes the transient behavior of the polarization. In most cases (see Table 1.1.6) T_2 is much smaller than T_1, and the transient oscillations of the electrons can be neglected. In (1.1.48a) the polarization is replaced by its steady-state value (1.1.50)/(1.1.51) and the rate equations are obtained. They have to be completed by the time-dependent pump term, here labeled as Δn_0. It depends on the specific pump scheme (see Sect. 1.1.5.3). The rate equations are widely used in laser design to evaluate output power, spiking behavior and Q-switching dynamics. The spontaneous emission contributes to the intensity of the interacting field, but only with a very small amount and is neglected here. Nevertheless it is important, because the laser is started by spontaneous emission and in the lower limit it determines the laser band width (Chap. 5.1).

With these approximations the field equations (1.1.48a)/(1.1.48b)/(1.1.48c) for the interaction with a monochromatic field reduce to one equation for the inversion density and a transport equation for the intensity:

$$\frac{\partial \Delta n}{\partial t} = -\frac{J f(\omega)}{J_s T_1} \Delta n - \frac{(\Delta n - \Delta n_0)}{T_1} \, , \tag{1.1.67}$$

$$\left(\frac{\partial}{\partial z} + \frac{1}{c} \frac{\partial}{\partial t} \right) J = (\Delta n \, \sigma_0 f(\omega)) \, J \tag{1.1.68}$$

(rate equations for a homogeneously broadened two-level system and a plane monochromatic wave)

with

$J(z,t)$: local intensity,
J_s: saturation intensity, depends on the level system (2,3, or 4 levels), see Sects. 1.1.4.1/1.1.5.3,
$\Delta n(z,t)$: local inversion density.

1.1.5.2 Thermodynamic considerations

So far the interaction with a monochromatic field of intensity $J(\omega)$ was discussed. Now the intensity is replaced by the spectral energy density ρ_ω of black-body radiation, providing the Einstein coefficients of spontaneous and induced emission.

Einstein published in 1917 [17Ein] his famous work on the quantum theory of radiation, where for the first time induced emission was introduced, the cornerstone of laser physics. He discussed the two-level system in equilibrium with thermal radiation of spectral energy density ρ_ω (energy per volume and spectral range $\mathrm{d}\omega$). The density is given by Planck's law [61Mor]:

$$\rho_\omega = \frac{\hbar \omega^3}{\pi^2 c^3} \frac{1}{\exp[\hbar \omega / \kappa T] - 1} \qquad \left[\frac{\mathrm{VAs}^2}{\mathrm{m}^3} \right] \tag{1.1.69}$$

with

$\kappa = 1.38 \times 10^{-23}$ VAs/K: Boltzmann's constant.

In thermal equilibrium the levels $|\varphi_1\rangle$, $|\varphi_2\rangle$ are populated according to Boltzmann's law [61Mor]:

$$\frac{n_2}{n_1} = \exp\left[-\hbar \omega_\mathrm{A}/\kappa T\right] \, . \tag{1.1.70}$$

These two fundamental laws can only be fulfilled, if induced emission is introduced, and Einstein postulated the following equation in steady state for the interaction of thermal radiation with a two-level system:

$$B_{12} \, \rho_\omega \, n_1 = B_{21} \, \rho_\omega \, n_2 \qquad + A_{21} \, n_2 \tag{1.1.71}$$

(absorption = induced emission + spontaneous emission)

with

B_{12}, B_{21}, A_{21}: Einstein coefficients of induced and spontaneous emission.

The transition of atoms from the lower level to the upper level by absorption of radiation must be balanced by induced emission and spontaneous emission from the upper level. This equation was derived by thermodynamical considerations. The quantum-mechanical equation (1.1.67) delivers in steady state, replacing Δn by $n_2 - n_1$ and n_0 by $n_1 + n_2$, and furthermore taking into account that for steady state without interaction holds $\Delta n_0 = -n_0$:

$$J\frac{\sigma}{\hbar\omega_A}n_1 = J\frac{\sigma}{\hbar\omega_A}n_2 + \frac{n_2}{T_1} \,. \tag{1.1.72}$$

This equation has the same structure as the Einstein equation. If the monochromatic intensity $J(\omega)$ is replaced by the spectral density ρ_ω and integration over the full spectral range is performed, a relation between the Einstein coefficients and the atomic parameters is obtained. These relations read in general for degenerated levels with weighting factors g_1, g_2 (degeneracies) [92Koe, 81Ver, 00Dav]:

$$B_{12} = \frac{g_2}{12\pi}\frac{|\mu_A|^2}{\hbar^2\varepsilon\varepsilon_0} \,, \tag{1.1.73a}$$

$$B_{21} = \frac{g_1}{12\pi}\frac{|\mu_A|^2}{\hbar^2\varepsilon\varepsilon_0} \,, \tag{1.1.73b}$$

$$A_{21} = \frac{1}{T_1} = \frac{g_1}{3}|\mu_A|^2 \frac{\omega_A^3}{\pi\varepsilon\varepsilon_0\hbar c^3} \,, \tag{1.1.74}$$

$$\mu_A = \mu_{12} = \mu_{21} \,,$$

$$\sigma_{21}(\omega) = \frac{\lambda^2}{4}A_{21}h(\omega) \,, \tag{1.1.75}$$

$$\sigma_{12}(\omega) = \frac{g_2}{g_1}\sigma_{21}(\omega) \,, \tag{1.1.76}$$

$$\sigma_{21}(\omega_A) = \frac{\lambda^2}{4\pi}\frac{T_2}{T_1} \leq \frac{\lambda^2}{4} \qquad \text{(holds for Lorentzian line shape),} \tag{1.1.77}$$

$$B_{12}g_1 = B_{21}g_2 \,, \tag{1.1.78}$$

$$\frac{A_{21}}{B_{21}} = \frac{2\hbar\omega_A^3}{\pi c^3} \,. \tag{1.1.79}$$

The above relations were derived for isotropic media. Anisotropic media are discussed in [86Sie]. Equation (1.1.80) holds for all dipole transitions, as long as the quantum system is coupled to a large number of modes (free space or a resonator with dimensions large compared with the wavelength). With these equations the gain coefficient can be related to the Einstein coefficient of spontaneous emission [92Koe]:

$$g(\omega) = \frac{\lambda^2}{4}h(\omega,\omega_A)\left[n_2 - \frac{g_2}{g_1}n_1\right]A_{21} \tag{1.1.80}$$

with

$h(\omega,\omega_A)$: the spectral line shape, depending on the type of broadening (see Sect. 1.1.6).

1.1.5.3 Pumping schemes and complete rate equations

The fundamental methods to obtain inversion are presented, discussing the idealized 3- and 4-level system.

Till now a two-level system was discussed, assuming a steady-state inversion Δn_0, which is always negative. To obtain positive inversion $\Delta n = n_2 - n_1 > 0$ and gain, additional levels are necessary.

$\Delta n > 0$ is a state of non-equilibrium. To support this state, energy has to be pumped into the system. This pumping energy can be incoherent light, kinetic energy of electrons/ions, chemical energy or electric energy. The pumping schemes can become very complicated, and in most cases many energy levels are involved. To understand the principal process for the generation of inversion, two idealized pumping schemes will be discussed.

1.1.5.3.1 The three-level system

The simplified diagram of the three-level system is shown in Fig. 1.1.8. The level E_3 is excited by absorption of light or by electron collisions, depending on the specific system. The decay from E_3 to E_2, the upper laser level, is very fast. Nearly all excited atoms are transferred into this level, which has a very long life time. If the pumping power is sufficiently high to overcome the decay of level E_2, atoms will be accumulated and finally n_2 is larger than n_1. The adiabatic rate equations give for the upper-level population without induced emission between the two levels ($J = 0$):

$$\frac{d n_2}{d t} = W(n_0 - n_2) - \frac{n_2}{T_1}. \tag{1.1.81}$$

W is the pumping rate, the product of the cross-section σ_{13} and the specific pump parameters. T_1 is the upper laser-level lifetime. This holds under the assumption that the population of level E_3 is zero and that $n_1 + n_2 = n_0$. Equation (1.1.81) reads with the inversion density $\Delta n = n_2 - n_1$:

$$\frac{d\Delta n}{d t} = W(n_0 - \Delta n) - \frac{n_0 - \Delta n}{T_1} \tag{1.1.82}$$

and in steady state one obtains:

$$\frac{\Delta n_{\text{steady},3}}{n_0} = \frac{W T_1 - 1}{W T_1 + 1}. \tag{1.1.83}$$

The relation between the inversion density and the pump rate is shown in Fig. 1.1.9. Inversion occurs for $W T_1 > 1$. With increasing pump rate the inversion increases also and approaches finally one, all atoms are in the upper level. To obtain $\Delta n_{\text{steady},3} > 0$ requires at least 50 % of the active atoms to be pumped into the upper level, high pump rates are necessary and the efficiency is low. Equation (1.1.82) has to be completed by the coherent interaction term of (1.1.67). The complete rate equation for the three-level system with pump rate W, interacting with a monochromatic field of intensity J is given in (1.1.84). For the intensity (1.1.48c), (1.1.48d) hold, depending on the type of line-broadening (Sect. 1.1.6).

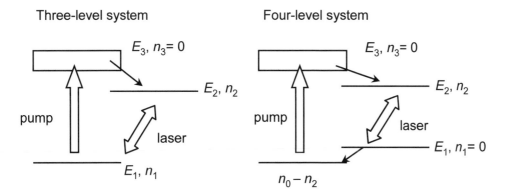

Fig. 1.1.8. The idealized three- and four-level system.

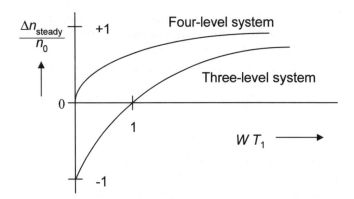

Fig. 1.1.9. Inversion density vs. pump rate for a three- and four-level system.

$$\frac{\partial \Delta n}{\partial t} = -\frac{J}{J_s}\frac{f(\omega)}{T_1}\Delta n + W(n_0 - \Delta n) - \frac{n_0 + \Delta n}{T_1} \qquad (1.1.84)$$

(rate equation of a three-level system).

1.1.5.3.2 The four-level system

The commonly used pump scheme, due to its high efficiency, is the four-level system as shown in Fig. 1.1.8. The two laser levels are E_2 and E_1, where the lower level E_1 has a very short lifetime and its population n_1 is nearly zero. This requires that the energy $E_1 - E_0$ is much larger than the thermal energy κT. The pump level E_3 decays very rapidly to the upper laser level E_2 and its population is again nearly zero. The inversion density now is $\Delta n = n_2 - n_1 \approx n_2$. Then the rate equation for the pump process reads:

$$\frac{\partial \Delta n}{\partial t} = W(n_0 - \Delta n) - \frac{\Delta n}{T_1} \qquad (1.1.85)$$

with the steady-state solution (without coherent interaction):

$$\frac{\Delta n_{\text{steady},4}}{n_0} = \frac{WT_1}{1 + WT_1} \, . \qquad (1.1.86)$$

Inversion is reached now at very small pump-power levels as shown in Fig. 1.1.9. The efficiency of such systems is much higher than of three-level systems. The complete rate equation for pumping and interaction with a field of intensity J is obtained by taking into account the corresponding term of (1.1.67). It has to be considered that $n_1 = 0$, and therefore the saturation intensity is higher by a factor of 2.

$$\frac{\partial \Delta n}{\partial t} = -\frac{J}{J_{s,4}}\frac{f(\omega)}{T_1}\Delta n + W(n_0 - \Delta n) - \frac{\Delta n}{T_1} \qquad (1.1.87)$$

(rate equation of a four-level system)

with

$$J_{s,4} = \frac{\hbar \omega_A}{\sigma_0 T_1} \, : \quad \text{saturation intensity of the four-level system.}$$

1.1.5.4 Adiabatic pulse amplification

The amplification and shaping of light pulses by saturable two-level systems is presented.

The pulse is adiabatic if its width τ is small compared with T_1 and large compared with T_2. Then the variation of the upper-level population due to spontaneous emission and pump can be neglected and this term can be neglected. If such a pulse travels through an active medium of length ℓ, it depletes the upper level, is amplified and shaped as depicted in Fig. 1.1.10. The initial conditions at $t = -\infty$ are:

Inversion density:	$\Delta n(z) = \Delta n_0$,	$0 \leq z \leq \ell$.
Input intensity:	J_0,	$z = 0$.
Input energy:	E_{in},	$z = 0$.

The equations (1.1.67)/(1.1.68) can be solved for a loss-free-medium ($\alpha = 0$) with a four-level system and yield for the output intensity [63Fra]:

$$J_{\text{out}}(t) = J_{\text{in}}(t - \ell/c) \frac{G_0}{G_0 - (G_0 - 1)\exp\left[-\dfrac{1}{E_s}\displaystyle\int_{-\infty}^{t-\ell/c} J_{\text{in}}(t')\mathrm{d}t'\right]}. \qquad (1.1.88)$$

The total output energy density E_{out} of the pulse is

$$E_{\text{out}} = E_s \ln\left[1 + G_0\left(\exp\left(E_{\text{in}}/E_s\right) - 1\right)\right] \qquad (1.1.89)$$

with the two limiting cases

$$E_{\text{out}} = \begin{cases} G_0 E_{\text{in}}, & E_{\text{in}} \ll E_s, \\ E_{\text{in}} + E_s \ln G_0 = E_{\text{in}} + \dfrac{\Delta n_0 \ell \, \hbar \omega_A}{2}, & E_{\text{in}} \gg E_s \end{cases} \qquad (1.1.90)$$

with

G_0: small-signal gain factor, (1.1.63),
$E_s = J_{s,4} T_1$: saturation energy density,
$E_{\text{in,out}}$: input/output energy density.

Equations (1.1.88)–(1.1.90) also hold for saturable absorbers with $G_0 < 1$. The pulse will be shaped in any case and the peak velocity will differ from the phase- and group velocities.

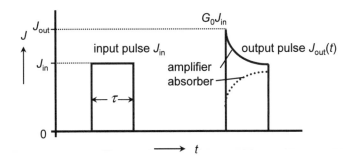

Fig. 1.1.10. Pulse amplification and shaping by a saturable amplifier/absorber.

1.1.5.5 Rate equations for steady-state laser oscillators

In the oscillator system, two counter-propagating traveling waves J^+, J^- appear, see Fig. 1.1.11, which are amplified by an intensity- and z-dependent gain coefficient according to (1.1.58a), (1.1.58b):

$$\frac{\mathrm{d}J^+}{\mathrm{d}z} = [g(J) - \alpha]\, J^+ \,, \tag{1.1.91a}$$

$$\frac{\mathrm{d}J^-}{\mathrm{d}z} = -[g(J) - \alpha]\, J^- \,. \tag{1.1.91b}$$

For the two traveling waves the boundary conditions at the mirrors are:

$$J^+(z=0) = J^-(z=0) R_1 \,,$$

$$J^-(z=\ell) = J^+(z=\ell) R_2 \,.$$

The combination of (1.1.91a) and (1.1.91b) yields [81Ver]:

$$J^+(z) J^-(z) = \mathrm{const.} \,,$$

a useful relation for analytical solutions. The gain coefficient is saturated by both waves. In steady state (1.1.84)/(1.1.87) hold with $J = J^+ + J^-$, depending on the level system and on the type of broadening. For homogeneous broadening a solution is given in [81Ver]. In general, numerical calculations are necessary. For optimization a diagram is offered in [92Koe]. The intensity rate equations are very useful for laser design and optimization, but deliver no spectral effects such as line width [58Sch, 74Sar, 95Man], mode competition [86Sie, 00Dav], mode hopping [86Sie, 64Lam, 74Sar], or intensity-dependent frequency shifts (Lamb dip) [64Lam]. Multimode oscillation can be described by rate equations with restrictions [64Sta, 63Tan, 93Sve].

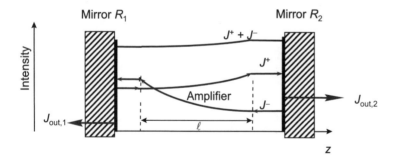

Fig. 1.1.11. The laser oscillator with two counter-propagating waves.

1.1.6 Line shape and line broadening

Shape and width of the spectral response of the two-level system depend on the special stochastic perturbation processes, in detail discussed by [81Ver, 86Eas]. An easy-to-read introduction is given by [86Sie, 00Dav].

1.1.6.1 Normalized shape functions

Normalized line shapes are introduced, which determine the relative strength of interaction.

The line shape depends on the specific interaction process. Two standard line shapes, easy to handle, are the Lorentzian and the Gaussian profiles [92Koe], shown in Fig. 1.1.12. They can be normalized differently.

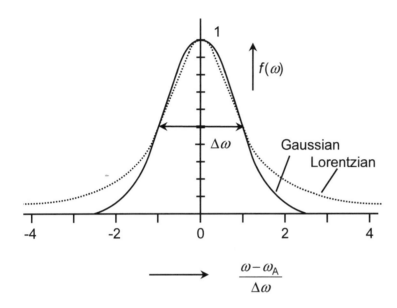

Fig. 1.1.12. Gaussian and Lorentzian line shape.

1.1.6.1.1 Lorentzian line shape

$$f_L(\omega, \omega_A) = \frac{(\Delta\omega_A/2)^2}{(\omega - \omega_A)^2 + (\Delta\omega_A/2)^2}, \qquad h_L(\omega, \omega_A) = \frac{2}{\pi\Delta\omega_A} f_L(\omega, \omega_A). \tag{1.1.92}$$

1.1.6.1.2 Gaussian line shape

$$f_G(\omega) = \exp\left[-\left(\frac{\omega - \omega_A}{\Delta\omega_A/2}\right)^2 \ln 2\right], \qquad h_G(\omega) = \sqrt{\frac{\ln 2}{\pi}} \frac{2}{\Delta\omega_A} f_G(\omega). \tag{1.1.93}$$

1.1.6.1.3 Normalization of line shapes

$$f_{G,L}(\omega = \omega_A) = 1, \quad f_{G,L}(\omega = \omega_A \pm \Delta\omega_A/2) = 0.5, \quad \int_{-\infty}^{+\infty} h_{G,L}(\omega, \omega_A) d\omega = 1. \tag{1.1.94}$$

1.1.6.2 Mechanisms of line broadening

1.1.6.2.1 Spontaneous emission

The spontaneous emission decay time T_{sp} of quantum dot lasers can be influenced by the geometry [97Scu], but for all macroscopic laser systems it is equal to the free-atom decay and related to the dipole moment (see Sect. 1.1.5.2). The line width of the power spectrum is $\Delta\omega = 1/T_{\text{sp}}$. The line shape is Lorentzian for undisturbed systems.

1.1.6.2.2 Doppler broadening

In thermal equilibrium the particles in a gas have a Maxwellian velocity distribution of the velocity v:

$$h(v) = \sqrt{\frac{m_{\text{A}}}{2\pi \kappa T}} \exp\left[-\frac{m_{\text{A}} v^2/2}{\kappa T}\right] \tag{1.1.95}$$

with

m_{A} : atomic mass,
κT : thermal energy of the particles.

The resonance frequency of a transition is shifted by the Doppler effect

$\Delta\omega = \omega_{\text{A}} v/c_0$.

Replacing the velocity in (1.1.76) by the frequency, delivers for the resulting spectral distribution a Gaussian line shape (1.1.74) with the width

$$\frac{\Delta\omega_{\text{D}}}{\omega_{\text{A}}} = \sqrt{\frac{8\,\kappa T \ln 2}{m_{\text{A}} c_0^2}} \ . \tag{1.1.96}$$

Some numbers are compiled in Table 1.1.5.

Table 1.1.5. Doppler and collision broadening for a thermal energy of $\kappa T = 1$ eV. The Doppler broadening refers to $\omega_{\text{A}} = 10^{15}$ s^{-1}, the collision broadening holds for a pressure of $p = 133$ Pa (1 torr) [81Ver, 01Men].

Gas	Doppler broadening $\Delta\omega_{\text{D}}$ [10^{10} s^{-1}]	Collision broadening $\Delta\omega_{\text{C}}$ [10^7 s^{-1}]
H_2	5.6	2.8
He	4	1.3
Ne	1.8	0.8
CO_2	1.2	1.2
Ar	1.5	9

1.1.6.2.3 Collision or pressure broadening

Elastic collisions between radiating atoms imply no energy loss, but a discontinuous jump in the phase of the emitted field. The average temporal length of the wave trains, in the undisturbed case given by the spontaneous life time T_{sp}, is reduced to the collision time τ. The Fourier transform of these shortened waves gives a Lorentzian line shape with the spectral width $\Delta\omega_{\mathrm{C}} = 2/\tau$ or

$$\Delta\omega_{\mathrm{C}} = \frac{32\,\sigma_{\mathrm{C}}\,p}{\sqrt{\pi m_{\mathrm{A}}\,\kappa T}} \qquad (1.1.97)$$

with

σ_{C}: collision cross section of the atom,
p: pressure of the gas.

The collision broadening is proportional to the gas pressure. Some numbers are given in Table 1.1.5.

1.1.6.2.4 Saturation broadening

A strong field of intensity J, comparable with the saturation intensity J_{s}, depletes the upper laser level. The gain is reduced according to (1.1.58a), (1.1.58b) and the gain profile becomes flatter and broader with the spectral width (see Fig. 1.1.13) [81Ver]:

$$\Delta\omega_{\mathrm{S}} = \Delta\omega_{\mathrm{A}}\sqrt{1 + J/J_{\mathrm{s}}}\,.$$

1.1.6.3 Types of broadening

The interaction of the field depends strongly on the type of broadening. Two idealized cases are the homogeneous and the inhomogeneous broadening [00Dav].

1.1.6.3.1 Homogeneous broadening

All transitions have the same resonance frequency ω_{A}. The gain is saturated for all atoms in the same way as given by (1.1.58a) and shown in Fig. 1.1.13. Examples for this type of broadening are:

- spontaneous emission,
- collision broadening,
- saturation broadening,
- thermal broadening in crystals by interaction with the lattice vibrations.

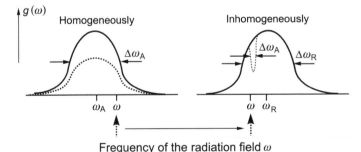

Fig. 1.1.13. Saturation of homogeneously and inhomogeneously broadened systems by a radiation field of frequency ω.

1.1.6.3.2 Inhomogeneous broadening

Groups of atoms with spectral density $h(\omega_R,\omega_A)$ and different frequencies ω_A produce a resulting line profile with center frequency ω_R and width $\Delta\omega_R$ as shown in Fig. 1.1.14. A strong monochromatic field of frequency ω interacts mainly with the group $\omega_A = \omega$ and saturates this particular group. A dip appears in the profile, which is called spectral hole-burning. Examples of inhomogeneous broadening are:

- Doppler broadening,
- Stark broadening in crystals due to statistical local crystalline fields.

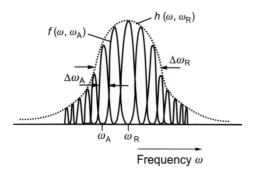

Fig. 1.1.14. An inhomogeneously broadened profile.

The resulting line profile is a convolution of the individual group profiles and the broadening process, which results in complicated integrals. The saturation process for inhomogeneously broadened lines is quite different, as will be shown by a simple example. In this case (1.1.58a) holds only for one group of atoms with the spectral density $h(\omega_A,\omega_R)$. Integration over all groups results in the total gain coefficient g_{inh}:

$$g_{\text{inh}}(\omega,\omega_R) = \int_{-\infty}^{+\infty} f(\omega,\omega_A) h(\omega_A,\omega_R) \, d\omega_A \; . \tag{1.1.98}$$

If the width $\Delta\omega_A$ is much smaller than the total width $\Delta\omega_R$, the function $h(\omega_A,\omega_R)$ can be taken outside of the integral at $\omega_A = \omega$. Assuming a Lorentzian profile for the single group, (1.1.98) becomes:

$$g_{\text{inh}}(\omega) = \Delta n_0 \sigma_0 h(\omega,\omega_R) \int \frac{f(\omega,\omega_A)}{1 + (J/J_s) f(\omega,\omega_A)} \, d\omega_A$$

and can be integrated:

$$g_{\text{inh}}(\omega) = \frac{\Delta n_0 \sigma_0}{\sqrt{1 + J/J_s}} h(\omega,\omega_R) \frac{\pi \Delta\omega_A}{2} = \frac{\Delta n_0 \sigma_0}{\sqrt{1 + J/J_s}} f(\omega) \frac{\Delta\omega_A}{\Delta\omega_R} \; . \tag{1.1.99}$$

The gain saturates slower than in the case of homogeneous broadening, but the maximum gain is lower by the ratio of the line widths. Inhomogeneous gain profiles can also be caused by spatial hole burning in solid-state laser systems. The standing waves between the mirrors produce an inversion grating and holes in the spectral gain profile [86Sie].

The spectral characteristics of lasers depend strongly on the type of broadening, see Fig. 1.1.15. In steady state the gain compensates losses and the gain profile saturates to fulfill the condition $GRV = 1$. A homogeneously broadened gain profile saturates till the steady-state condition is fulfilled for the central frequency. The bandwidth $\Delta\omega_{L,h}$ is very small and depends on the thermal and

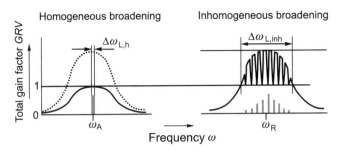

Fig. 1.1.15. Spectrum of an inhomogeneously and homogeneously broadened laser transition in steady state. Total gain factor GRV vs. frequency of the radiation field.

mechanical fluctuations [00Dav]. In the case of solid-state lasers spatial hole burning will influence the spectral behavior and can produce even for homogeneous transitions multi-mode oscillation [63Tan, 66Men]. In the case of inhomogeneous broadening each spectral group of atoms saturates separately and many modes will oscillate, which produces a large lasing bandwidth $\Delta\omega_{L,inh}$. If single-mode operation is enforced by suitable frequency selecting elements, the left → right and the right → left traveling waves produce two symmetric holes, due to the Doppler effect. This effect can be used for frequency stabilization (Lamb dip [64Lam]).

1.1.6.4 Time constants

The line profile of a real laser transition is in most cases a mixture of homogeneous and inhomogeneous profiles, depending on the temperature and the pressure. The following time constants are used in literature:

- T_{sp}: spontaneous life time,
- T_1: upper-laser-level life time (energy relaxation time, longitudinal relaxation time),
- T_2': Stochastic processes broaden the line homogeneously. The inverse of the line width is the dephasing time T_2'.
- T_2^*: The line is broadened inhomogeneously. The inverse of this line width $\Delta\omega_R$ is the dephasing time T_2^*.
- T_2: For the resulting dephasing time (transverse relaxation time, entropy time constant), approximately holds (depends on the line profiles):

$$\frac{1}{T_2^2} \approx \frac{1}{T_2^{*\,2}} + \frac{1}{T_2'^{\,2}}\ .$$

Some examples of decay times are given in Table 1.1.6.

1.1.7 Coherent interaction

Radiation field and two-level system are two coupled oscillators. Without stochastic perturbations the stored energy is permanently exchanged between these two systems.

If the interaction time of the radiation field with the two-level system is small compared with all relaxation times, including the pump term, the stochastic processes can be neglected and

Table 1.1.6. Spontaneous life time T_sp, upper-laser-level life time T_1, transverse relaxation time T_2, homogeneous relaxation time T_2' and inhomogeneous relaxation time T_2^* [01Iff, 92Koe, 86Sie], [01Men, Chap. 6].

	T_sp [s]	T_1 [s]	T_2 [s]	T_2' [s]	T_2^* [s]
Neon-atom (He/Ne-laser), $\lambda_0 = 632.8$ nm, He ($p = 130$ Pa), Ne ($p = 25$ Pa)	10^{-8}	10^{-8}	3×10^{-9}	10^{-8}	4×10^{-9}
Chromion-ion, $\lambda_0 = 694.3$ nm, R_1-transition in ruby					
$T = 300$ K	3×10^{-3}	3×10^{-3}	10^{-12}	10^{-12}	2×10^{-7}
$T = 4$ K	4×10^{-3}	4×10^{-3}	2×10^{-7}	3×10^{-3}	2×10^{-7}
SF$_6$-molecule, $\lambda_0 = 10.5$ μm, $p = 0.4$ Pa	10^{-3}	10^{-3}	6×10^{-9}	7×10^{-6}	6×10^{-9}
Rhodamin-molecule in ethanol, singlet-transition, $\lambda_0 = 570.0$ nm	5×10^{-9}	5×10^{-9}	10^{-12}	10^{-12}	–
Neodymium-ion in YAG-crystal, $\lambda_0 = 1060$ nm, $T = 300$ K	5×10^{-4}	2.3×10^{-4}	7×10^{-12}	–	–

(1.1.45a)/(1.1.45b) hold. This kind of coherent interaction is of strong interest in nonlinear spectroscopy [84She, 86Sie, 71Lam, 72Cou], [01Men, Chap. 7] and confirmed by many experiments. Examples of nonlinear coherent interaction are transient response of atoms, optical nutation, photon echoes, $n\pi$-pulses and quantum beats. Here only some very simple examples will be presented. A more detailed treatment is given in [95Man].

1.1.7.1 The Feynman representation of interaction

Feynman introduced a very elegant representation of interaction, which enables an easy-to-understand visualization.

A very compact description of the two-level interaction was given by Feynman [57Fey]. The real electric field is

$$\boldsymbol{E}_\text{real} = \frac{1}{2} \left\{ \boldsymbol{E}_0 \exp\left[\mathrm{i}\left(\omega t - kz\right)\right] + \boldsymbol{E}_0^* \exp\left[-\mathrm{i}\left(\omega t - kz\right)\right] \right\} .$$

It generates a real polarization, (1.1.23), shifted in phase against the field:

$$\begin{aligned}\boldsymbol{P}_\text{A,real} &= \frac{1}{2} \left\{ \boldsymbol{P}_\text{A0} \exp\left[\mathrm{i}(\omega t - kz)\right] + \boldsymbol{P}_\text{A0}^* \exp\left[-\mathrm{i}(\omega t - kz)\right] \right\} \\ &= \boldsymbol{C} \cos\left(\omega t - kz\right) + \boldsymbol{S} \sin\left(\omega t - kz\right)\end{aligned} \qquad (1.1.100)$$

with $\boldsymbol{C}$, $\boldsymbol{S}$ real vectors:

$$\boldsymbol{C} = \frac{1}{2} \left(\boldsymbol{P}_\text{A0} + \boldsymbol{P}_\text{A0}^*\right), \quad \boldsymbol{S} = \frac{1}{2} \mathrm{i} \left(\boldsymbol{P}_\text{A0} - \boldsymbol{P}_\text{A0}^*\right) .$$

In the following an isotropic medium is assumed. Then $\boldsymbol{\mu}_\text{A}$, $\boldsymbol{P}_\text{A}$ and $\boldsymbol{E}$ are parallel and can be treated as scalars. With these new real quantities the equations of interaction (1.1.45a), (1.1.45b) become:

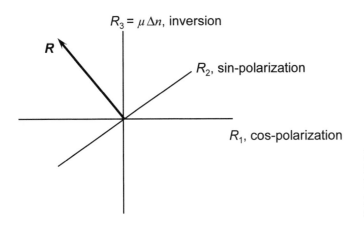

Fig. 1.1.16. In the case of coherent interaction, the system is characterized by its $\boldsymbol{R}$-vector which rotates in the polarization/inversion space with constant length.

$$\frac{\partial C}{\partial t} = -\delta S + i\mu_A \cdot \Delta n \left(\frac{\Lambda - \Lambda^*}{2}\right), \tag{1.1.101a}$$

$$\frac{\partial S}{\partial t} = \delta C - \mu_A \cdot \Delta n \left(\frac{\Lambda + \Lambda^*}{2}\right), \tag{1.1.101b}$$

$$\mu_A \frac{\partial \Delta n}{\partial t} = -i C \left(\frac{\Lambda - \Lambda^*}{2}\right) + S \left(\frac{\Lambda + \Lambda^*}{2}\right), \tag{1.1.101c}$$

where Λ is a complex quantity. Its modulus is called the Rabi frequency:

$$\Lambda(z,t) = \frac{\mu_A E_0}{\hbar}, \quad |\Lambda| : \text{Rabi frequency}. \tag{1.1.102}$$

Two vectors $\boldsymbol{R}$, $\boldsymbol{F}$ are introduced:

$$\boldsymbol{R} = (C, S, \mu_A \Delta n) = (R_1, R_2, R_3), \quad \boldsymbol{F} = \left(\frac{\Lambda + \Lambda^*}{2}, i\frac{\Lambda - \Lambda^*}{2}, \delta\right) = (F_1, F_2, F_3).$$

The $\boldsymbol{R}$-vector characterizes the state of the two-level system and can be depicted in an inversion/polarization space, as shown in Fig. 1.1.16. $\boldsymbol{R}$ corresponds to the Bloch vector of the spin-1/2 system [46Blo]. The equations (1.1.101a), (1.1.101b) of interaction can be condensed to:

$$\frac{\partial \boldsymbol{R}}{\partial t} = [\boldsymbol{F} \times \boldsymbol{R}] \quad \text{(coherent interaction)}. \tag{1.1.103}$$

Scalar multiplication of this equation with $\boldsymbol{R}$ results in:

$$\left\langle \boldsymbol{R} \frac{\partial \boldsymbol{R}}{\partial t} \right\rangle = \langle \boldsymbol{R} [\boldsymbol{F} \times \boldsymbol{R}] \rangle = 0,$$

which means that the length of the vector is constant during interaction:

$$|C|^2 + |S|^2 + |\mu_A \Delta n|^2 = |\boldsymbol{R}_0|^2. \tag{1.1.104}$$

The tip of the vector moves on a sphere in the inversion/polarization space with complicated trajectories [69McC, 74Sar, 69Ics]. The incoherent relaxation and pumping of the system can be included in (1.1.103) by an additional relaxation term [72Cou].

1.1.7.2 Constant local electric field

If the amplitude $\boldsymbol{E}_0$ of the electric field is assumed to be constant, a very simple solution of the rotating-wave equations is obtained with one main parameter, the Rabi frequency Λ.

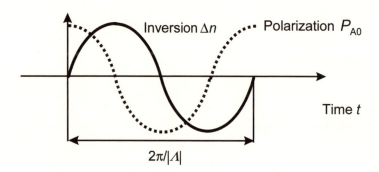

Fig. 1.1.17. Oscillation of inversion density Δn and polarization amplitude $\boldsymbol{P}_{A0}$ in resonance for a constant local electric field.

For a constant electric field at a fixed position z the rotating-wave approximation has a periodic solution. Inversion and polarization with the initial condition $t = 0$, $\Delta n = n_0$, $\boldsymbol{P}_{A0} = 0$ are:

$$\frac{\Delta n}{n_0} = \frac{\delta^2 + |\Lambda|^2 \cos \beta t}{\beta^2} \;, \quad \Lambda = \frac{\mu_A E_0}{\hbar} \;, \tag{1.1.105}$$

$$P_{A0} = n_0 \frac{\mu_A \Lambda}{\beta} \left[\frac{\delta}{\beta}(1 - \cos \alpha t) + \mathrm{i} \sin \alpha t \right] \;, \quad \beta = \sqrt{\delta^2 + |\Lambda|^2} \;. \tag{1.1.106}$$

In resonance $\omega = \omega_A$, $\delta = 0$, the inversion density Δn and the amplitude $\boldsymbol{P}_{A0}$ of the polarization oscillate with this frequency, see Fig. 1.1.17. The real polarization $\boldsymbol{P}_{A,\mathrm{real}}$ of (1.1.100) contains the frequencies $\omega_A \pm |\Lambda|$. Some values of dipole moments are given in Table 1.1.2 to estimate $|\Lambda|$. Off resonance the temporal behavior of inversion and polarization is more complicated (optical nutation) [72Cou]. If at $t = 0$ all atoms are in the lower level ($\Delta n_0 = -n_0$) a complete inversion is produced at $t = \pi/|\Lambda|$ by a coherent field. It is called pulse inversion [60Vuy]. At $t = 2/|\Lambda|$, all atoms are again in the lower level, no energy transfer has taken place.

1.1.7.3 Propagation of resonant coherent pulses

For short pulses, $\tau < T_2$, the perturbations can be neglected. The solution of the complete interaction equations (1.1.101a)–(1.1.101c) for a propagating resonant pulse is rather simple.

The propagation of pulses in a two-level system is described by the rotating-wave approximation, (1.1.45a)/(1.1.45b), and by the wave equation in the SVE approximation (1.1.28). The set of these three non-linear equations is difficult to solve, only special cases will be discussed here. At $t = 0$ the electric field E_0 is assumed to be real, $\Lambda = \Lambda^*$. In case of resonance, $\delta = 0$, (1.1.101a) delivers $C = 0$, $R_1 = 0$. The interaction equations (1.1.101b), (1.1.101c) reduce to

$$R_1 = 0 \;,$$

$$\frac{\partial R_2}{\partial t} = -\Lambda R_3 \;,$$

$$\frac{\partial R_3}{\partial t} = \Lambda R_2 \;.$$

The $\boldsymbol{R}$-vector moves in the R_2-R_3-plane, see Fig. 1.1.18. If the angle θ with the R_3-axis is introduced, one solution of the above equations is:

$$R_2 = R_0 \sin \theta \;,$$

$$R_3 = -R_0 \cos \theta$$

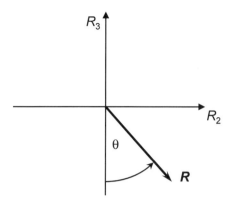

Fig. 1.1.18. In resonance, $\delta = 0$, the $\boldsymbol{R}$-vector of the two-level system rotates in the R_2-R_3-plane.

with

$$\Lambda = \frac{\partial \theta}{\partial t} = \frac{\mu_A E_0}{\hbar} . \qquad (1.1.107)$$

R_0 is given by the initial conditions at $t = 0$. The SVE-approximation of (1.1.28) then becomes:

$$\left(\frac{\partial^2}{\partial t\, \partial z} + \frac{1}{c} \frac{\partial^2}{\partial t^2} \right) \theta = -\frac{\alpha}{2} \frac{\partial \theta}{\partial t} + \frac{\gamma}{2} R_0 \sin \theta , \quad \gamma = \frac{\mu_A k_0}{n_r \varepsilon_0 \hbar} . \qquad (1.1.108)$$

From θ the amplitude E_0 of the electric field can be calculated with (1.1.107)/(1.1.105).

1.1.7.3.1 Steady-state propagation of $n\pi$-pulses

Steady state means that a pulse is propagating with velocity v and constant pulse envelope $\boldsymbol{E}_0(t, z) = \boldsymbol{E}_0(t - z/v)$. The amplitude depends on one parameter w only:

$$w = t - z/v$$

and (1.1.108) becomes:

$$\left(1 - \frac{c}{v} \right) \frac{\mathrm{d}^2 \theta}{\mathrm{d} w^2} + \frac{\alpha c}{2} \frac{\mathrm{d} \theta}{\mathrm{d} w} = c \frac{\gamma}{2} R_0 \sin \theta . \qquad (1.1.109)$$

This equation is equivalent to the equation of the pendulum with friction in a gravitational field. In the following examples two different initial conditions are assumed:

$$R_0 = \mu_A \Delta n_0 \begin{cases} > 0 & \text{(amplifier)} , \\ < 0 & \text{(absorber)} , \end{cases}$$

which corresponds to the pendulum up or down at $t = 0$.

1.1.7.3.1.1 2π-pulse in a loss-free medium

A medium without losses ($\alpha = 0$) interacts with a coherent pulse in resonance ($\delta = 0$). The initial condition is $\Delta n_0(t = -\infty) = +\Delta n_0$ ($\Delta n_0 < 0$, absorber). One steady-state solution is the 2π-pulse, see Fig. 1.1.19, which corresponds to a local field of duration $\tau = 2\pi/\Lambda$. The leading edge of the pulse produces an inversion and energy is transferred to the atomic system, the amplitude is reduced. The trailing part of the pulse is then amplified by this inversion. In total the pulse

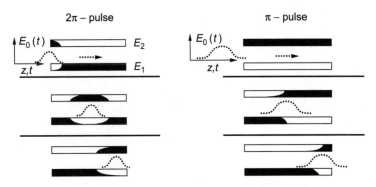

Fig. 1.1.19. Propagation of 2π- and π-pulses in a two-level system.

has lost no energy, but is delayed in time. Such a pulse is only stable, if the broadband losses are negligible and if the initial inversion is negative. The steady-state solution is:

$$E = E_{\text{peak}} \frac{\exp\left[i\omega(t - z/c)\right]}{\cosh\left[(t - z/v)/\tau\right]} \quad \text{(field)}, \tag{1.1.110}$$

$$E_{\text{peak}} = 2\sqrt{\hbar\omega}\sqrt{\frac{\Delta n_0}{\varepsilon_0(1 - c/v)}} \quad \text{(peak amplitude)}, \tag{1.1.111}$$

$$J_{\text{peak}} = \frac{2\hbar\omega\Delta n_0 c}{1 - c/v} \quad \text{(peak intensity)}, \tag{1.1.112}$$

$$T_{2\pi} = 2\tau = 2\sqrt{\frac{(1 - c/v)T_2}{g_0 c}} \quad \text{(pulse duration)}, \tag{1.1.113}$$

$$v = \frac{c}{1 - g_0 c\tau^2/T_2} \quad \text{(pulse peak velocity)} \tag{1.1.114}$$

with

$g_0 = \Delta n_0 \sigma < 0$: small-signal absorption coefficient,
c: phase velocity in the medium,
v: pulse peak velocity.

This two-level system is the most simple model of a saturable absorber, which in the case of incoherent interaction absorbs the radiation. But the coherent 2π-pulse transmits the absorber without losing energy. Therefore this effect is called self-induced transparency [75Kri]. The pulse is characterized by three parameters: peak velocity v, peak amplitude E_{peak} and the width $T_{2\pi}$. One of these parameters can be chosen arbitrarily, the other two result from (1.1.112)/(1.1.113)/(1.1.114). But the interaction is coherent only as long as $T_{2\pi} \ll T_2$.

1.1.7.3.1.2 π-pulse in an amplifying medium

A steady-state solution in an amplifying medium, initial condition $\Delta n(t = -\infty) = \Delta n_0 > 0$, with broadband losses ($\alpha \neq 0$) is the π-pulse [74Loy], see Fig. 1.1.19:

$$E = E_{\text{peak}} \frac{\exp\left[i\omega(t - z/c)\right]}{\cosh\left[(t - z/c)/\tau\right]} \quad \text{(field)}, \tag{1.1.115}$$

$$E_{\text{peak}} = \frac{\hbar}{\tau\mu} \quad \text{(peak amplitude)}, \tag{1.1.116}$$

$$J_{\text{peak}} = \frac{\hbar\omega}{2\sigma_0 T_2}\left[\frac{g_0}{\alpha}\right]^2 \quad \text{(peak intensity)}, \tag{1.1.117}$$

$$T_\pi = 2\tau = 2\,T_2\,\frac{\alpha}{g_0} \quad \text{(pulse duration)}. \tag{1.1.118}$$

The pulse propagates approximately with c, depletes at each position the upper level, and converts this energy via the broadband losses α into heat. The saturated gain just compensates the losses. The pulse is only stable for $\alpha > 0$ and $g_0 > 0$.

So far solutions of the steady-state SVE-equation were presented, assuming resonance and a homogeneously broadened two-level system. Off-resonance interaction and inhomogeneously broadened systems are much more complicated and are discussed in detail in the literature [74Sar, 69Ics, 72Cou]. Moreover, the stability of the pulses with respect to small perturbations was not yet mentioned. It is controlled by the area theorem [67McC, 74Sar].

1.1.7.3.2 Superradiance

The spontaneous emission was neglected in the coherent interaction. An initial state, $\boldsymbol{R} = (0, 0, \mu\,\Delta n)$, complete inversion, without external field $\boldsymbol{F}$ would be stable according to the interaction equations (1.1.103). But due to spontaneous emission and amplified spontaneous emission, the $\boldsymbol{R}$-vector will be pushed a bit out of equilibrium and decay into the stable position $\boldsymbol{R} = (0, 0, -\mu\,\Delta n)$. This phenomenon is called superradiance and discussed in detail in Chap. 6.2.

1.1.8 Notations

Symbol	Unit	Meaning
A_{21}	s^{-1}	Einstein coefficient of spontaneous emission
$\boldsymbol{B}$	Vs/m^2	magnetic induction
B_{12}, B_{21}	m^3/VAs^3	Einstein coefficient of induced emission
$\boldsymbol{C}$	As/m^2	component of the Feynman vector $\boldsymbol{R}$
c_0	m/s	vacuum velocity of a plane wave
c	m/s	phase velocity of light in a medium
$c_{1,2}$	–	coefficients of the eigenvector
$\boldsymbol{D}$	As/m^2	electric displacement
$\boldsymbol{E}$	V/m	electric field
$\boldsymbol{E}_0$	V/m	electric-field amplitude
$E_{1,2}$	VAs	energy eigenstates of the two-level system
E_{in}	VAs	amplifier input energy
E_{out}	VAs	amplifier output energy
E_{S}	VAs/m^2	amplifier saturation energy density
$f(\omega, \omega_{\text{A}})$	–	line shape factor
G	–	gain factor
G_0	–	small-signal gain factor
g	m^{-1}	gain coefficient
g_0	m^{-1}	small-signal gain coefficient
$g_{1,2}$	–	degeneracies of lower/upper laser level

Symbol	Unit	Description
g_h	m^{-1}	gain coefficient of a homogeneously broadened transition
g_{inh}	m^{-1}	gain coefficient of an inhomogeneously broadened transition
$\boldsymbol{H}$	A/m	magnetic field
$\boldsymbol{H}_0$	A/m	magnetic-field amplitude
H_0	VAs	Hamilton operator of the undisturbed transition
H_{int}	VAs	Hamilton operator of interaction
$h(\omega, \omega_A)$	s	line shape factor
$\boldsymbol{j}$	A/m^2	current density
$\boldsymbol{J}$	Vs/m^2	magnetic polarization
J	VA/m^2	intensity
J^+, J^-	VA/m^2	intensity inside the resonator
J_s, J_{s4}	VA/m^2	saturation intensity of 2-, 3- and 4-level system
k	m^{-1}	wave number
$\boldsymbol{k}$	m^{-1}	wave vector inside the medium
$\boldsymbol{k}_0$	m^{-1}	wave vector in vacuum
ℓ	m	geometrical length of the active medium
n	–	complex refractive index
n_r	–	real refractive index
n_0	m^{-3}	density of active atoms
$n_{1,2}$	m^{-3}	density of lower/upper population
$\boldsymbol{P}_{A,real}$	As/m^2	real polarization of the active atoms
$\boldsymbol{P}_A$	As/m^2	complex polarization of the active atoms
$\boldsymbol{P}_{A0}$	As/m^2	amplitude of the complex polarization
$\boldsymbol{P}_H$	As/m^2	complex polarization of the host material
$\boldsymbol{R}$	As/m^2	Feynman vector
R	–	$= \sqrt{R_1 R_2}$, average mirror reflectivity
$R_{1,2}$	–	reflectivity of mirror 1, 2
$\boldsymbol{r}$	m	position vector
$\boldsymbol{S}$	VA/m^2	Poynting vector
T_1	s	upper-laser-level life time
T_2'	s	dephasing time due to homogeneous broadening
T_2^*	s	dephasing time due to inhomogeneous broadening
T_2	s	resulting dephasing time
T_{sp}	s	spontaneous decay time
$T_\pi, T_{2\pi}$	s	pulse duration of π-, 2π-pulses
V	–	resonator loss factor per transit
v	m/s	pulse peak velocity
Z	V/A	impedance
Z_0	V/A	vacuum impedance
α	m^{-1}	absorption coefficient
χ_A	–	susceptibility of the active atoms
χ_e	–	electric susceptibility
χ_H	–	susceptibility of the host material
χ_m	–	magnetic susceptibility
δ	s^{-1}	detuning
Δn	m^{-3}	inversion density
Δ_{tr}	m^{-2}	transverse delta-operator
$\Delta \omega_A$	s^{-1}	line width of homogeneous broadening
$\Delta \omega_C$	s^{-1}	line width of collision broadening

Symbol	Unit	Description
$\Delta\omega_R$	s^{-1}	line width of inhomogeneous broadening
$\Delta\omega_S$	s^{-1}	line width of saturation broadening
$\Delta\omega_{L,inh}, \Delta\omega_{L,h}$	s^{-1}	lasing bandwidth of inhomogeneous/homogeneous transitions
ε	–	permittivity
ε_0	8.8542×10^{-12} As/Vm	electric constant
$|\varphi\rangle$	–	state vector of the two-level system
$|\varphi_{1,2}\rangle$	–	eigenfunctions of the two-level system
κ	1.38×10^{-23} VAs^2/K	Boltzmann's constant
λ_0	m	vacuum wavelength
Λ	s^{-1}	Rabi frequency
μ	–	permeability
μ_0	$4\pi \times 10^{-7}$ Vs/Am	magnetic constant
$\boldsymbol{\mu}_{12}, \boldsymbol{\mu}_{21}$	Asm	$= \boldsymbol{\mu}_A$, dipole moment of the two-level transition
$\boldsymbol{\mu}_A$	Asm	dipole moment of the two-level transition
θ	–	beam divergence, slope of the Feynman vector
ρ_ω	VAs^2/m^3	spectral energy density (per $d\omega$)
$\sigma(\omega)$	m^2	cross section of the two-level system
σ_e	A/Vm	electric conductivity
σ_0	m^2	cross section of the two-level system in resonance
τ	s	pulse width
ω	s^{-1}	frequency of the radiation field
ω_A	s^{-1}	resonance frequency of the homogeneously broadened transition
ω_R	s^{-1}	resonance frequency of the inhomogeneously broadened transition

References for 1.1

17Ein   Einstein, A.: Phys. Z. **18** (1917) 121.

19Mei   Meissner, A.: Patentschrift Reichspatentamt, Deutsches Reich, No. 291604, 1919.

46Blo   Bloch, F.: Phys. Rev. **70** (1946) 460.

47Lam   Lamb, W.E., Retherford, R.C.: Phys. Rev. **72** (1947) 241.

54Bas   Basov, N.G., Prokhorov, A.M.: Zh. Eksp. Teor. Fiz. **28** (1954) 249.
54Max   Maxwell, J.C.: Treatise on electricity and magnetism, New York: Dover Publications Inc., 1954 (reprint of the 3. edition 1891).

57Fey   Feynman, R.P., Vernon, F.L., Helwarth, R.W.: J.Appl. Phys. **28** (1957) 49.

58Sch   Schawlow, A.L., Townes, C.H.: Phys. Rev. **112** (1958) 1940.

60Mai   Maiman, T.H.: Nature (London) **187** (1960) 493.
60Vuy   Vuylsteke, A.A.: Elements of maser theory, N.Y.: v. Nostrand Comp., 1960.

61Mes   Messiah A.: Quantum mechanics, Vol. I–II, Amsterdam: North Holland Publ. Comp., 1961–1962.
61Mor   Morse, P.M.: Thermal physics, New York: W.A. Benjamin Inc, 1961.

63Fra   Frantz, L.M., Nodvik, I.S.: J. Appl. Phys. **34** (1963) 2346.
63Tan   Tang, C.L., Statz, H., de Mars, G.: Phys. Rev. **34** (1963) 2289.

64Lam   Lamb, W.E.: Phys. Rev. A **134** (1964) 1429.
64Sta   Statz, H., Tang, C.L.: Appl. Phys. **35** (1964) 1377.

66Men   Menne, T.J.: IEEE J. Quantum Electron. **2** (1966) 47.
66War   Ward, J.F.: Phys. Rev. **143** (1966) 569.

67McC   McCall, S.L., Hahn, E.L.: Phys. Rev. Lett. **18** (1967) 908.

68Sch   Schiff, L.I.: Quantum mechanics, New York.: McGraw Hill, 1968.

69Are   Arecchi, F.T., Masserini, G.L., Schwendimann, P.: Riv. Nuovo Cimento Soc. Ital. Fis. **1** (1969) 181.
69Ics   Icsevgi, A., Lamb, W.E.: Propagation of light pulses in a Laser amplifier; Phys. Rev. **185** (1969) 517.
69McC   McCall, L., Hahn, E.L.: Phys. Rev. **183** (1969) 457.

70Hak   Haken, H.: Laser theory, Handbuch der Physik, Vol.XXV/2c, Flügge, S. (ed.), Berlin: Springer-Verlag, 1970.

71Lam   Lamb, G.L.: Rev. Mod. Phys. **43** (1971) 99.

72Cou   Courtens, E.: Nonlinear coherent resonant phenomena, in: Laser handbook Vol. 2, Arecchi, F.T., Schulz-Dubois, E.O. (eds.), Amsterdam, New York, Oxford: North Holland Publ. Comp, 1972, p. 1259.

74Loy	Loy, M.M.T.: Phys. Rev. Lett. **32** (1974) 814.
74Sar	Sargent III, M., Scully, O.M., Lamb jr, W.E.: Laser physics, Reading (MA): Addison-Wesley Publ. Comp., 1974.
75All	Allen, L., Eberly, J.H.: Optical resonance and two level systems, New York: J.Wiley & Sons, 1975.
75Kri	Krieger, W., Toschek, P.E.: Phys. Rev. A **11** (1975) 276.
77Coh	Cohen-Tannoudji, C., Diu, B., Laloe, F.: Quantum mechanics, Vol. 1, 2, New York: John Wiley, 1977.
78Dri	Driscoll, W.G., Vaugham, W. (eds.): Handbook of optics, New York: McGraw Hill Publ. Comp., 1978.
81Ver	Verdeyen, J.T.: Laser Electronics, Englewood Cliffs, N.J.: Prentice Hall, 1981, p. 175.
82Fer	Ferguson, T.R.: Vector modes in cylindrical resonators; J. Opt. Soc. Am. 72 (1982) 1328–1334.
82Gra	Gray, D.E. (ed.): American institute of physics handbook, New York: McGraw-Hill Book Company, 1982.
84She	Shen, Y.R.: The principles of nonlinear optics, New York: John Wiley & Sons, 1984, 563 pp.
85Pal	Palik, E.D. (ed.): Handbook of optical constants of solids, New York: Academic Press, 1985.
86Eas	Eastham, D.A.: Atomic physics of lasers, London: Taylor & Francis, 1986, 230 pp.
86Sie	Siegman, A.E.: Lasers, Mill Valley (Ca): University Science Books, 1986.
88Her	Hertz, H.: Pogg. Ann. Phys. Chem. (2) **34** (1888) 551.
89Yar	Yariv, A.: Quantum electronics, New York: John Wiley & Sons, 1989.
91Wei	Weiss, C.O., Vilaseca, R.: Dynamics of lasers, Weinheim, Germany: VCH Verlagsgesellschaft, 1991.
92Koe	Koechner, W.: Solid state Laser engineering, Berlin: Springer-Verlag, 1992.
93Sve	Svelto, O., Hanna, D.C.: Principles of Lasers, New York, London: Plenum Press, 1993, p. 223.
93Wit	Wittrock, U., Kumkar, M., Weber, H.: Coherent radiation fields with pure radial or azimuthal polarisation, Proc. of the 1st international workshop on laser beam characterisation, Mejias, P.M., Weber, H., Martinez-Herrero, R., Gonzáles-Urena, A. (eds.), Madrid: Sociedad Espanola de Optica, 1963, p. 41.
95Man	Mandel, L., Wolf, E.: Optical coherence and quantum optics, Cambridge: Cambridge University Press, 1995.
95Wal	Walls, D.F., Milburn, G.J.: Quantum optics, Berlin: Springer Verlag, 1995.
97Scu	Scully, O.M., Zubairy, M.S.: Quantum optics, Cambridge: Cambridge University Press, 1997.

99Ber	Bergmann/Schäfer: Lehrbuch der Experimentalphysik, Vol. 3, Optics, Niedrig, H. (ed.), New York: W. de Gruyter, 1999.
99Bor	Born, M., Wolf, E.: Principles of optics, Cambridge: Cambridge University Press, 1999.
99Jac	Jackson, J.D.: Classical electrodynamics, New York: John Wiley & Sons, 1999, p. 410.
00Dav	Davies, C.C.: Laser and electro-optics, Cambridge: Cambridge Univ. Press, 2000.
01Iff	Iffländer, R.: Solid State Lasers for Material Processing, Berlin: Springer-Verlag, 2001, p. 257–318.
01Men	Menzel, R.: Photonics, Berlin: Springer-Verlag, 2001.
01Vog	Vogel, W., Welsch, D.G., Wallentowitz, S.: Quantum optics – an introduction, Berlin: Wiley-VCH, 2001.

Part 2

Radiometry

2.1 Definition and measurement of radiometric quantities

B. WENDE, J. FISCHER

2.1.1 Introduction

Radiometry is the science and technology of the measurement of electromagnetic energy. Here we confine ourselves on the subfield of optical radiometry which covers the measurement of electromagnetic radiation in the wavelength range from about 0.01 µm to 1000 µm. Radiometric quantities are derived from the quantity energy. The corresponding photometric quantities on the other hand involve the additional evaluation of the radiant energy in terms of a defined weighting function, usually the standard photometric observer. In the following only the definitions of the radiometric quantities are explained in detail. Starting from the radiant energy the other fundamental radiometric quantities radiant power, radiant excitance, irradiance, radiant intensity, and radiance are derived by considering the additional physical quantities time, area, and solid angle.

The radiometric quantities defined in abstract terms are practically embodied by radiometric standards. Radiometry is based on primary detector standards and primary source standards. Primary detector standards are mostly electrical-substitution thermal detectors whereas for primary source standards the emitted radiant power is accurately calculable. For the radiometric measurement of cw laser emission radiation detectors or radiometers calibrated against primary detector standards are the preferred secondary standards. The detection principle of the radiometers could be thermal (thermopiles, bolometers, and pyroelectric detectors) or photoelectric (semiconductors). As secondary standards for pulsed laser radiation mostly thermally absorbing glass-disk calorimeters are used. These standards are derived from the cw standards using accurately measured shuttering of the laser radiation to produce pulses of known radiant energy.

2.1.2 Definition of radiometric quantities

Radiometric and photometric quantities are represented by the same principal symbol and may be distinguished by their subscripts. While radiometric quantities either have the subscript "e" or no subscript (as in the whole Chap. 2.1), photometric quantities have the subscript "v", where "e" stands for "energetic" and "v" for "visible". The most frequently used radiometric quantities are listed in Table 2.1.1 together with their symbols, defining equations, and units. The additional physical quantities applied in Table 2.1.1 are the time t, the element of solid angle $d\omega$, and the angle θ between the line of sight and the normal of the radiating or receiving surface with the area element dA, see Fig.2.1.1.

In the case that the quantities are functions of wavelength their designations must be preceded by the adjective "spectral". For example, the symbol for spectral radiance is $L(\lambda)$. This has to be well distinguished from the convention for the spectral concentration of a quantity, which is also preceded by the adjective "spectral". In that case, however, the symbol has the subscript λ, i.e. $dL/d\lambda = L_\lambda$.

Table 2.1.1. Radiometric quantities, their defining equations and units.

Quantity	Symbol	Defining equation	Unit
Radiant energy	Q		J
Radiant power	Φ	$\Phi = \mathrm{d}Q/\mathrm{d}t$	W
Radiant excitance	M	$M = \mathrm{d}\Phi/\mathrm{d}A$	W m^{-2}
Irradiance	E	$E = \mathrm{d}\Phi/\mathrm{d}A$	W m^{-2}
Radiant intensity	I	$I = \mathrm{d}\Phi/\mathrm{d}\omega$	W sr^{-1}
Radiance	L	$L = \mathrm{d}^2\Phi/(\cos\theta\,\mathrm{d}A\,\mathrm{d}\omega)$	W m^{-2} sr^{-1}

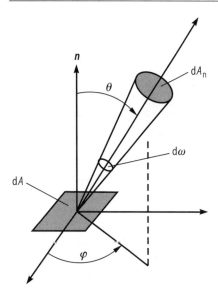

Fig. 2.1.1. Geometry for definition of the radiance.

To explain the defining equations given in Table 2.1.1 a radiation source of finite extent is considered. If we surround the radiation source with a closed surface and calculate the radiant energy Q penetrating the surface per unit time we get the total radiant power Φ emitted by the source. For clarity, the above mentioned symbols for the spectral properties of the radiation are omitted in this chapter. The radiant power per unit area of the radiation source associated with the emission into the hemispheric space above $\mathrm{d}A$ is defined as the radiant excitance M. At this point it is appropriate to introduce the radiation incident from all directions in the hemispheric space above the surface of a detector. The irradiance E is defined as the radiant power incident on a surface per unit area of the surface. The irradiance represents also the energy which propagates per unit time through the unit area perpendicular to the direction of energy transport. This is known as the density of energy flow identical to the magnitude of the Poynting vector averaged over time.

Coming back to the source-based radiometric quantities we consider now the radiant power proceeding from a point source per unit solid angle $\mathrm{d}\omega$ in a specified direction. The corresponding quantity appropriate especially for nearly point-shaped sources is denoted as radiant intensity I. If we generalize and consider again a source of finite extent the directional nature of radiation has to be taken into account accurately. From Fig. 2.1.1 we formally define as radiance L the radiant power emitted in the (θ, φ) direction, per unit area of the surface normal to this direction and per unit solid angle. Note that the area $\mathrm{d}A_\mathrm{n}$ used to define the radiance is the component of $\mathrm{d}A$ perpendicular to the direction of the radiation. This projected area is equal to $\cos\theta\,\mathrm{d}A$ and in effect, this is how $\mathrm{d}A$ would appear to an observer situated on the surface in the (θ, φ) direction.

Although the directional distribution of surface emission varies according to the nature of the surface, there is a special case which provides a reasonable approximation for many surfaces. For an isotropically diffuse emitter the radiance is independent of direction:

$$L(\theta, \varphi) = L . \tag{2.1.1}$$

Such an emitter is denoted as a lambertian radiator which emits in accordance with Lambert's cosine law:

$$I(\theta) = I(0) \cos \theta . \tag{2.1.2}$$

The radiant intensity of a perfectly diffuse surface element in any direction varies as the cosine of the angle between that direction and the normal to the surface element. It is noted that this law is consistent with the definitions of radiance and radiant intensity given in Table 2.1.1. It may be helpful to derive the relationship between radiance and radiant excitance for a lambertian radiator. The radiant excitance into the hemispheric space above $\mathrm{d}A$ is calculated from the radiance by integration over the solid angle $\mathrm{d}\omega = \sin\theta\, \mathrm{d}\theta\, \mathrm{d}\varphi$:

$$M = \frac{\mathrm{d}\Phi}{\mathrm{d}A} = \int_0^{2\pi} \int_0^{\pi/2} L(\theta, \varphi) \cos\theta \sin\theta\, \mathrm{d}\theta\, \mathrm{d}\varphi . \tag{2.1.3}$$

By removing $L(\theta, \varphi)$ from the integrand according to (2.1.1) and performing the integration we get

$$M = \pi L . \tag{2.1.4}$$

Note that the constant appearing in the above expression is π, not 2π, and has the unit steradian (sr).

2.1.3 Radiometric standards

2.1.3.1 Primary standards

Depending on the application primary source and primary detector standards are used to establish radiometric scales. Black-body radiators of known temperature with calculable spectral radiance are operated as primary source standards at temperatures up to about 3200 K [96Sap]. Due to the steep decrease of their Planckian radiation spectrum in the UV spectral range radiometry with black-body radiators is limited to wavelengths above 200 nm. In comparison with a black-body radiator, the maximum of the synchrotron radiation spectrum emitted by an electron storage ring is shifted to shorter wavelengths by several orders of magnitude [96Wen]. In a storage ring electrons move with nearly the velocity of light along a circular trajectory and emit a calculable radiant power through an aperture stop situated near the orbital plane. Radiometry can thus be extended into the X-ray region up to photon energies of 100 keV.

Electrical-substitution thermal detectors operated at ambient temperature have been the most frequently used primary detector standards. However, their performance is limited by the thermal properties of materials at room temperature resulting in complicated corrections that have to be applied. Hence, their uncertainties remain near 0.1 % to 0.3 % [89Fro, 79Wil]. Cryogenic radiometers have been developed to satisfy the increasing demands for more accurate detector standards from users especially in new and expanding fields of optical fibers, laser technology, and space science. Today, these instruments with absorption cavities at nearly the temperature of liquid helium

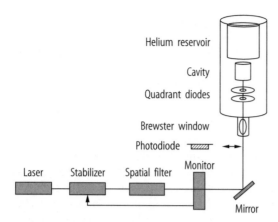

Fig. 2.1.2. Cryogenic radiometer for the calibration of photodiodes. The stabilized laser beam enters the cryostat via a Brewster window and is aligned by quadrant photodiodes. In the cavity the laser radiation is absorbed and electrically substituted.

are the most accurate among all primary standards, with relative uncertainties of less than 0.01 % [85Qui, 96Fox]. The principle of operation of both the cryogenic radiometers and the instruments at ambient temperature is that a thermometer measures the temperature rise of an absorption cavity, relative to a constant-temperature heat sink, during radiant and electrical heating cycles. By adjusting the electrical power so that the absorption cavity temperature rise is the same for both types of heating, the radiant power can be equated to the easily measured quantity of electrical power. For cryogenic radiometers the corrections due to the limited absorptance of the cavity, the lead heating of electrical connections, the radiative heat loss, and the background radiation can be made sufficiently small to reach very accurate equivalence of optical and electrical heating [96Fox]. Today, high-precision calibrations of laser radiometry secondary standards are mostly traceable to cryogenic radiometers. In Fig. 2.1.2 a typical experimental arrangement for the calibration of transfer photodiodes is shown [93Fu].

2.1.3.2 Secondary standards

Secondary standards serve to disseminate a metrologic scale or quantity to the user in science and industry. In this section, first, the common detectors used in the secondary standards for laser radiometry are shortly described, and second, some examples for laser radiometers and calorimeters are given. The detection principle of the secondary standards is usually thermal or photoelectric. The thermal detectors have the remarkable advantage of a flat spectral responsivity function which makes the calibration for different laser wavelengths not necessary or at least easier compared to that of photoelectric detectors. Among the thermal detectors we distinguish between thermopile detectors, bolometric and pyroelectric detectors.

A thermopile consists of a number of thermocouples in series to provide a thermoelectric voltage proportional to the temperature difference between the receiver and its thermal environment. Its optimization in detector applications has received considerable attention [68Smi, 58Sch, 70Ste]. At this point the term responsivity s is introduced which is the ratio of the detector output to the detector input. Whereas the detector input is a radiometric quantity, the detector output is usually an electrical quantity, for example current, voltage, or change in resistance. In order to optimize the responsivity of a thermopile one has to maximize the Seebeck coefficient of the two materials used for each thermocouple, the thermal resistance between the receiver and the environment, and the absorptance of the surface. The materials used for thermocouples are either metals, alloys, or semiconductors, for examples see [89Hen].

A bolometer is a temperature transducer based on the change of electrical resistance with temperature. The important quantity is the temperature difference between the receiver and its

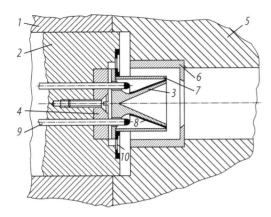

Fig. 2.1.3. Cross section of a cone-shaped laser radiometer. *3*: blackened cone, *6*: aperture, *7*: heat protection tube, *8*: electrical heater, *9*: electrical connections, *10*: thermopile; *1, 2, 4, 5*: parts of the heat sink.

thermal environment. Therefore one resistance element is needed to measure the temperature of the receiver and one to measure that of the thermal environment. AC and DC bridge techniques are applied for the comparison, the most common employing Wheatstone bridge configurations. The second resistance element should be physically close to the radiation-measuring element to compensate for convective disturbances, pressure fluctuations, changes in temperature of the housing, and instabilities in the bridge supply. The resistors are preferably made of metal wires or films of nickel, platinum, or gold [65Ble]. Thermistors are also used which have a larger temperature coefficient of the resistance. At lower operation temperatures the signal-to-noise ratio of bolometers can be increased considerably [82Mat, 87McD].

Pyroelectric detectors produce a current proportional to the rate of temperature change. The detection mechanism is based on the temperature dependence of the electrical polarization in ferroelectric crystals. Since pyroelectric detectors respond to modulated radiant power only, their use in laser radiometers for measuring cw radiation requires chopping of the incident beam. This can provide considerable drift immunity and allows for the use of drift-free AC amplification techniques [70Put, 75Tif].

Beside the thermal detectors also photoelectric devices or quantum detectors are used in laser radiometry. Photoelectric detectors for laser radiometric applications are either photoconductors or photodiodes. In a photoconductor made of a thin film of a semiconductor material the incident radiation generates additional carriers. These intrinsic band-to-band transitions or extrinsic transitions involving forbidden-gap energy levels result in an increase of conductivity [81Sze]. For sensitive infrared detection, the photoconductor must be cooled in order to reduce thermal ionization of the energy levels. In photodiodes the carriers are mainly generated in the depletion layer of the diode junction. The electron-hole pairs separated by an internal or external electric field recombine by driving an external current. Photodiodes are operated in two different modes: In the photovoltaic mode no bias voltage is applied and the photodiode can be considered as current source. In contrast, in the reverse-bias mode the photocurrent generates a voltage drop at an external load resistance which is used as measuring quantity. The reverse-bias mode is preferred for the detection of pulsed laser radiation.

A practical example of a radiometer for *cw laser radiation* is shown in Fig. 2.1.3. It measures radiant power in the range from 1 mW to 10 W, whereas the lower limit is set by detector and amplifier noise and the upper by the load limit of the electrical heater [89Moe]. The radiation absorber is a polished hollow cone electro-plated with a nearly specular reflecting black nickel layer. The temperature difference between the absorber cone and the heat sink is measured by a thermopile. The electric heater for moderate-accuracy in-situ calibrations of the instrument is wound around the cone. Another design of a thermopile-type radiometer with an integral alignment module can be found in [88Ino]. Further similar systems are described in [77Gun, 91Rad]. A commercial version of a laser radiometer based on a pyroelectric lithium tantalate crystal is described in [89Hen]. For higher radiant power levels of up to 1 kW cavity absorbers cooled by a surrounding jacket of flowing water are employed. The difference in temperature between the outflowing and inflowing water is measured and serves as quantity for the absorbed laser radiant power [96Bra]. A special design of the surface geometry of the cavity reduces the irradiance of the laser beam, thus improving the protection from damaging the surface.

The preferred instruments for *pulsed laser radiation* are thermally absorbing devices such as calorimeters. The receiver element is often a glass-disk, where the radiation is absorbed in the volume instead of on the surface. The absorptance exhibits an excellent stability under chemical and mechanical stress. This type of calorimeter is described in [70Edw, 74Gun]. The radiative load can be reduced by using glass with a low absorption coefficient which increases the length of the absorption path. On the other hand the heat capacity increases linearly with the thickness of the glass-disk which, in conjunction with the poor thermal conductivity of glass, results in long response and cooling times of these detectors. The radiometric scale for laser radiant energy is usually derived from the scale for cw laser radiant power. In [91Moe] a fast electromechanical shutter is used to produce pulses of known laser radiant energy of up to 5 J. The influence of the pulse duration has to be corrected in the calibration procedure. A laser energy meter not depending on a cw laser radiant power scale is described in [90Yua]. In this instrument the light pressure of the laser beam sensed by two mirrors is converted by a moving coil to an electrical signal. The main advantages of this system are fast response and no interruption of the laser beam. The device has been investigated for single laser pulses of radiant energies between 10 mJ and 6 J. Another method not interrupting the laser beam is the photoacoustic calorimetry [86Kim]. There, the radiant energy incident upon a mirror is absorbed at the mirror surface. The absorbed energy generates elastic strain waves which propagate through the mirror substrate. The strain waves eventually pass through a piezoelectric transducer attached to the back of the mirror substrate. The voltage of the piezoelectric crystal gives a direct indication of the amount of energy absorbed at the mirror surface. Since a priori the absorptance of the mirror is not known the instrument has to be calibrated against a standard energy meter.

2.1.4 Outlook – State of the art and trends

Although optical radiometry has been developed for 100 years, measurements of the various radiometric quantities only recently have achieved the required small uncertainties. Today the most accurate detector-based primary radiometric standard is the electrically calibrated cryogenic radiometer. In this instrument the radiant power of – preferably – a laser beam is measured by substituting the absorbed optical power of the laser beam by the electrical power of a heating system. Cryogenic radiometers operate at liquid helium temperatures and have a measurement uncertainty of a few parts in 10^4, a significant improvement over earlier room-temperature radiometers.

Accurate characterization of laser sources is crucial to the effective development and use of industrial technologies such as light-wave telecommunications, laser-based medical instrumentation, materials processing, photolithography, data storage, and laser safety equipment. Traceable measurement standards are essential both for users to have confidence in their measurements and to support quality assurance in the manufacture of lasers and laser systems. Because lasers present a potential safety hazard, it is also important to have measurement standards to satisfy nationally and internationally agreed safety limits. The traceability for laser radiometric measurements in Germany is maintained by the Physikalisch-Technische Bundesanstalt. It meets the requirements for calibration and testing laboratories, certification and accreditation bodies defined in the ISO/IEC Guide 17025 and the DIN/EN 45000 and DIN/EN/ISO 9000 series of standards, see http://www.ptb.de/en/org/q/q3/q33/_index.htm.

References for 2.1

58Sch	Schley, U., Hoffmann, F.: Optik (Stuttgart) **15** (1958) 358.
65Ble	Blevin, W.R., Brown, W.J: J. Sci. Instrum. **42** (1965) 19.
68Smi	Smith, R.A., Jones, F.E., Chasmar, R.P.: The detection and measurement of infrared radiation, London and New York: Oxford University Press, 1968.
70Edw	Edwards, J.G.: J. Phys. E: Sci. Instrum. **3** (1970) 452.
70Put	Putley, E.H.: Semiconductors and semimetals, Vol. 5, New York: Academic Press, 1970, p. 259.
70Ste	Stevens, N.B.: Semiconductors and semimetals, Vol. 5, New York: Academic Press, 1970, p. 287.
74Gun	Gunn, S.R.: Rev. Sci. Instrum. **45** (1974) 936.
75Tif	Tiffany, W.B.: Proc. SPIE (Int. Soc. Opt. Eng.) **62** (1975) 153.
77Gun	Gunn, S.R., Rupert, V.: Rev. Sci. Instrum. **48** (1977) 1375.
79Wil	Willson, R.C.: Appl. Opt. **18** (1979) 179.
81Sze	Sze, S.M.: Physics of semiconductor devices, New York: Wiley, 1981, p. 743.
82Mat	Mather, J.C.: Appl. Opt. **21** (1982) 1125.
85Qui	Quinn, T.J., Martin, J.E.: Philos. Trans. R. Soc. (London) A **316** (1985) 85.
86Kim	Kimura, W.D., Ford, D.H.: Rev. Sci. Instrum. **57** (1986) 2754.
87McD	McDonald, D.G.: Appl. Phys. Lett. **50** (1987) 775.
88Ino	Inoue, T., Endo, M., Yokoshima, I., Kawahara, K.: Rev. Sci. Instrum. **59** (1988) 2384.
89Fro	Fröhlich, C.: Inst. Phys. Conf. Ser. **92** (1989) 73.
89Hen	Hengstberger, F.: Absolute radiometry, San Diego: Academic Press, 1989.
89Moe	Möstl, K.: Inst. Phys. Conf. Ser. **92** (1989) 11.
90Yua	Yuan, Y.: Rev. Sci. Instrum. **61** (1990) 1743.
91Moe	Möstl, K., Brandt, F.: Metrologia **28** (1991) 121.
91Rad	Radak, Bo.B., Radak, Br.B.: Rev. Sci. Instrum. **62** (1991) 318.
93Fu	Fu Lei, Fischer, J.: Metrologia **30** (1993) 297–303.
96Bra	Brandt, F., Möstl, K.: Laser in Forschung und Technik, Berlin: Springer-Verlag, 1996, p. 730.
96Fox	Fox, N.P.: Metrologia **32** (1995/96) 535–543.
96Sap	Sapritsky, V.I.: Metrologia **32** (1995/96) 411–417.
96Wen	Wende, B.: Metrologia **32** (1995/96) 419–424.

2.2 Beam characterization

B. EPPICH

2.2.1 Introduction

The success of almost any laser application depends mainly on the power density distributions in a certain area of the laser beam, usually the focal region. It is the aim of laser beam characterization to describe and predict the profiles a beam takes on under free-space propagation or behind optical systems.

The attributes of a power density distribution in a plane transverse to the direction of propagation can be divided into *size* and *shape*. Under free-space propagation the size of the power density profile is always changing with the distance from the source, whereas the shape of the profile may vary or not. Examples for shape-invariant laser beams are the well-known Gaussian, Laguerre-Gaussian, Hermite–Gaussian, and Gauss-Schell model beams.

A complete characterization of laser beams would allow the prediction of power density distributions, including size and shape, behind arbitrary optical systems as far as they are sufficiently known. Admittedly for such detailed characterization a huge amount of data and sophisticated measurement procedures are necessary. But for many applications the knowledge and prediction of the transverse extent of the laser beam profile might be sufficient. Restriction to nearly aberration-free optical systems then enables beam characterization by only ten or less parameters.

In the following the validity of the paraxial approximation will be presumed. In practical this means that the full divergence angle of the beam should not exceed 30 degrees. Furthermore, any polarization effects are neglected. Beam characterization methods based on the considerations presented in this chapter have recently become an international standard, published as ISO 11146 [99ISO].

2.2.2 The Wigner distribution

A complete description of partially coherent radiation fields (within the restrictions stated above) can be given by a two-point-correlation integral of the field in a transverse plane at location z [99Bor]:

$$\tilde{\Gamma}(\boldsymbol{r}_1, \boldsymbol{r}_2, z, \tau) = \frac{1}{T} \int_{t_0}^{t_0+T} E^*(\boldsymbol{r}_1, z, t)\, E(\boldsymbol{r}_2, z, t+\tau)\, \mathrm{d}t\,, \qquad (2.2.1)$$

where $E(\boldsymbol{r}, z, t)$ is the electrical field, z the coordinate along the direction of propagation, $\boldsymbol{r} = (x, y)^{\mathrm{T}}$ a transverse spatial vector (see Fig. 2.2.1), and T the integration time which shall be large enough to ensure that the integration results are independent of the starting time t_0. The temporal Fourier transform of this correlation integral is known as the cross-spectral density or the (mutual) power spectrum:

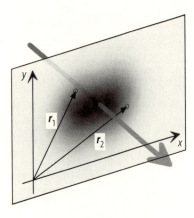

Fig. 2.2.1. Spatial coordinates r_1 and r_2 of a pair of points in a plane transverse to the direction of propagation.

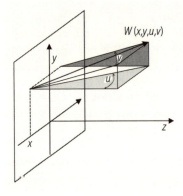

Fig. 2.2.2. The phase space coordinates of the Wigner distribution. x and y are spatial transverse coordinates, u and v are the corresponding angular coordinates.

$$\Gamma(\boldsymbol{r}_1, \boldsymbol{r}_2, z, \omega) = \int \tilde{\Gamma}(\boldsymbol{r}_1, \boldsymbol{r}_2, z, \tau) \, e^{i\omega\tau} \, d\tau \, . \tag{2.2.2}$$

Since laser beams in general can be considered as quasi-monochromatic, the frequency dependency will be dropped in the following:

$$\Gamma(\boldsymbol{r}_1, \boldsymbol{r}_2, z, \omega_0) \to \Gamma(\boldsymbol{r}_1, \boldsymbol{r}_2, z) \, . \tag{2.2.3}$$

From the cross-spectral density in a transverse plane at location z the power density in that plane can easily be obtained by

$$I(\boldsymbol{r}, z) = \Gamma(\boldsymbol{r}, \boldsymbol{r}, z) \, . \tag{2.2.4}$$

Given the cross-spectral density at an entry plane the further propagation through arbitrary, but well-defined optical systems can be calculated by several methods and hence the power density distribution in the output plane of the systems predicted [99Bor].

The Wigner distribution $W(\boldsymbol{r}, \boldsymbol{q}, z)$ of partially coherent beams is defined as the Fourier transform of the cross spectral density with respect to the separation vector $\boldsymbol{s}$ [78Bas]:

$$W(\boldsymbol{r}, \boldsymbol{q}, z) = \int \Gamma\left(\boldsymbol{r} + \tfrac{1}{2}\boldsymbol{s}, \boldsymbol{r} - \tfrac{1}{2}\boldsymbol{s}, z\right) e^{-ik\boldsymbol{q}\boldsymbol{s}} \, d\boldsymbol{s} \, . \tag{2.2.5}$$

The Wigner distribution contains the same information as the cross-spectral density, but in a different, more descriptive manner. Considering $\boldsymbol{q} = (u, v)^{\mathsf{T}}$ as an angular vector with respect to the z-axis (Fig. 2.2.2), the Wigner distribution gives the part (amount) of the radiation power which passes the plane at z through the point $\boldsymbol{r}$ in the direction given by $\boldsymbol{q}$. Within this picture the Wigner distribution might be considered as a generalization of the geometric optical radiance, although this analogy is limited. E.g. the Wigner distribution may take on negative values.

The power density distribution in a transverse plane is obtained by integration over the angles of direction,

$$I(\mathbf{r}, z) = \int W(\mathbf{r}, \mathbf{q}, z) \, \mathrm{d}\mathbf{q} \,, \tag{2.2.6}$$

and the far-field power density distribution by integration over the spatial coordinates,

$$I_\mathrm{F}(\mathbf{q}) = \int W(\mathbf{r}, \mathbf{q}, z) \, \mathrm{d}\mathbf{r} \,. \tag{2.2.7}$$

The Wigner distribution represents the beam in a transverse plane at location z. As the beam propagates in free space or through an optical system the Wigner distribution changes. This is reflected in the z-dependency of the Wigner distribution in the equations above. In the following equations this z-dependency will be dropped wherever appropriate.

The propagation of the Wigner distribution through aberration-free first-order optical systems (combinations of parabolic elements and free-space propagation) is very similar to that of geometric-optical rays. Such rays are specified by their position $\mathbf{r}$ and direction $\mathbf{q}$. After propagation through an aberration-free optical system position and direction will change according to

$$\begin{pmatrix} \mathbf{r}_\mathrm{out} \\ \mathbf{q}_\mathrm{out} \end{pmatrix} = \mathsf{S} \cdot \begin{pmatrix} \mathbf{r}_\mathrm{in} \\ \mathbf{q}_\mathrm{in} \end{pmatrix} \,, \tag{2.2.8}$$

where S is a 4×4-matrix representing the optical system, the system matrix (see Chap. 3.1). Considering the Wigner distribution as a density distribution of geometric optical rays, its propagation law is given by ray tracing [78Bas]:

$$W_\mathrm{out}(\mathbf{r}_\mathrm{out}, \mathbf{q}_\mathrm{out}) = W_\mathrm{in}(\mathbf{r}_\mathrm{in}, \mathbf{q}_\mathrm{in}) \quad \text{with} \quad \begin{pmatrix} \mathbf{r}_\mathrm{in} \\ \mathbf{q}_\mathrm{in} \end{pmatrix} = \mathsf{S}^{-1} \cdot \begin{pmatrix} \mathbf{r}_\mathrm{out} \\ \mathbf{q}_\mathrm{out} \end{pmatrix} \,. \tag{2.2.9}$$

2.2.3 The second-order moments of the Wigner distribution

From the Wigner distribution smaller sets of data can be derived, which can be associated to certain physical properties of the beams. These sets of data are the so-called moments of the Wigner distribution [86Bas]:

$$\langle x^k y^\ell u^m v^n \rangle = \frac{\int W(x,y,u,v) \, x^k y^\ell u^m v^n \, \mathrm{d}x \, \mathrm{d}y \, \mathrm{d}u \, \mathrm{d}v}{\int W(x,y,u,v) \, \mathrm{d}x \, \mathrm{d}y \, \mathrm{d}u \, \mathrm{d}v} \quad \text{with} \quad k,\ell,m,n \geq 0 \,, \tag{2.2.10}$$

where

$$W(x,y,u,v) = W(\mathbf{r}, \mathbf{q}) \quad \text{with} \quad \mathbf{r} = (x,y)^\mathsf{T} \,, \quad \mathbf{q} = (u,v)^\mathsf{T} \,. \tag{2.2.11}$$

The order of the moments is defined by the sum of the exponents, $k + \ell + m + n$. There are four first-order moments, $\langle x \rangle$, $\langle y \rangle$, $\langle u \rangle$, and $\langle v \rangle$, which together specify position and direction of propagation of the beam profile's centroids within the given coordinate system.

The *centered* moments of the Wigner distribution are defined to be independent of the coordinate system:

$$\langle x^k y^\ell u^m v^n \rangle_\mathrm{c} = \\ \frac{\int W(x,y,u,v) \, (x - \langle x \rangle)^k (y - \langle y \rangle)^\ell (u - \langle u \rangle)^m (v - \langle v \rangle)^n \, \mathrm{d}x \, \mathrm{d}y \, \mathrm{d}u \, \mathrm{d}v}{\int W(x,y,u,v) \, \mathrm{d}x \, \mathrm{d}y \, \mathrm{d}u \, \mathrm{d}v} \,. \tag{2.2.12}$$

There are ten centered second-order moments, specified by $k + \ell + m + n = 2$. Three pure spatial moments, $\langle x^2 \rangle_c$, $\langle y^2 \rangle_c$, $\langle xy \rangle_c$, three pure angular moments, $\langle u^2 \rangle_c$, $\langle v^2 \rangle_c$, $\langle uv \rangle_c$, and four mixed moments, $\langle x u \rangle_c$, $\langle y v \rangle_c$, $\langle x v \rangle_c$, and $\langle y u \rangle_c$. The centered second-order moments are associated to the beam extents in the near and far field and to the propagation of beam widths as will be discussed in the next section.

Only the three pure spatial moments can directly be measured since they can be obtained from the power density distribution in the observation plane by

$$\langle x^k y^\ell \rangle_c = \frac{1}{P} \int I(x,y) \, (x - \langle x \rangle)^k \, (y - \langle y \rangle)^\ell \, \mathrm{d}x \, \mathrm{d}y \tag{2.2.13}$$

with

$$\langle x \rangle = \frac{1}{P} \int I(x,y) \, x \, \mathrm{d}x \, \mathrm{d}y , \tag{2.2.14}$$

$$\langle y \rangle = \frac{1}{P} \int I(x,y) \, y \, \mathrm{d}x \, \mathrm{d}y , \tag{2.2.15}$$

and

$$P = \int I(x,y) \, \mathrm{d}x \, \mathrm{d}y . \tag{2.2.16}$$

As the beam propagates through optical systems the Wigner distribution changes and consequently the moments change, too. A simple propagation law for the centered second-order moments through aberration-free optical systems can be derived from the propagation law of the Wigner distribution (2.2.9). Combining the ten moments in a symmetric 4×4-matrix, the variance matrix

$$\mathsf{P} = \begin{pmatrix} \langle x^2 \rangle_c & \langle xy \rangle_c & \langle xu \rangle_c & \langle xv \rangle_c \\ \langle xy \rangle_c & \langle y^2 \rangle_c & \langle yu \rangle_c & \langle yv \rangle_c \\ \langle xu \rangle_c & \langle yu \rangle_c & \langle u^2 \rangle_c & \langle uv \rangle_c \\ \langle xv \rangle_c & \langle yv \rangle_c & \langle uv \rangle_c & \langle v^2 \rangle_c \end{pmatrix} , \tag{2.2.17}$$

delivers the propagation law

$$\mathsf{P}_{\text{out}} = \mathsf{S} \cdot \mathsf{P}_{\text{in}} \cdot \mathsf{S}^\mathsf{T} , \tag{2.2.18}$$

where P_{in} and P_{out} are the variance matrices in the input and output planes of the optical system, respectively, and S is the system matrix.

2.2.4 The second-order moments and related physical properties

In this section the relations between the centered second-order moments and some more physical properties are discussed.

2.2.4.1 Near field

The three spatial-centered second-order moments are related to the spatial extent of the power density in the reference plane as can be derived from (2.2.13). For example, the centered second-order moments $\langle x^2 \rangle_c$, defined by

$$\langle x^2 \rangle_c = \frac{1}{P} \int I(x,y) \, (x - \langle x \rangle)^2 \, dx \, dy \,, \tag{2.2.19}$$

can be considered as the intensity-weighted average of the squared distances in x-direction of all points in the plane from the beam-profile center. Obviously, this quantity increases with increasing beam extent in x-direction. A beam width in x-direction can be defined as

$$d_x = 4\sqrt{\langle x^2 \rangle_c} \,. \tag{2.2.20}$$

The factor of 4 in this equation has been chosen by convention to adapt this beam-width definition to the former $1/e^2$-definition for the beam radius of Gaussian beams. For an aligned elliptical Gaussian beam profile,

$$I(x,y) \propto e^{-2\frac{x^2}{w_x^2}} \cdot e^{-2\frac{y^2}{w_y^2}} \,, \tag{2.2.21}$$

where w_x and w_y are the $1/e^2$-beam radii in x- and y-direction, respectively, the relation

$$d_x = 2\,w_x$$

holds. Similar, a beam width in y-direction can be defined as

$$d_y = 4\sqrt{\langle y^2 \rangle_c} \,. \tag{2.2.22}$$

The beam width along an arbitrary azimuthal direction enclosing an angle of α with the x-axis can be derived from a rotation of the coordinate system delivering

$$d_\alpha = 4\sqrt{\langle x^2 \rangle_c \cos^2 \alpha + 2 \langle xy \rangle_c \sin \alpha \cos \alpha + \langle y^2 \rangle_c \sin^2 \alpha} \,. \tag{2.2.23}$$

In general, the beam width considered as a function of the azimuthal direction α has unique maximum and minimum. The related directions are orthogonal to each other and define the principal axes of the beam. The signed angle between the x-axis and that principal axis which is closer to the x-axis is given by

$$\varphi = \frac{1}{2} \operatorname{atan} \left(\frac{2 \langle xy \rangle_c}{\langle x^2 \rangle_c - \langle y^2 \rangle_c} \right) \,. \tag{2.2.24}$$

The beam width along that principal axis which is closer to the x-axis is determined by

$$d'_x = 2\sqrt{2} \left\{ (\langle x^2 \rangle_c + \langle y^2 \rangle_c) + \varepsilon \left[(\langle x^2 \rangle_c - \langle y^2 \rangle_c)^2 + 4 \langle xy \rangle_c^2 \right]^{\frac{1}{2}} \right\}^{\frac{1}{2}} \tag{2.2.25}$$

with

$$\varepsilon = \operatorname{sgn} \left(\langle x^2 \rangle_c - \langle y^2 \rangle_c \right) \,. \tag{2.2.26}$$

Correspondingly, the beam width along the principal axis closer to the y-axis is given by

$$d'_y = 2\sqrt{2} \left\{ (\langle x^2 \rangle_c + \langle y^2 \rangle_c) - \varepsilon \left[(\langle x^2 \rangle_c - \langle y^2 \rangle_c)^2 + 4 \langle xy \rangle_c^2 \right]^{\frac{1}{2}} \right\}^{\frac{1}{2}} \,. \tag{2.2.27}$$

Hence, the three spatial-centered second-order moments define the size and orientation of the so-called variance ellipse as the representation of a beam profile's extent (Fig. 2.2.3).

Beam profiles having approximately equal beam widths in both principal planes, $d'_x \approx d'_y$, may be considered as circular and a beam diameter may be defined by

$$d = 2\sqrt{2} \sqrt{\langle x^2 \rangle + \langle y^2 \rangle} \,. \tag{2.2.28}$$

Sometimes this is an useful definition even for non-circular beam profiles, denoted then as "generalized beam diameter".

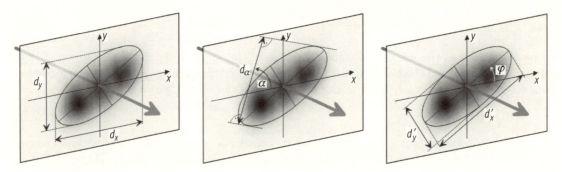

Fig. 2.2.3. Widths and variance ellipse of a power density profile. Left: widths d_x and d_y along the coordinate axes, middle: width d_α along an arbitrary direction, right: widths d'_x and d'_y along the principal axes.

2.2.4.2 Far field

The three angular-centered second-order moments are related to the beam-profile extent in the far field, far away from the reference plane, or in the focal plane of a focusing lens. From the propagation law of the second-order moments, (2.2.18), the dependency of the spatial moments on the propagation distance z from the reference plane can be derived:

$$\begin{aligned}
\langle x^2 \rangle_{\rm c}(z) &= \langle x^2 \rangle_{\rm c,0} + 2z\, \langle xu \rangle_{\rm c,0} + z^2 \langle u^2 \rangle_{\rm c,0}\,, \\
\langle xy \rangle_{\rm c}(z) &= \langle xy \rangle_{\rm c,0} + z\left(\langle xv \rangle_{\rm c,0} + \langle yu \rangle_{\rm c,0}\right) + z^2 \langle uv \rangle_{\rm c,0}\,, \\
\langle y^2 \rangle_{\rm c}(z) &= \langle y^2 \rangle_{\rm c,0} + 2z\, \langle yv \rangle_{\rm c,0} + z^2 \langle v^2 \rangle_{\rm c,0}\,.
\end{aligned} \qquad (2.2.29)$$

For large distances z the spatial moments depend only on the angular moments in the reference plane:

$$\begin{aligned}
\langle x^2 \rangle_{\rm c}(z) &\approx z^2 \langle u^2 \rangle\,, \\
\langle xy \rangle_{\rm c}(z) &\approx z^2 \langle uv \rangle\,, \\
\langle y^2 \rangle_{\rm c}(z) &\approx z^2 \langle v^2 \rangle\,.
\end{aligned} \qquad (2.2.30)$$

The azimuthal angle $\varphi_{\rm F}$ of that principal axis in the far field, which is closer to the x-axis is then obtained by

$$\varphi_{\rm F} = \lim_{z \to \infty} \frac{1}{2} \operatorname{atan}\left(\frac{2\langle xy \rangle_{\rm c}(z)}{\langle x^2 \rangle_{\rm c}(z) - \langle y^2 \rangle_{\rm c}(z)}\right) = \frac{1}{2} \operatorname{atan}\left(\frac{2\langle uv \rangle_{\rm c}}{\langle u^2 \rangle_{\rm c} - \langle v^2 \rangle_{\rm c}}\right)\,, \qquad (2.2.31)$$

and the (full) divergence angles along the principal axes of the far field might be defined as

$$\theta'_x = \lim_{z \to \infty} \frac{d'_x(z)}{z} = 2\sqrt{2}\left\{\left(\langle u^2 \rangle_{\rm c} + \langle v^2 \rangle_{\rm c}\right) + \eta\left[\left(\langle u^2 \rangle_{\rm c} - \langle v^2 \rangle_{\rm c}\right)^2 + 4\langle uv \rangle_{\rm c}^2\right]^{\frac{1}{2}}\right\}^{\frac{1}{2}}\,, \qquad (2.2.32)$$

$$\theta'_y = \lim_{z \to \infty} \frac{d'_y(z)}{z} = 2\sqrt{2}\left\{\left(\langle u^2 \rangle_{\rm c} + \langle v^2 \rangle_{\rm c}\right) - \eta\left[\left(\langle u^2 \rangle_{\rm c} - \langle v^2 \rangle_{\rm c}\right)^2 + 4\langle uv \rangle_{\rm c}^2\right]^{\frac{1}{2}}\right\}^{\frac{1}{2}} \qquad (2.2.33)$$

with

$$\eta = \operatorname{sgn}\left(\langle x^2 \rangle_{\rm c} - \langle y^2 \rangle_{\rm c}\right)\,. \qquad (2.2.34)$$

The generalized beam divergence angle might be defined as

$$\theta = 2\sqrt{2}\sqrt{\langle u^2 \rangle_{\rm c} + \langle v^2 \rangle_{\rm c}}\,. \qquad (2.2.35)$$

The azimuthal orientation of the far field may differ from the orientation of the near field.

2.2.4.3 Phase paraboloid and twist

The four mixed moments $\langle xu \rangle_c$, $\langle xv \rangle_c$, $\langle yu \rangle_c$, and $\langle yv \rangle_c$ are closely related to the phase properties of the beam in the reference plane. Together with the three spatial moments they determine the radii of curvature and azimuthal orientation of the best-fitting phase paraboloid. Although the phase properties of partially coherent beams might be quite complicated, it is always possible to find a best-fitting phase function being quadratic (bilinear) in x and y:

$$\Phi(x,y) = k \left(a x^2 + 2 b x y + c y^2 \right) . \tag{2.2.36}$$

The best-fitting parameters a, b, c are defined by minimizing the generalized divergence angle, (2.2.35), if a phase function according to (2.2.36) would be subtracted from the actual phase distribution in the reference plane (e.g. by introducing a cylindrical lens) resulting in

$$a = \frac{\langle y^2 \rangle \langle xu \rangle \left(\langle x^2 \rangle + \langle y^2 \rangle \right) - \langle xy \rangle^2 \left(\langle xu \rangle - \langle yv \rangle \right) - \langle xy \rangle \langle y^2 \rangle \left(\langle xv \rangle + \langle yu \rangle \right)}{\left(\langle x^2 \rangle + \langle y^2 \rangle \right) \left(\langle x^2 \rangle \langle y^2 \rangle - \langle xy \rangle^2 \right)} , \tag{2.2.37}$$

$$b = \frac{\langle x^2 \rangle \langle y^2 \rangle \left(\langle xv \rangle + \langle yu \rangle \right) - \langle xy \rangle \left(\langle x^2 \rangle \langle yv \rangle + \langle y^2 \rangle \langle xu \rangle \right)}{\left(\langle x^2 \rangle + \langle y^2 \rangle \right) \left(\langle x^2 \rangle \langle y^2 \rangle - \langle xy \rangle^2 \right)} , \tag{2.2.38}$$

$$c = \frac{\langle x^2 \rangle \langle yv \rangle \left(\langle x^2 \rangle + \langle y^2 \rangle \right) + \langle xy \rangle^2 \left(\langle xu \rangle - \langle yv \rangle \right) - \langle xy \rangle \langle x^2 \rangle \left(\langle xv \rangle + \langle yu \rangle \right)}{\left(\langle x^2 \rangle + \langle y^2 \rangle \right) \left(\langle x^2 \rangle \langle y^2 \rangle - \langle xy \rangle^2 \right)} . \tag{2.2.39}$$

A phase distribution as given in (2.2.36) can be considered as a rotated phase paraboloid, with

$$\varphi_P = \frac{1}{2} \operatorname{atan} \left(\frac{2b}{a-c} \right) \tag{2.2.40}$$

as the signed angle between the x-axis and that principal axis of the phase paraboloid, which is closer to the x-axis, and with

$$R'_x = \frac{2}{(a+c) + \mu \sqrt{(a-c)^2 + 4 b^2}} \tag{2.2.41}$$

and

$$R'_y = \frac{2}{(a+c) - \mu \sqrt{(a-c)^2 + 4 b^2}} \tag{2.2.42}$$

with

$$\mu = \operatorname{sgn}(a-c) \tag{2.2.43}$$

as the radii of curvature along that principal axis of the phase paraboloid, which is closer to the x- and y-axis, respectively. The radii of curvature R'_x and R'_x independently may be positive or negative or infinite, the later indicating a plane phase front along that azimuthal direction. The azimuthal orientation of the phase paraboloid's principal axes may differ from the orientation of the near field and/or far field.

If the radii of phase curvature along both principal axes are approximately equal, $R'_x \approx R'_y$, a generalized phase curvature of the best-fitting rotational symmetric phase paraboloid is defined by

$$R = \frac{\langle x^2 \rangle_c + \langle y^2 \rangle_c}{\langle xu \rangle_c + \langle yv \rangle_c} . \tag{2.2.44}$$

Another phase-related parameter is the so-called twist, defined as

$$t_{\mathrm{w}} = \langle xv \rangle - \langle yu \rangle \,. \tag{2.2.45}$$

The twist parameter is proportional to the orbital angular momentum transferred by the beam [93Sim].

2.2.4.4 Invariants

From the ten centered second-order moments two basic quantities can be derived, that are invariant under propagation through aberration-free first-order optics [03Nem].

The effective beam propagation ratio is defined as

$$M_{\mathrm{eff}}^2 = \frac{4\pi}{\lambda} \left(\det (\mathrm{P}) \right)^{\frac{1}{4}} \geq 1 \tag{2.2.46}$$

and can be considered as a measure of the focusability of a beam. The lower limit holds only for coherent Gaussian beams.

The intrinsic astigmatism a, given by

$$a = \frac{8\pi^2}{\lambda^2} \Big[\left(\langle x^2 \rangle_{\mathrm{c}} \langle u^2 \rangle_{\mathrm{c}} - \langle xu \rangle_{\mathrm{c}}^2 \right) + \left(\langle y^2 \rangle_{\mathrm{c}} \langle v^2 \rangle_{\mathrm{c}} - \langle yv \rangle_{\mathrm{c}}^2 \right)$$
$$+ 2 \left(\langle xy \rangle_{\mathrm{c}} \langle uv \rangle_{\mathrm{c}} - \langle xv \rangle_{\mathrm{c}} \langle yu \rangle_{\mathrm{c}} \right) \Big] - \left(M_{\mathrm{eff}}^2 \right)^2 \geq 0 \,, \tag{2.2.47}$$

is related to the visible and hidden astigmatism of the beam (see below).

2.2.4.5 Propagation of beam widths and beam propagation ratios

Under free-space propagation any directional beam width d_α, as well as the generalized beam diameter d, obeys an hyperbolic propagation law:

$$d_\alpha (z) = d_{0,\alpha} \sqrt{1 + \left(\frac{z - z_{0,\alpha}}{z_{\mathrm{R},\alpha}} \right)^2} = \sqrt{d_{0,\alpha}^2 + \theta_\alpha^2 \left(z - z_{0,\alpha} \right)^2} \,, \tag{2.2.48}$$

where $z_{0,\alpha}$ is the z-position of the smallest width, the waist position, $d_{0,\alpha}$ is the waist width, θ_α the divergence angle, and $z_{\mathrm{R},\alpha}$ the Rayleigh length, i.e. the distance from the waist position, where the width has grown by factor of $\sqrt{2}$. For the width along the x-direction, $\alpha = 0$, see Fig. 2.2.4, the parameters can be obtained by

$$z_0 = -\frac{\langle xu \rangle_{\mathrm{c}}}{\langle u^2 \rangle_{\mathrm{c}}} \,, \tag{2.2.49}$$

$$d_0 = 4 \sqrt{\langle x^2 \rangle_{\mathrm{c}} - \frac{\langle xu \rangle_{\mathrm{c}}^2}{\langle u^2 \rangle_{\mathrm{c}}}} \,, \tag{2.2.50}$$

and

$$z_{\mathrm{R}} = \sqrt{\frac{\langle x^2 \rangle_{\mathrm{c}}}{\langle u^2 \rangle_{\mathrm{c}}} - \frac{\langle xu \rangle_{\mathrm{c}}^2}{\langle u^2 \rangle_{\mathrm{c}}^2}} \,. \tag{2.2.51}$$

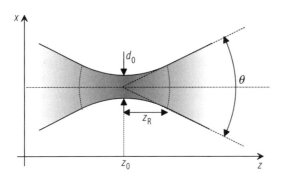

Fig. 2.2.4. Free-space propagation of beam widths with the beam waist position z_0, the beam waist width d_0, the Rayleigh length z_R, and the full divergence angle θ.

For other azimuthal directions α the same equations apply with the following substitutions:
$$\begin{aligned} \langle x^2 \rangle_c &\to \langle x^2 \rangle_c \cos^2 \alpha + 2 \langle xy \rangle_c \cos\alpha \sin\alpha + \langle y^2 \rangle_c \sin^2 \alpha \,, \\ \langle xu \rangle_c &\to \langle xu \rangle_c \cos^2 \alpha + 2 \left(\langle xv \rangle_c + \langle yu \rangle_c \right) \cos\alpha \sin\alpha + \langle yv \rangle_c \sin^2 \alpha \,, \\ \langle u^2 \rangle_c &\to \langle u^2 \rangle_c \cos^2 \alpha + 2 \langle uv \rangle_c \cos\alpha \sin\alpha + \langle v^2 \rangle_c \sin^2 \alpha \,. \end{aligned} \qquad (2.2.52)$$

For the generalized diameter d the propagation parameters are obtained by
$$z_0 = -\frac{\langle xu \rangle_c + \langle yv \rangle_c}{\langle u^2 \rangle_c + \langle v^2 \rangle_c}\,, \qquad (2.2.53)$$

$$d_0 = 2\sqrt{2} \sqrt{\left(\langle x^2 \rangle_c + \langle y^2 \rangle_c \right) - \frac{\left(\langle xu \rangle_c + \langle yv \rangle_c \right)^2}{\langle u^2 \rangle_c + \langle v^2 \rangle_c}}\,, \qquad (2.2.54)$$

and
$$z_R = \sqrt{\frac{\langle x^2 \rangle_c + \langle y^2 \rangle_c}{\langle u^2 \rangle_c + \langle v^2 \rangle_c} - \left(\frac{\langle xu \rangle_c + \langle yv \rangle_c}{\langle u^2 \rangle_c + \langle v^2 \rangle_c} \right)^2}\,. \qquad (2.2.55)$$

It should be noted that beam widths along the principal axes, d'_x and d'_y, do *not* obey the hyperbolic propagation law in the case of a general astigmatic beam with rotating variance ellipse (see next section).

The product of the (directional) beam waist diameter d, d_α and the corresponding far-field divergence angle θ, θ_α is called the beam parameter product. Due to diffraction the beam parameter product has a lower limit given by
$$d_0 \cdot \theta = \frac{d_0^2}{z_R} \geq 4\frac{\lambda}{\pi}\,, \qquad d_{0,\alpha} \cdot \theta_\alpha = \frac{d_{0,\alpha}^2}{z_{R,\alpha}} \geq 4\frac{\lambda}{\pi}\,. \qquad (2.2.56)$$

Normalization to this lower limit delivers the so-called beam parameter ratios
$$M^2 = \frac{\pi}{\lambda}\frac{d_0 \cdot \theta}{4}\,, \qquad M_\alpha^2 = \frac{\pi}{\lambda}\frac{d_{0,\alpha} \cdot \theta_\alpha}{4}\,. \qquad (2.2.57)$$

The beam parameter ratios M^2 and M_α^2 are invariant in stigmatic aberration-free first-order optical systems (combinations of perfect spherical lenses). In contrast to the effective beam parameter ratio M_{eff}^2, they may change under propagation through cylindrical lenses.

2.2.5 Beam classification

Lasers beams can be classified according to their propagation behavior. The classification is based on the discrimination between circular and non-circular power density profiles and the azimuthal orientation of the non-circular profiles. A beam profile is considered circular if the beam widths along both principal axes are approximately equal, or, in practice, if

$$\frac{\min\left(d'_x, d'_y\right)}{\max\left(d'_x, d'_y\right)} > 0.87 \; . \tag{2.2.58}$$

In this sense a homogeneous profile with square footprint is regarded circular, see Fig. 2.2.5.

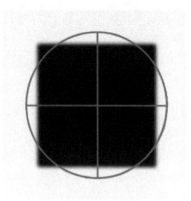

Fig. 2.2.5. Within the concept of second-order-moment beam characterization a square top-hat profile is considered circular: Its width is independent of the azimuthal direction.

2.2.5.1 Stigmatic beams

A laser beam is considered stigmatic if all its profiles under free-space propagation are circular and if all non-circular profiles behind an arbitrary cylindrical lens, inserted somewhere in the beam, have the same azimuthal orientation as the lens. The system matrix P_{st} of a perfectly stigmatic beam has only three independent parameters:

$$\mathsf{P}_{\mathrm{st}} = \begin{pmatrix} \langle x^2 \rangle_c & 0 & \langle xu \rangle_c & 0 \\ 0 & \langle x^2 \rangle_c & 0 & \langle xu \rangle_c \\ \langle xu \rangle_c & 0 & \langle u^2 \rangle_c & 0 \\ 0 & \langle xu \rangle_c & 0 & \langle u^2 \rangle_c \end{pmatrix} \; . \tag{2.2.59}$$

Physical parameters of a stigmatic beam are the beam diameter in the reference plane

$$d = 4\sqrt{\langle x^2 \rangle_c} \tag{2.2.60}$$

and the full divergence angle

$$\theta = 4\sqrt{\langle u^2 \rangle} \; . \tag{2.2.61}$$

Since the properties of a stigmatic beam are independent of the azimuthal direction, it has a unique waist position

$$z_0 = -\frac{\langle xu \rangle_c}{\langle u^2 \rangle_c} \tag{2.2.62}$$

with a waist diameter of

$$d_0 = 4\sqrt{\langle x^2\rangle_c - \frac{\langle xu\rangle_c^2}{\langle u^2\rangle_c}}\ . \tag{2.2.63}$$

The Rayleigh length z_R is the distance from the waist position where the diameter has grown by a factor of $\sqrt{2}$, given by

$$z_R = \sqrt{\frac{\langle x^2\rangle_c}{\langle u^2\rangle_c} - \frac{\langle xu\rangle_c^2}{\langle u^2\rangle_c^2}}\ . \tag{2.2.64}$$

Finally, the phase paraboloid is of rotational symmetry with the radius of curvature being

$$R = \frac{\langle x^2\rangle_c}{\langle xu\rangle_c}\ . \tag{2.2.65}$$

2.2.5.2 Simple astigmatic beams

A laser beam is classified as simple astigmatic if at least some of the power density profiles the beam takes on under free-space propagation are non-circular, but all non-circular profiles have the same azimuthal orientation. In practice, the orientations of two non-circular beam profiles are regarded as equal, if the azimuthal angles differ by less than 10 degrees. A simple astigmatic beam whose principal axes are parallel to the x- and y-axis is called aligned simple astigmatic. The variance matrix P_{asa} of a perfect aligned simple astigmatic beam has six independent parameters:

$$P_{asa} = \begin{pmatrix} \langle x^2\rangle_c & 0 & \langle xu\rangle_c & 0 \\ 0 & \langle y^2\rangle_c & 0 & \langle yv\rangle_c \\ \langle xu\rangle_c & 0 & \langle u^2\rangle_c & 0 \\ 0 & \langle yv\rangle_c & 0 & \langle v^2\rangle_c \end{pmatrix}\ . \tag{2.2.66}$$

All the physical parameters given for stigmatic beams can be assigned separately for each principal axis of a simple astigmatic beam. The diameters in x- and y-direction are

$$d_x = 4\sqrt{\langle x^2\rangle_c}\ ,\quad d_y = 4\sqrt{\langle y^2\rangle_c} \tag{2.2.67}$$

and the according full divergence angle

$$\theta_x = 4\sqrt{\langle u^2\rangle}\ ,\quad \theta_y = 4\sqrt{\langle v^2\rangle}\ . \tag{2.2.68}$$

Aligned simple astigmatic beams have in general two different waist positions for each principal axis:

$$z_{0,x} = -\frac{\langle xu\rangle_c}{\langle u^2\rangle_c}\ ,\quad z_{0,y} = -\frac{\langle yv\rangle_c}{\langle v^2\rangle_c} \tag{2.2.69}$$

with the associated waist diameters

$$d_{0,x} = 4\sqrt{\langle x^2\rangle_c - \frac{\langle xu\rangle_c^2}{\langle u^2\rangle_c}}\ ,\quad d_{0,y} = 4\sqrt{\langle y^2\rangle_c - \frac{\langle yv\rangle_c^2}{\langle v^2\rangle_c}}\ . \tag{2.2.70}$$

Similarly, two Rayleigh lengths are defined by

$$z_{R,x} = \sqrt{\frac{\langle x^2\rangle_c}{\langle u^2\rangle_c} - \frac{\langle xu\rangle_c^2}{\langle u^2\rangle_c^2}}\ ,\quad z_{R,y} = \sqrt{\frac{\langle y^2\rangle_c}{\langle v^2\rangle_c} - \frac{\langle yv\rangle_c^2}{\langle v^2\rangle_c^2}}\ , \tag{2.2.71}$$

and the radii of phase curvature are

$$R_x = \frac{\langle x^2 \rangle_c}{\langle xu \rangle_c}, \quad R_y = \frac{\langle y^2 \rangle_c}{\langle yv \rangle_c}. \tag{2.2.72}$$

The propagation laws for the beam diameters along both principal axes are:

$$d_x(z) = d_{0,x} \sqrt{1 + \left(\frac{z - z_{0,x}}{z_{R,x}}\right)^2} = \sqrt{d_{0,x}^2 + \theta_x^2 (z - z_{0,x})^2} \tag{2.2.73}$$

and

$$d_y(z) = d_{0,y} \sqrt{1 + \left(\frac{z - z_{0,y}}{z_{R,y}}\right)^2} = \sqrt{d_{0,y}^2 + \theta_y^2 (z - z_{0,y})^2}. \tag{2.2.74}$$

For non-aligned simple astigmatic beams similar relations hold.

2.2.5.3 General astigmatic beams

All other beams are classified as general astigmatic. Usually all ten second-order moments are necessary to describe a general astigmatic beam.

2.2.5.4 Pseudo-symmetric beams

Pseudo-symmetric beams are general astigmatic but "look like" stigmatic or simple astigmatic under free-space propagation. They possess an inner astigmatism which is hidden under free propagation and propagation through stigmatic (isotropic) optical systems (i.e. combinations of spherical lenses). Pseudo-symmetric beams differ from real stigmatic or simple astigmatic beams by a non-vanishing twist parameter, $t_w \neq 0$.

The variance matrix $\mathsf{P}_{\mathrm{pst}}$ of pseudo-stigmatic beams is therefore

$$\mathsf{P}_{\mathrm{pst}} = \begin{pmatrix} \langle x^2 \rangle_c & 0 & \langle xu \rangle_c & \frac{t}{2} \\ 0 & \langle x^2 \rangle_c & -\frac{t}{2} & \langle xu \rangle_c \\ \langle xu \rangle_c & -\frac{t}{2} & \langle u^2 \rangle_c & 0 \\ \frac{t}{2} & \langle xu \rangle_c & 0 & \langle u^2 \rangle_c \end{pmatrix}. \tag{2.2.75}$$

Under free-space propagation there is no difference between a real stigmatic beam, $t_w = 0$, and the corresponding pseudo-stigmatic one, $t_w \neq 0$, (2.2.29). The difference can be uncovered by inserting an arbitrary cylindrical lens somewhere in the beam path. The stigmatic beam is converted into a simple astigmatic beam with non-rotating variance ellipse while the pseudo-stigmatic one is turned into a general astigmatic beam with rotating variance ellipse. Figure 2.2.6 illustrates the different behaviors.

The variance matrix $\mathsf{P}_{\mathrm{psa}}$ of aligned pseudo-simple astigmatic beams is given by

$$\mathsf{P}_{\mathrm{psa}} = \begin{pmatrix} \langle x^2 \rangle_c & 0 & \langle xu \rangle_c & \frac{t}{2} \\ 0 & \langle y^2 \rangle_c & -\frac{t}{2} & \langle yv \rangle_c \\ \langle xu \rangle_c & -\frac{t}{2} & \langle u^2 \rangle_c & 0 \\ \frac{t}{2} & \langle yv \rangle_c & 0 & \langle v^2 \rangle_c \end{pmatrix}. \tag{2.2.76}$$

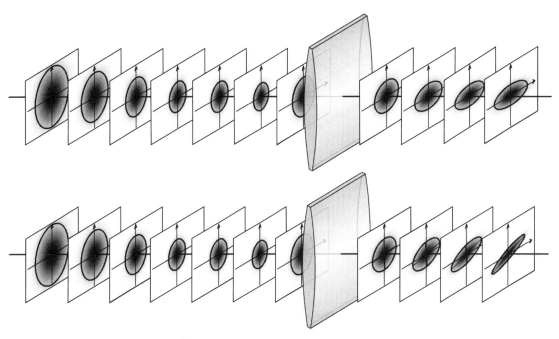

Fig. 2.2.6. Propagation of a stigmatic (top) and pseudo-stigmatic (bottom) laser beam. In free-space propagation both beams are indistinguishable. But a cylindrical lens transforms the stigmatic beam into a simple astigmatic one, whereas the pseudo-stigmatic beam becomes general astigmatic with rotating variance ellipse.

Again, under free-space propagation there is no difference between a real simple astigmatic beam, $t_w = 0$, and the corresponding pseudo-simple astigmatic one, $t_w \neq 0$, (2.2.29). Inserting an aligned cylindrical lens somewhere in the beam pass unveils the difference. The simple astigmatic beam keeps being simple astigmatic while the pseudo-simple astigmatic one is turned into a general astigmatic beam with rotating variance ellipse. Figure 2.2.7 illustrates the different behaviors.

2.2.5.5 Intrinsic astigmatism and beam conversion

Applying astigmatic (anisotropic) optical systems (including cylindrical lenses) may convert beams from one class to another. But only beams with vanishing intrinsic astigmatism a, (2.2.47), can be converted into stigmatic ones [94Mor]. In practice, beams with

$$\frac{a}{\left(M_{\text{eff}}^2\right)^2} < 0.039 \tag{2.2.77}$$

are considered intrinsic stigmatic, all others intrinsic astigmatic (the limit of 0.039 is a consequence of (2.2.58)). Intrinsic astigmatic beams can always be converted into pseudo-stigmatic or simple astigmatic ones.

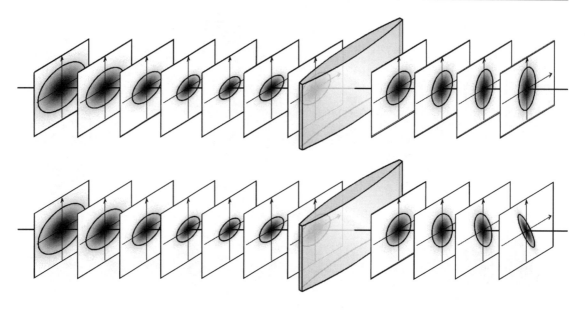

Fig. 2.2.7. Propagation of a simple astigmatic (top) and a pseudo-simple astigmatic (bottom) laser beam. In free-space propagation both beams are indistinguishable. But an aligned cylindrical lens transforms the simple astigmatic beam into a simple astigmatic one, whereas the pseudo-simple astigmatic beam becomes general astigmatic with rotating variance ellipse.

2.2.6 Measurement procedures

Only the three pure spatial moments out of the ten second-order moments are accessible for direct measurement. The other seven moments are retrieved indirectly based on the propagation law of the spatial moments (2.2.29).

The measurement method is based on the acquisition of a couple of power density profiles at different z-locations near the generalized beam waist, (2.2.53), e.g. by means of CCD cameras or similar devices (Fig. 2.2.8, left). From the measured profiles the spatial moments at each measurement plane are calculated. Fitting parabolas with three free parameters to the curve of each spatial moment delivers nine independent quantities: the moments $\langle x^2 \rangle_{c,0}$, $\langle xy \rangle_{c,0}$, $\langle y^2 \rangle_{c,0}$, $\langle xu \rangle_{c,0}$, $\langle yv \rangle_{c,0}$, $\langle u^2 \rangle_{c,0}$, $\langle uv \rangle_{c,0}$, $\langle v^2 \rangle_{c,0}$ and the sum of the crossed mixed moments $\langle xv \rangle_{c,0} + \langle yu \rangle_{c,0}$. If the waist of the beam is not accessible, an artificial waist has to be created by inserting an almost aberration-free focusing lens into the beam path. Approximately half of the profiles should be acquired close to the waist within one generalized Rayleigh length, the rest outside two Rayleigh lengths. This ensures balanced accuracy for all parameters of the fitting process.

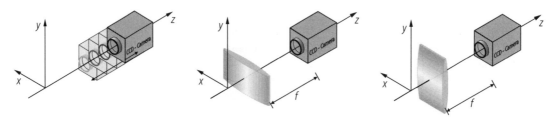

Fig. 2.2.8. Determination of the ten second-order moments in three steps. First step is a z-scan measurement (left), in the second step the CCD camera is placed in the focal plane behind a horizontally oriented cylindrical lens (middle), in the third step the lens is rotated by 90 degrees (right).

At least one cylindrical lens is needed for the measurement of the missing difference of the crossed mixed moments $\langle xv\rangle_{c,0} - \langle yu\rangle_{c,0}$. To retrieve it, a cylindrical lens with focal length f is inserted into the beam path at an arbitrary position in the beam waist region. Firstly, this cylindrical lens shall be aligned with the x-axis and the spatial moment $\langle xy\rangle_1$ is measured in the focal distance behind the lens (Fig. 2.2.8, middle). Next, the lens is rotated by 90 degrees and the spatial moment $\langle xy\rangle_2$ is again measured in the focal distance from the lens (Fig. 2.2.8, right). The missing difference of the crossed mixed moments of the reference plane is then given by

$$\langle xv\rangle_{c,0} - \langle yu\rangle_{c,0} = \frac{\langle xy\rangle_2 - \langle xy\rangle_1}{f} \,. \tag{2.2.78}$$

2.2.7 Beam positional stability

2.2.7.1 Absolute fluctuations

For various reasons a laser beam may fluctuate in position and/or direction. The positional fluctuations in a transverse plane may be measured by the variance of the first-order spatial moments of the beam profile:

$$\langle x^2\rangle_s = \frac{1}{N}\sum_{i=1}^{N}\langle x\rangle_i^2 - \left(\frac{1}{N}\sum_{i=1}^{N}\langle x\rangle_i\right)^2, \tag{2.2.79}$$

$$\langle xy\rangle_s = \frac{1}{N}\sum_{i=1}^{N}\langle x\rangle_i\langle y\rangle_i - \frac{1}{N}\sum_{i=1}^{N}\langle x\rangle_i \frac{1}{N}\sum_{i=1}^{N}\langle y\rangle_i, \tag{2.2.80}$$

$$\langle y^2\rangle_s = \frac{1}{N}\sum_{i=1}^{N}\langle y\rangle_i - \left(\frac{1}{N}\sum_{i=1}^{N}\langle y\rangle_i\right)^2, \tag{2.2.81}$$

where $\langle x\rangle_i$ and $\langle y\rangle_i$ are the first-order moments determined in N individual measurements and $\bar{x} = \frac{1}{N}\sum_{i=1}^{N}\langle x\rangle_i$, $\bar{y} = \frac{1}{N}\sum_{i=1}^{N}\langle y\rangle_i$ define the long-term average beam position. Obviously, the positional fluctuations are different from plane to plane. It can be shown that, under some reasonable assumptions, the positional fluctuations can be characterized closely analogous to the characterization of the beam extent based on the second-order moments of the Wigner distribution [94Mor, 96Mor]. Within this concept, the fluctuation properties of a laser beam are completely determined by ten different parameters, arranged in a symmetric 4×4 matrix

$$\mathsf{P}_s = \begin{pmatrix} \langle x^2\rangle_s & \langle xy\rangle_s & \langle xu\rangle_s & \langle xv\rangle_s \\ \langle xy\rangle_s & \langle y^2\rangle_s & \langle yu\rangle_s & \langle yv\rangle_s \\ \langle xu\rangle_s & \langle yu\rangle_s & \langle u^2\rangle_s & \langle uv\rangle_s \\ \langle xv\rangle_s & \langle yv\rangle_s & \langle uv\rangle_s & \langle v^2\rangle_s \end{pmatrix}, \tag{2.2.82}$$

obeying the same simple propagation law as the centered second-order moments:

$$\mathsf{P}_{s,\mathrm{out}} = \mathsf{S}\cdot\mathsf{P}_{s,\mathrm{in}}\cdot\mathsf{S}^\mathsf{T}\,. \tag{2.2.83}$$

The elements of the beam fluctuation matrix may be considered as the centered second-order moments of a probability distribution $p(x,y,u,v)$ giving the probability that the fluctuation beam

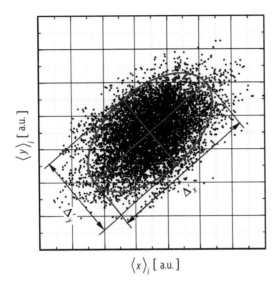

Fig. 2.2.9. Centroid coordinates of fluctuating beam and corresponding variance ellipse characterizing the fluctuations.

has a position (x, y) and direction (u, v) at a random measurement. Similar to the second-order moments of the Wigner distribution, only the three spatial moments are directly measurable. The complete set can be obtained from a z-scan measurement as described in the section above, by acquiring a couple of power density distributions in any measurement plane, calculating the first-order spatial moments from each profile, derive the three variances according to (2.2.79)–(2.2.81), and obtaining the second-order fluctuation moments in the reference plane from a fitting process. Again, measurements behind a cylindrical lens are necessary to achieve all ten parameters.

Fluctuation widths can be derived from the second-order fluctuation moments. In analogy to the beam width definitions, the fluctuation widths are

$$\Delta'_x = 2\sqrt{2} \left\{ \left(\langle x^2 \rangle_s + \langle y^2 \rangle_s\right) + \tau \left[\left(\langle x^2 \rangle_s - \langle y^2 \rangle_s\right)^2 + 4 \langle xy \rangle_s^2\right]^{\frac{1}{2}} \right\}^{\frac{1}{2}}, \qquad (2.2.84)$$

$$\Delta'_y = 2\sqrt{2} \left\{ \left(\langle x^2 \rangle_s + \langle y^2 \rangle_s\right) - \tau \left[\left(\langle x^2 \rangle_s - \langle y^2 \rangle_s\right)^2 + 4 \langle xy \rangle_s^2\right]^{\frac{1}{2}} \right\}^{\frac{1}{2}} \qquad (2.2.85)$$

with

$$\tau = \operatorname{sgn}\left(\langle x^2 \rangle_s - \langle y^2 \rangle_s\right), \qquad (2.2.86)$$

where Δ'_x and Δ'_y are the beam fluctuation widths along the principal axes of the beam positional fluctuations and where

$$\beta = \frac{1}{2} \operatorname{atan} \left(\frac{2 \langle xy \rangle_s}{\langle x^2 \rangle_s - \langle y^2 \rangle_s} \right) \qquad (2.2.87)$$

is the signed angle between the x-axis and that principal axis of the beam fluctuation which is closer to the x-axis (Fig. 2.2.9). The principal axes of the beam positional fluctuations may not coincide with the principal axes of the power density distribution.

The width of the positional fluctuations along an arbitrary direction, given by the azimuthal angle α, is given by

$$\Delta_\alpha = 4 \sqrt{\langle x^2 \rangle_s \cos^2 \alpha + 2 \langle xy \rangle_s \sin \alpha \cos \alpha + \langle y^2 \rangle_s \sin^2 \alpha} . \qquad (2.2.88)$$

2.2.7.2 Relative fluctuations

For many applications the widths of the positional fluctuations compared to the momentary beam profile width might be more relevant than the absolute fluctuation widths. The relative fluctuation along an arbitrary direction, given by the azimuthal angle α, is defined by

$$\Delta_{\text{rel},\alpha} = \sqrt{\frac{\langle x^2\rangle_s \cos^2\alpha + 2\langle xy\rangle_s \sin\alpha\cos\alpha + \langle y^2\rangle_s \sin^2\alpha}{\langle x^2\rangle_c \cos^2\alpha + 2\langle xy\rangle_c \sin\alpha\cos\alpha + \langle y^2\rangle_c \sin^2\alpha}} \ . \tag{2.2.89}$$

The effective relative fluctuation may by specified by

$$\Delta_{\text{rel}} = \sqrt{\frac{\langle x^2\rangle_s + \langle y^2\rangle_s}{\langle x^2\rangle_c + \langle y^2\rangle_c}} \ . \tag{2.2.90}$$

2.2.7.3 Effective long-term beam widths

For applications with response times much longer than the typical fluctuation durations the time-averaged intensity distribution rather than the momentary beam profile determines the process results:

$$\bar{I}(x,y) = \frac{1}{T}\int_{t_0}^{t_0+T} I(x,y,t)\,\mathrm{d}t \ . \tag{2.2.91}$$

The effective width of the time-averaged power density profile along an azimuthal direction enclosing an angle of α with the x-axis can be obtained from the widths of the momentary beam profile and the fluctuation width by

$$d_{\text{eff},\alpha} = \sqrt{d_\alpha^2 + \Delta_\alpha^2} \ . \tag{2.2.92}$$

References for 2.2

78Bas   Bastiaans, M.J.: Wigner distribution function applied to optical signals; Opt. Commun. **25** (1978) 26.

86Bas   Bastiaans, M.J.: Propagation laws for the second-order moments of the Wigner distribution function in first-order optical systems; J. Opt. Soc. Am. A **3** (1986) 1227.

93Sim   Simon, R., Mukunda, N.: Twisted Gaussian Schell-model beams; J. Opt. Soc. Am. A **10** (1993) 95.

94Mor   Morin, M., Bernard, P., Galarneau, P.: Moment definition of the pointing stability of a laser beam; Opt. Lett. **19** (1994) 1379.

96Mor   Morin, M., Levesque, M., Mailloux, A., Galarneau, P.: Moment characterization of the position stability of laser beams; Proc. SPIE (Int. Soc. Opt. Eng.) **2870** (1996) 206.

99Bor   Born, M., Wolf, E.: Principles of optics, Cambridge: Cambridge University Press, 1999.
99ISO   ISO 11146, Lasers and laser-related equipment – test methods for laser beam widths, divergence angles and beam propagation ratios, 1999 (new revised edition 2005).

03Nem   Nemes, G.: Intrinsic and geometrical beam classification, and the beam identification after measurement; Proc. SPIE (Int. Opt. Soc. Eng.) **4932** (2003) 624.

Part 3

Linear optics

3.1 Linear optics

R. GÜTHER

The propagation of light and its interaction with matter is completely described by *Maxwell's equations* (1.1.4)–(1.1.7) and the material equations (1.1.8) and (1.1.9), see Chap. 1.1.

In this chapter the propagation of light in dielectric homogeneous and nonmagnetic media is discussed. Furthermore, monochromatic waves are assumed and linear interaction. The implications thereof for the medium are:

- *Relative permittivity*: ε_r ($\varepsilon(\boldsymbol{E}, \boldsymbol{H})$ in (1.1.8)) is a complex tensor, which in most cases depends on the frequency only, but in special cases also on the spatial coordinate.
- *Relative permeability*: $\mu_r = 1$ ($\mu(\boldsymbol{E}, \boldsymbol{H})$ in (1.1.9)).
- *Electrical charge density*: $\rho = 0$.
- *Current density*: $j = 0$.

3.1.1 Wave equations

Maxwell's equations together with the material equations and the above assumptions result in the time-dependent wave equation for the electric field

$$\Delta \boldsymbol{E}(\boldsymbol{r}, t) - \frac{\varepsilon_r}{c_0^2} \frac{\partial^2}{\partial t^2} \boldsymbol{E}(\boldsymbol{r}, t) = 0 \tag{3.1.1}$$

with

$c_0 = 2.99792458 \times 10^8$ m/s: vacuum velocity of light,

$$\Delta = \frac{\partial^2}{\partial x^2} + \frac{\partial^2}{\partial y^2} + \frac{\partial^2}{\partial z^2} : \quad \text{delta operator.}$$

An identical equation holds for the magnetic field $\boldsymbol{H}(\boldsymbol{r}, t)$.

For the following discussion we assume monochromatic fields, so that

$$\boldsymbol{E}(\boldsymbol{r}, t) = \boldsymbol{E}(\boldsymbol{r}) \, \mathrm{e}^{\mathrm{i}\omega t} \tag{3.1.2}$$

with

ω: angular temporal frequency.

The magnetic field is related to $\boldsymbol{E}$ by the corresponding Maxwell equation (1.1.7)

$$\operatorname{curl} \boldsymbol{E}(\boldsymbol{r}) = -\mathrm{i}\,\omega\,\mu_0 \boldsymbol{H}(\boldsymbol{r}) \,. \tag{3.1.3}$$

Together with the ansatz (3.1.2), for isotropic media (ε_r is a complex scalar) (3.1.1) results in

$$\Delta \boldsymbol{E}(\boldsymbol{r}) + k_0^2 \, \hat{n}^2 \, \boldsymbol{E}(\boldsymbol{r}) = 0 \quad (\text{wave equation}) \tag{3.1.4}$$

with

$k_0 = 2\pi/\lambda_0$: *wave number*,
λ_0 : wavelength in vacuum,
$\hat{n}$: complex refractive index, see (1.1.20).

For isotropic media and fields with uniform polarization the vector property of the field can be neglected. This results in

$$\Delta E(\boldsymbol{r}) + k_0^2\, \hat{n}^2\, E(\boldsymbol{r}) = 0 \quad (\textit{Helmholtz equation}). \tag{3.1.5}$$

In most cases the field can be approximated by a quasiplane wave, propagating in z-direction

$$\boldsymbol{E} = \boldsymbol{E}_0(\boldsymbol{r})\, \mathrm{e}^{\mathrm{i}(\omega t - k_0\, \hat{n}\, z)}\,. \tag{3.1.6}$$

Remark: There are different *conventions* for writing the complex wave (3.1.6):

1. Electrical engineering and most books on quantum electronics:

    $\boldsymbol{E} \propto \exp(\mathrm{i}\,\omega\,t - \mathrm{i}\,k_0\,\hat{n}\,z)\,,$

    for example [96Yar, 86Sie, 66Kog2, 84Hau, 91Sal, 98Sve, 96Die] and this chapter, Chap. 3.1.
2. Physical optics:

    $\boldsymbol{E} \propto \exp(\mathrm{i}\,k_0\,\hat{n}\,z - \mathrm{i}\,\omega\,t)\,,$

    for example [99Bor, 92Lan, 75Jac, 05Hod, 98Hec, 70Col].

[94Fol] discusses both cases.

Consequences of the convention: shape of results on phases of wave propagation, diffraction, interferences, Jones matrix, Collins integral, Gaussian beam propagation, absorption, and gain.

With

$$\left|\frac{\partial \boldsymbol{E}_0}{\partial z}\right| \ll |k_0\,\hat{n}\,\boldsymbol{E}_0|$$

(3.1.4) can be reduced to

$$\Delta_\mathrm{t} \boldsymbol{E}_0 + 2\,\mathrm{i}\,k_0 \hat{n}\, \frac{\partial\,\boldsymbol{E}_0}{\partial\,z} = 0 \quad (\textit{Slowly Varying Envelope (SVE) equation})\,, \tag{3.1.7}$$

with

$$\Delta_\mathrm{t} = \frac{\partial^2}{\partial\,x^2} + \frac{\partial^2}{\partial\,y^2} \; : \quad \text{transverse delta operator (rectangular symmetry)},$$

see Chap. 1.1, (1.1.24a). *Other names for SVE* are: paraxial wave equation [86Sie], paraxial Helmholtz equation [96Ped, 78Gra].

The analogue approximation with respect to time t instead of the spatial coordinate z is used in ultrashort laser pulse physics [96Die, 86Sie].

3.1.2 Polarization

Restriction of (3.1.2) to a plane wave along the z-axis, see Fig. 3.1.1, results in

$$\begin{bmatrix} E_x \\ E_y \end{bmatrix} = \begin{bmatrix} E_{0x} \cos(\omega t - kz + \delta_x) \\ E_{0y} \cos(\omega t - kz + \delta_y) \end{bmatrix} \Rightarrow$$

$$\begin{bmatrix} E_{0x} \exp(\mathrm{i}\,\delta_x) \\ E_{0y} \exp(\mathrm{i}\,\delta_y) \end{bmatrix} \exp[\mathrm{i}(\omega t - kz)] \equiv E_0\, \boldsymbol{J} \exp[\mathrm{i}(\omega t - kz)] \ . \tag{3.1.8}$$

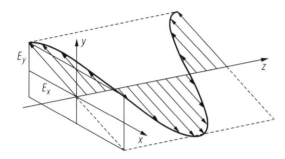

Fig. 3.1.1. Electric field of a linear polarized wave with propagation along the z-axis.

Definitions:

$$E_0 = \sqrt{E_{0x}^2 + E_{0y}^2} \ ,$$

$$\boldsymbol{J} = \frac{1}{E_0} \begin{bmatrix} E_{0x} \exp(\mathrm{i}\,\delta_x) \\ E_{0y} \exp(\mathrm{i}\,\delta_y) \end{bmatrix} \ : \quad (\textit{normalized})\ \textit{Jones vector}\ ,$$

δ_x and δ_y : phase angles ,

$\Rightarrow$: transition to the complex representation ,

$\varepsilon_0 n c_0 E_0^2 \boldsymbol{J}\boldsymbol{J}^*/2$: light intensity [W/m^2] .

Different *conventions* for *right*-hand polarization:

1. Looking against the direction of light propagation the light vector moves clockwise in the x-y-plane of Fig. 3.1.1 ([99Bor, 91Sal, 96Ped, 98Hec, 88Kle, 87Nau]).
2. The clockwise case occurs looking with the propagation direction (right-hand screw, elementary particle physics) ([84Yar, 88Yeh, 05Hod] and *in this chapter*).

Remark: $\boldsymbol{J}$ without normalization is also called Jones vector in [84Yar, 88Yeh, 90Roe, 77Azz, 86Sol], [95Bas, Vol. II, Chap. 27].

Jones Calculus [41Jon, 97Hua, 88Yeh, 90Roe, 75Ger]:

$$\mathsf{J}_2 = \mathsf{M}\,\mathsf{J}_1 \tag{3.1.9}$$

with

J_1 : Jones matrix for the initial polarization state,
M : Jones matrix describing an optical element or system,
J_2 : Jones matrix of the polarization state after light has passed the element or system.

In Table 3.1.1 the characterization of the polarization states of light with the *Jones vector* is given, in Table 3.1.2 the characterization of optical elements with the *Jones matrix*.

Table 3.1.1. Characterization of the polarization states of light with the *Jones vector*.

Jones vector J	$\begin{bmatrix} \cos\psi \\ \sin\psi \end{bmatrix}$	$\frac{1}{\sqrt{2}}\begin{bmatrix} 1 \\ i \end{bmatrix}$	$\frac{1}{\sqrt{2}}\begin{bmatrix} i \\ 1 \end{bmatrix}$	$\begin{bmatrix} a \cdot i \\ b \end{bmatrix}$ and $a^2 + b^2 = 1$
State of polarization	Linear polarization	Left circular polarization	Right circular polarization	Right elliptical polarization
Projection of the vector E onto the x-y-plane viewed along the propagation direction z				

Table 3.1.2. Characterization of optical elements with the *Jones matrix*.

Opt. Element	Polarizer along the x-direction	Polarizer along the y-direction	Quarter-wave plate	Half-wave plate	Brewster-angle-tilted plate: index n	Faraday rotator (angle β)	Coordinate rotation by an angle α: $M(\alpha)$
Jones Matrix	$\begin{bmatrix} 1 & 0 \\ 0 & 0 \end{bmatrix}$	$\begin{bmatrix} 0 & 0 \\ 0 & 1 \end{bmatrix}$	$\begin{bmatrix} 1 & 0 \\ 0 & \pm i \end{bmatrix}$	$\begin{bmatrix} 1 & 0 \\ 0 & -1 \end{bmatrix}$	$\begin{bmatrix} \left[\frac{2n}{n^2+1}\right]^2 & 0 \\ 0 & 1 \end{bmatrix}$	$\begin{bmatrix} \cos\beta & -\sin\beta \\ \sin\beta & \cos\beta \end{bmatrix}$	$\begin{bmatrix} \cos\alpha & \sin\alpha \\ -\sin\alpha & \cos\alpha \end{bmatrix}$ Rotated element: $M(\alpha)\,M\,M(-\alpha)$

Example 3.1.1.
$$M = M_3 \cdot M_2 \cdot M_1 ,\qquad(3.1.10)$$
M: Jones matrix of the system which consists of elements with the matrices M_1, M_2, M_3. Light passes first the element with M_1 and last the element with M_3.

Example 3.1.2. $\boldsymbol{J}_1 = \frac{1}{\sqrt{2}} \begin{bmatrix} 1 \\ 1 \end{bmatrix}$ (linear 45°-polarization), $M = \begin{bmatrix} 1 & 0 \\ 0 & \pm i \end{bmatrix}$ ($\begin{Bmatrix} \text{right} \\ \text{left} \end{Bmatrix}$ quarter-wave plate), $\boldsymbol{J}_2 = M \cdot \boldsymbol{J}_1 = \frac{1}{\sqrt{2}} \begin{bmatrix} 1 \\ \pm i \end{bmatrix}$ ($\begin{Bmatrix} \text{right} \\ \text{left} \end{Bmatrix}$ circular polarization).

Development: Any Jones vector can be developed into a superposition of two orthogonal Jones vectors:
$$\boldsymbol{J} = a_1 \boldsymbol{J}_1 + a_2 \boldsymbol{J}_2 \qquad(3.1.11)$$
with $\boldsymbol{J}_1 \boldsymbol{J}_2^* = 0$.

Example 3.1.3. linearly polarized light = left polarized light + right polarized light.

Partially polarized light: If parts of both coefficients of the $\boldsymbol{E}$-vector are uncorrelated, there is a mixing of polarized and *non*polarized light. It is described by the four components of the *Stokes vector* $\{s_0, s_1, s_2, s_3\}$, using $\langle \ldots \rangle$ to signify averaging by detection:

$$s_0 = \langle E_x^2 \rangle + \langle E_y^2 \rangle \quad \Rightarrow \quad E_{0x}^2 + E_{0y}^2 ,\qquad(3.1.12)$$
$$s_1 = \langle E_x^2 \rangle - \langle E_y^2 \rangle \quad \Rightarrow \quad E_{0x}^2 - E_{0y}^2 ,\qquad(3.1.13)$$
$$s_2 = 2 \langle E_x E_y \cos[\delta_y - \delta_x] \rangle \quad \Rightarrow \quad 2 E_{0x} E_{0y} \cos(\delta_y - \delta_x) ,\qquad(3.1.14)$$
$$s_3 = 2 \langle E_x E_y \sin[\delta_y - \delta_x] \rangle \quad \Rightarrow \quad 2 E_{0x} E_{0y} \sin(\delta_y - \delta_x)\qquad(3.1.15)$$

with
$$s_0^2 > s_1^2 + s_2^2 + s_3^2 \quad \Rightarrow \quad s_0^2 = s_1^2 + s_2^2 + s_3^2 ,\qquad(3.1.16)$$

where $\Rightarrow$ means the transition from partially polarized light to completely polarized light, shown with the terms of Fig. 3.1.1.

Meaning of the s_i:
s_0: power flux,
$\sqrt{s_1^2 + s_2^2 + s_3^2}/s_0$: degree of polarization,
$\sqrt{s_1^2 + s_2^2}/s_0$: degree of linear polarization,
s_3/s_0: degree of circular polarization.

Mueller calculus ([75Ger, 77Azz, 90Roe, 95Bas]): extension of the *Jones calculus* for *partial-coherent light*, where the four dimensional Stokes vector replaces the Jones vector and the real 4×4 Mueller matrices the complex 2×2 Jones matrices. The Jones calculus is usually sufficient to describe coherent laser radiation.

Measurement of the polarization state:

– *Partially polarized light*: [87Nau], [76Jen, Chap. 27.6], [77Azz, Chap. 3], [61Ram, Sect. 14–25], [95Bas, Vol. 2, Chap. 22.15], [75Ger]. Result: Stokes vector.
– *Pure coherent light*: see [05Hod]. There are commercial systems for this task.

Eigenstates of polarized light are those two polarization states (Jones vectors) which reproduce themselves, multiplied with a complex factor (eigenvalue), if monochromatic light passes an optical element or system.
Calculation: see [97Hua, 77Azz], application: decoupling of the polarization mixing during round trips in resonators [74Jun].

3.1.3 Solutions of the wave equation in free space

Following (3.1.2), each of the wave solutions given in this section must be multiplied with the factor $e^{i\omega t}$ to obtain the propagating wave of (3.1.1).

3.1.3.1 Wave equation

The solutions of the wave equation (3.1.4) are vector fields.

3.1.3.1.1 Monochromatic plane wave

$$\boldsymbol{E} = \boldsymbol{E}_0 \exp\{-i k_0 \hat{n}\, \boldsymbol{e}\boldsymbol{r} + i\varphi\}, \tag{3.1.17}$$

$$\boldsymbol{H} = \frac{\hat{n}}{c_0 \mu_0} (\boldsymbol{e} \times \boldsymbol{E}_0) \exp\{-i k_0 \hat{n}\, \boldsymbol{e}\boldsymbol{r} + i\varphi\} \tag{3.1.18}$$

with

$\boldsymbol{r}$: position vector,
$\boldsymbol{e}$: unit vector normal to the wave fronts,
$k_0 = 2\pi/\lambda_0$: wave number,
$\hat{n}$: complex refractive index,
φ : phase.

For the phase velocity and the wave group velocity see Sect. 3.1.5.3.

3.1.3.1.2 Cylindrical vector wave

$$\boldsymbol{E} = E_0\, \boldsymbol{e}_z\, H_0^{(2)}(k_0 \rho), \tag{3.1.19}$$

$$\boldsymbol{H} = i\frac{E_0}{c_0 \mu_0} \left(\boldsymbol{e}_z \times \frac{\boldsymbol{\rho}}{\rho}\right) H_1^{(2)}(k_0 \rho) \quad (\rho > \lambda) \tag{3.1.20}$$

for time-harmonic electric source current density on the z-axis of a cylindrical coordinate system with the coordinates (ρ, φ, z) : (radial distance, azimuthal angle, z-axis) [94Fel, Chap. 5].

$H_m^{(2)}$: m^{th} order Hankel function of the second kind [70Abr];
  the change of convention in Sect. 3.1.1 includes: $H_m^{(2)} \Rightarrow H_m^{(1)}$ [94Fel, p. 487];
$\boldsymbol{\rho}$: radial position vector,
$\boldsymbol{e}_z$: unit vector along the z-axis.

3.1.3.1.3 Spherical vector wave

$$\boldsymbol{E} = E_0 \cdot (\boldsymbol{n} \times \boldsymbol{p}) \times \boldsymbol{n} \cdot \frac{\exp(-i k_0 \hat{n}\, r)}{r}, \tag{3.1.21}$$

$$\boldsymbol{H} = \frac{E_0}{c_0 \mu_0} \cdot (\boldsymbol{n} \times \boldsymbol{p}) \cdot \frac{\exp(-i k_0 \hat{n}\, r)}{r} \quad (r \gg \lambda_0) \tag{3.1.22}$$

is the *far field* ($1/r^2$ and higher inverse power terms $\ll 1/r$-term) of an oscillating electric dipole ([99Bor, 94Leh, 75Jac]) with

E_0 : amplitude [V],
$\boldsymbol{p}$: unit vector of the dipole moment,
$\boldsymbol{n}$: unit vector pointing from dipole to spatial position,
r : radial distance.

3.1.3.2 Helmholtz equation

The approximative transition from the *vectorial* wave equation (3.1.4) to the Helmholtz equation (3.1.5) ([99Bor]) results in *scalar* solutions. E is called: "field" [72Mar], "complex displacement" or "scalar wave function" [99Bor], "disturbance" [95Bas, Vol. I].

3.1.3.2.1 Plane wave

$$E = E_0 \exp\{-\mathrm{i}\, k_0\, \hat{n}\, \boldsymbol{er} + \mathrm{i}\,\varphi\,.\} \tag{3.1.23}$$

For the parameters see (3.1.18).

3.1.3.2.2 Cylindrical wave

$$E = E_0\, H_0^{(2)}(k_0\, \hat{n}\, \rho) \quad (\rho > \lambda_0) \tag{3.1.24}$$

is the diverging field of a homogeneous line source [41Str, Chap. IV], [94Fel, Chap. 5]. For the parameters see (3.1.19).

3.1.3.2.3 Spherical wave

$$E = E_0 \cdot \frac{\exp(-\mathrm{i}\, k_0\, \hat{n}\, r)}{r} \quad (r > \lambda_0)\,, \tag{3.1.25}$$

parameters see (3.1.21).

3.1.3.2.4 Diffraction-free beams

3.1.3.2.4.1 Diffraction-free Bessel beams

Diffraction-free Bessel beams without transversal limitation are discussed in [05Hod, 91Nie, 88Mil].

$$E(x,y,z) = E_0 \cdot J_0(a\,\rho) \cdot \exp\{-\mathrm{i}\cos(\theta_B)\, k_0 z\} \tag{3.1.26}$$

with

E_0 : amplitude vector [V/m],
J_0 : zero-order Bessel function of the first kind [70Abr]; higher-order Bessel beams see [96Hal];
$\rho = \sqrt{x^2 + y^2}$: radial distance from the z-axis,
$a = k_0 \sin \Theta_B$ [m^{-1}],
Θ_B : convergence angle of the conus of the plane wave normal to the z-axis, see Fig. 3.1.2.

3.1.3.2.4.2 Real Bessel beams

Real Bessel beams are limited by a finite aperture D of the optical elements needed or Gaussian beam illumination (Gaussian Bessel beams [87Gor]).

Methods of generation: axicons [85Bic] (Fig. 3.1.2), annular aperture in the focus of a lens [87Dur, 91Nie], holographic [91Lee] or diffractive [96Don] elements. Because of finite aperture diffraction the latter display approximately the shape of (3.1.26) with cutoff at a geometric determined radius r_N, which includes N maxima (Fig. 3.1.3) and different amplitude patterns in dependence on z.

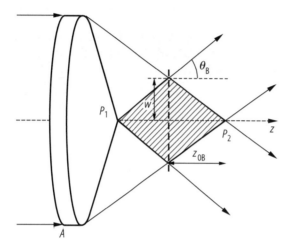
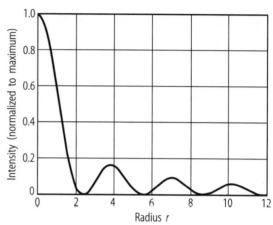

Fig. 3.1.2. Generation of a Bessel beam with help of an axicon A by a conus of plane-waves propagation directions.

Fig. 3.1.3. Transversal intensity structure of a Bessel beam ($\propto J_0^2(r)$).

Advantage of Bessel beams: Large depth of focus $2\,z_{0B}$ between $P1$ and $P2$ in Fig. 3.1.2 (thin "needle of light") for measurement purposes.

Disadvantage: Every maximum in Fig. 3.1.3 contains in the corresponding circular ring nearly the same power as the central peak. High power loss occurs if the central part is used only [05Hod].

3.1.3.2.4.3 Vectorial Bessel beams

Vectorial Bessel beams are discussed in [96Hal].

3.1.3.3 Solutions of the slowly varying envelope equation

Gaussian beams are solutions of the *SVE-equation* (3.1.7) [91Sal, 96Ped, 86Sie, 78Gra], which is equivalent to paraxial approximation or Fresnel's approximation, see Sect. 3.1.4.

The *transition* from SVE-approximated Gaussian beams towards an exact solution of the wave equation in the non-paraxial range is given in a Lax-Wünsche series [75Lax, 79Agr, 92Wue]. For contour plots of the relative errors in the Gaussian beam volume see [97For, 97Zen].

The *vectorial field* of Gaussian beams is discussed in [79Dav, 95Gou], containing a Lax-Wünsche series; Gaussian beam in elliptical cylinder coordinates are given in [94Soi, 00Gou].

3.1.3.3.1 Gauss-Hermite beams (rectangular symmetry)

Elliptical higher-order Gauss-Hermite beam:

$$E_{mn}(x,y,z) = E_0\, U_m(x,z)\, U_n(y,z)\, \exp\{-i\,k_0 z\}\,, \tag{3.1.27}$$

$$U_m(x,z) = \sqrt{\frac{w_{0x}}{w_x(z)}}\, H_m\!\left(\frac{\sqrt{2}\,x}{w_x(z)}\right) \exp\!\left\{-\frac{x^2}{w_x^2(z)} - i\,\frac{k_0\, x^2}{2\,R_x(z)}\right\} \exp\{i\,\varphi_m(z)\}\,, \tag{3.1.28}$$

$$U_n(y,z) = U_{m \Rightarrow n}(x \Rightarrow y, z) \tag{3.1.29}$$

with

w_{0x} : the $1/e^2$-intensity waist radius,

$z_{0x} = \dfrac{\pi\, w_{0x}^2}{\lambda}$: the Rayleigh distance (half depth of focus),

$w_x(z) = w_{0x}\sqrt{1 + \dfrac{z^2}{z_0^2}}$: the E_{00}-beam $1/e^2$-intensity radius,

$R_x(z) = z\sqrt{1 + \dfrac{z_0^2}{z^2}}$: the radius of curvature of the wavefront at position z,

$\varphi_m(z) = \left(\tfrac{1}{2} + m\right) \arctan\!\left(\dfrac{z}{z_0}\right)$: Gouy's phase, changing sign for the transition through $z=0$,

$H_m\!\left(\dfrac{\sqrt{2}}{w_x(z)}\right)$: the Hermite polynomial of order m [70Abr],

$H_0(\xi) = 1,\ H_1(\xi) = 2\xi,\ H_2(\xi) = 4\xi^2 - 2,\ H_3(\xi) = 8\xi^3 - 12\xi,\ H_4(\xi) = 16\xi^4 - 48\xi^2 + 12,\ \dots\,,$

$$\int_{-\infty}^{\infty} d\xi \left\{\frac{\exp(-\xi^2/2)}{\sqrt{\sqrt{\pi}\, m!\, 2^m}} H_m(\xi)\right\} \left\{\frac{\exp(-\xi^2/2)}{\sqrt{\sqrt{\pi}\, n!\, 2^n}} H_n(\xi)\right\} = \delta_{mn}\,,$$

$$\delta_{mn} = \begin{cases} 1 & \text{for } m = n \\ 0 & \text{for } m \ne n \end{cases} \quad (\textit{orthogonality relation})\,. \tag{3.1.30}$$

Example 3.1.4. Rotational symmetrical Gaussian fundamental mode (Gaussian beam):

Specialization of (3.1.27): $m = n = 0$, $w_{0x} = w_{0y} = w_0$, $r = \sqrt{x^2 + y^2}$.

$$E_{00}(r,z) = E_0\, \frac{w_0}{w(z)}\, \exp\!\left\{-\frac{r^2}{w^2(z)} - i\,\frac{k r^2}{2R(z)}\right\} \exp\!\left\{i\,\frac{1}{2}\arctan\frac{z}{z_0}\right\} \exp\{-i\,kz\}\,, \tag{3.1.31}$$

$$w(z) = w_0\sqrt{1 + \frac{z^2}{z_0^2}}\,,\quad R(z) = z\sqrt{1 + \frac{z_0^2}{z^2}}\,.$$

Properties of E_{00} (fundamental mode): The shape of the Gaussian E_{00}-beam is depicted in Fig. 3.1.4. *Parameters* of E_{00} in Fig. 3.1.4 are:

C: curves with constant amplitude decrease as $E(r,z) = E(0,z)/e$
 or constant intensity decrease as $I(r,z) = I(0,z)/e^2$,
P: phase fronts with radius of curvature $R(z)$,

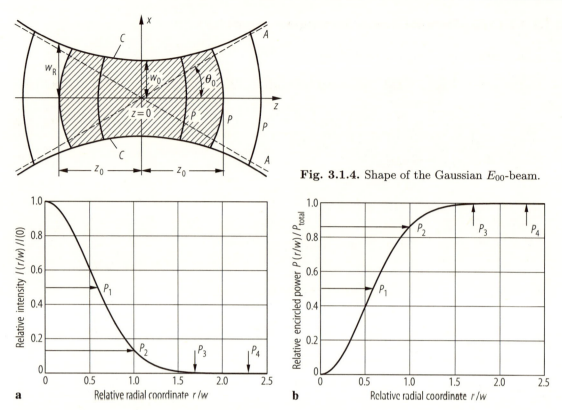

Fig. 3.1.4. Shape of the Gaussian E_{00}-beam.

Fig. 3.1.5. (a) Cross section of a Gaussian beam perpendicular to the z-axis. (b) Power transmitted by a circular aperture with the relative radius r/w in a cross section.

Table 3.1.3. Characteristic points in Fig. 3.1.5.

Point in Fig. 3.1.5a, b	Relative abscissa r/w	Relative intensity, Fig. 3.1.5a	Relative transmission, Fig. 3.1.5b	Characterization
P_1	0.588	0.5	0.5	FWHM [a]
P_2	1	0.135	0.865	$1/e^2$-int. [b]
P_3	1.57	0.01	0.99	trunc. [c]
P_4	2.3	0.001	0.999	trunc. [d]

[a] Full width half maximum/2.
[b] $1/e^2$-intensity or $1/e$-amplitude.
[c] Diffraction of E_{00}-beam by circular aperture $\Rightarrow$ 17 % intensity ripple [86Sie, p. 667].
[d] Diffraction of E_{00}-beam by circular aperture $\Rightarrow$ 1 % intensity ripple [86Sie, p. 667] (no essential effect of truncation).

w_0 : *beam waist*,

z_0 : *Rayleigh distance*, half of the *confocal parameter* $b = 2z_0$ (similarly to depth of focus in usual optics), that z-value, where the cross section $\pi w_R^2 = 2\pi w_0^2$ of the Gaussian beam has doubled in comparison with the waist,

$\Theta_0 = \lambda/(\pi w_0)$: $1/e^2$-intensity divergence angle toward the asymptotes A.

In Fig. 3.1.5a the cross section of a Gaussian beam perpendicular to the z-axis is given, in Fig. 3.1.5b the power transmitted by a circular aperture with the relative radius r/w in a cross section. Characteristic points in Fig. 3.1.5 are listed in Table 3.1.3.

Astigmatic and general astigmatic generalizations of the elliptical Gaussian beam: see Sect. 3.1.7.

3.1.3.3.2 Gauss-Laguerre beams (circular symmetry)

$$E_{lp}(r,\psi,z) = E_0 \exp\{-\mathrm{i}\,[kz - \varphi_{lp}(z)]\}\,\frac{w_0}{w(z)}\left(\frac{\sqrt{2}\,r}{w(z)}\right)^l L_p^l\left(\frac{2r^2}{w^2(z)}\right)$$
$$\times \exp\left\{-\frac{r^2}{w^2(z)} - \mathrm{i}\,\frac{k\,x^2}{2\,R(z)}\right\}\begin{Bmatrix}\cos(l\psi)\\ \sin(l\psi)\end{Bmatrix} \qquad (3.1.32)$$

with

z : propagation direction,

r,φ : polar coordinates in the plane $\perp$ z-axis,

$z_0 = \dfrac{\pi w_0^2}{\lambda}$: the Rayleigh distance (half depth of focus),

$w(z) = w_0 \sqrt{1 + \left(\dfrac{z}{z_0}\right)^2}$: the E_{00}-beam $1/e^2$-intensity radius,

$R(z) = z\left\{1 + \left(\dfrac{z_0}{z}\right)^2\right\}$: the radius of curvature of the wavefront at position z,

$\varphi_{lp} = (2p+l+1)\arctan\left(\dfrac{z}{z_0}\right)$: Gouy's phase,

L_p^l : Laguerre polynomial of degree p and order l [70Abr]:

$$L_0^l(\xi) = 1,\quad L_1^l(\xi) = (l+1) - \xi,\quad L_2^l(\xi) = \frac{(l+1)(l+2)}{2} - (l+2)\xi - \frac{1}{2}\xi^2,$$

$$L_3^l(\xi) = \frac{(l+3)(l+2)(l+1)}{6} - \frac{(l+3)(l+2)}{2}\xi + \frac{(l+3)}{2}\xi^2 - \frac{1}{6}\xi^3 \ldots,$$

$$\int_0^\infty \mathrm{d}\xi\,\xi^l\,\exp(-\xi)\,L_p^l(\xi)\,L_q^l(\xi) = \delta_{pq}\,\frac{(l+p)!}{p!} \quad (\textit{orthogonality relation}), \qquad (3.1.33)$$

$p!$: the factorial p.

- *Two degenerate mode patterns* are formed by the cos- and sin-terms in (3.1.32).
- $l = p = 0$ means the rotational symmetrical Gaussian beam E_{00}.
- The *symmetry* determines what system of Gauss-Laguerre polynomials or Gauss-Hermite polynomials is more appropriate for a wave field development.

3.1.3.3.3 Cross-sectional shapes of the Gaussian modes

In Fig. 3.1.6 intensity distributions of Gauss-Hermite modes E_{mn} are given (rectangular symmetry), in Fig. 3.1.7 intensity distributions of Gauss-Laguerre modes E_{pl} (circular symmetry).

Rectangular symmetry (Gauss-Hermite modes)

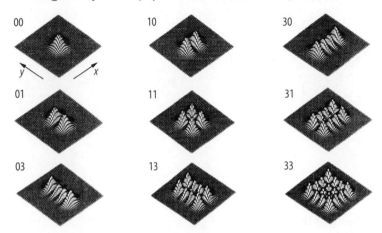

Fig. 3.1.6. Intensity distributions of Gauss-Hermite modes E_{mn}. The two digits at each distribution are m and n.

Circular symmetry (Gauss-Laguerre modes)

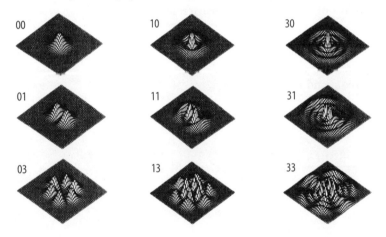

Fig. 3.1.7. Intensity distributions of Gauss-Laguerre modes E_{pl}. The two digits at each distribution are p and l. .

3.1.4 Diffraction

Diffraction of light by aperture rims or amplitude and phase modifications inside the aperture:

- Solutions of Maxwell's equations taking into account the material properties of the aperture:
    - special cases: exact solutions [99Bor, 86Sta],
    - mostly: numerical solutions.
- Starting with a field near the aperture with reasonable assumptions for this field or its measurement: large variety of methods for different ranges of validity [99Bor, 86Sta, 61Hoe].

3.1.4.1 Vector theory of diffraction

- *Vectorial generalization* of Kirchhoff's theory: Given $\boldsymbol{E}$ and $\boldsymbol{H}$ in an aperture $\Rightarrow$ $\boldsymbol{E}$ and $\boldsymbol{H}$ in the volume by Stratton-Chu Green's function representation [23Kot, 41Str, 86Sol, 91Ish].
- *Two-dimensional problem* and meridional incidence of light [61Hoe]: Separation of the polarizations $\boldsymbol{E}$ parallel and $\boldsymbol{E}$ perpendicular to the plane of incidence for half plane [99Bor], slit [99Bor], gratings [80Pet], and volume gratings [69Kog, 81Sol, 81Rus].

3.1.4.2 Scalar diffraction theory

Two sources of scalar diffraction theory are:

- Transition from *vectorial* theory to *scalar* theory: [99Bor, 86Sol]. The information about the polarization is lost.
- Mathematical formulation and generalization of *Huygens' principle*: Each point on a wavefront may be regarded as a source of secondary waves, and the position of the wavefront at a later time is determined by the envelope of these secondary waves.

In Table 3.1.4 *diffraction formulae* with fields given near the diffraction aperture are listed. Figures 3.1.8 and 3.1.9 are related to Table 3.1.4.

Remarks on the formulae of Table 3.1.4:

(3.1.37): Approximation of (3.1.34): Huygens' principle with an additional directional factor (Fresnel).

(3.1.38): Approximation of (3.1.36): Huygens' principle with a modified directional factor.

(3.1.39): *Fresnel's approximation* (= paraxial approximation). The approximation conditions from (3.1.34) to (3.1.39) resp. (3.1.40) are explained in [96For, 86Sta, 87Ree].

Fresnel's approximation: The condition $N_\mathrm{F}(a/d)^2/4 \ll 1$ [91Sal] is valid for sharp-edged apertures A, but it is weakened for the transmission of Gaussian-beam-like fields [86Sie, p. 635] or Gaussian-like soft apertures. Fresnel's approximation describes the propagation of the field from plane $z = 0$ to plane $z = z$. This transformation can be cascaded to describe complex systems and is an often used *tool in paraxial propagation* of radiation (Sect. 3.1.4.5.2).

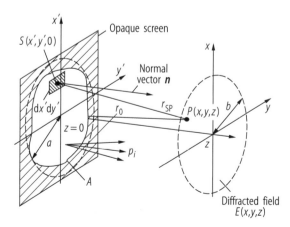

Fig. 3.1.8. Diffraction at an aperture A with source terms $E(x',y',0)$ and/or $\frac{\partial}{\partial z}E(x',y',z)\big|_{z=0}$, respectively, and a or b the maximum radial distances of source S or image point P, respectively. p_i symbolizes different plane waves for (3.1.41)–(3.1.43).

Table 3.1.4. *Diffraction formulae* with fields given near the diffraction aperture (r_{SP}: see Fig. 3.1.8).

Integrals	Formula		Restrictions	Ref.
Rayleigh-Sommerfeld of 1st kind	$E_{RS1}(x,y,z) = -\dfrac{1}{4\pi}\iint\limits_A E(x',y',0)\dfrac{\partial}{\partial z}\left(\dfrac{\exp(-\mathrm{i}kr_{SP})}{r_{SP}}\right)\mathrm{d}x'\,\mathrm{d}y'$	(3.1.34)	$r_{SP} > \lambda_0$, plane aperture	[99Bor] [86Sta]
Rayleigh-Sommerfeld of 2nd kind	$E_{RS2}(x,y,z) = -\dfrac{1}{2\pi}\iint\limits_A \left[\dfrac{\partial E(x',y',z')}{\partial z}\right]_{z'=0}\dfrac{\exp(-\mathrm{i}kr_{SP})}{r_{SP}}\mathrm{d}x'\,\mathrm{d}y'$	(3.1.35)	$r_{SP} > \lambda_0$, plane aperture	
Fresnel-Kirchhoff	$E_{FK}(x,y,z) = \dfrac{1}{2}\left[E_{RS1}(x,y,z) + E_{RS2}(x,y,z)\right]$	(3.1.36)	$r_{SP} > \lambda_0$, curved aperture	
Rayleigh-Sommerfeld 1st kind approx.	$E_{RS1a}(x,y,z) = \dfrac{1}{\mathrm{i}\lambda}\iint\limits_A E(x',y',0)\dfrac{\exp(-\mathrm{i}kr_{SP})}{r_{SP}}\cos(n,r_{SP})\,\mathrm{d}x'\,\mathrm{d}y'$	(3.1.37)	$r_{SP} \gg \lambda_0$	
Fresnel-Kirchhoff approximation, refers to Fig. 3.1.8	$E_{FKa}(x,y,z) = \dfrac{1}{\mathrm{i}\lambda}\iint\limits_A E(x',y',0)\cdot\dfrac{\exp(-\mathrm{i}kr_{SP})}{r_{SP}}\cdot\dfrac{1+\cos(n,r_{SP})}{2}\mathrm{d}x'\,\mathrm{d}y'$	(3.1.38)	$r_{SP} \gg \lambda_0$	
Fresnel's approximation, refers to Fig. 3.1.8	$E_{Fre}(x,y,z) = \dfrac{\mathrm{i}\exp(-\mathrm{i}kz)}{\lambda d}\iint\limits_A E(x',y',0)\exp\left\{-\mathrm{i}\pi\dfrac{(x-x')^2+(y-y')^2}{\lambda z}\right\}\mathrm{d}x'\,\mathrm{d}y'$	(3.1.39)	$z \gg \lambda_0$	[99Bor] [96For] [97For] [87Ree] [86Sta]

(continued)

Table 3.1.4 continued.

Integrals	Formula		Restrictions	Ref.
Fraunhofer far-field approximation, refers to Fig. 3.1.8	$E_{\text{Fra}}(x,y,z) = \dfrac{\mathrm{i}\exp(-\mathrm{i}kz)\, p}{\lambda z} \iint\limits_{A} E(x',y',0) \exp\left\{\mathrm{i}2\pi \dfrac{xx'+yy'}{\lambda z}\right\} \mathrm{d}x'\,\mathrm{d}y'$ with the additional phase term $p = \begin{cases} 1 & \text{for } \dfrac{b^2}{\lambda z} \ll 1 \\ \exp\left\{-\mathrm{i}\pi \dfrac{x^2+y^2}{\lambda z}\right\} & \text{otherwise} \end{cases}$	(3.1.40)	$\dfrac{a^2}{\lambda d} \ll 1$	[99Bor] [68Goo] [96For] [97For] [86Sta]
Plane-wave representation (also: angular-spectrum representation), refers to Figs. 3.1.8 and 3.1.9	2-D Fourier transform (see remark on (3.1.40)) of the source distribution E_s in plane $z=0$: $A_0(f_x, f_y) = \displaystyle\int\limits_{-\infty}^{\infty}\int\limits_{-\infty}^{\infty} E_s(x',y',0) \exp\{\mathrm{i}2\pi(f_x x' + f_y y')\}\,\mathrm{d}x'\,\mathrm{d}y'$, propagation of plane waves with the spatial frequencies f_x and f_y along the z-direction by distance z: $\exp\{-\mathrm{i}2\pi(f_x x + f_y y)\} \Rightarrow \exp\left\{-\mathrm{i}2\pi\left(f_x x + f_y y + \sqrt{1/\lambda^2 - f_x^2 - f_y^2}\, z\right)\right\}$, addition of plane waves at distance z: $E(x,y,z) = \displaystyle\iint\limits_{f_x^2+f_y^2 < 1/\lambda^2} A_0(f_x, f_y) \exp\left\{-\mathrm{i}2\pi\left(f_x x + f_y y + \sqrt{1/\lambda^2 - f_x^2 - f_y^2}\, z\right)\right\} \mathrm{d}f_x\,\mathrm{d}f_y$, equivalent to (3.1.34) [97For]	(3.1.41) (3.1.42) (3.1.43)	$r_{\text{SP}} > \lambda_0$	[91Sal] [78Loh] [86Sta] [97For] [99Bor]
Far field in the focal plane of a lens, refers to Fig. 3.1.9	$E_{\text{P}}(x,y) = \dfrac{\mathrm{i}p}{\lambda f} \displaystyle\int\limits_{-\infty}^{\infty}\int\limits_{-\infty}^{\infty} E_{\text{S}}(x',y') \exp\left\{\mathrm{i}2\pi\left(\dfrac{x}{\lambda f}x' + \dfrac{y}{\lambda f}y'\right)\right\} \mathrm{d}x'\,\mathrm{d}y'$, $p = \exp\left(\mathrm{i}\pi\dfrac{(x^2+y^2)(d-f)}{\lambda f^2}\right)$	(3.1.44) (3.1.45)	$d, f \gg \lambda$	[91Sal]

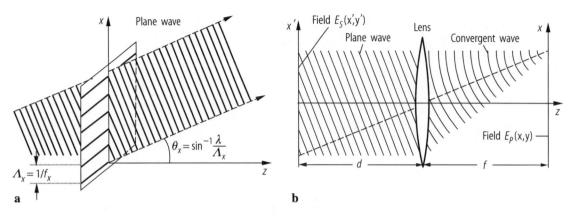

Fig. 3.1.9. (a) *Spatial frequencies* of a plane wave with propagation direction Θ_x with respect to the plane $x = 0$ (and Θ_y analogously) are f_x and f_y with $\Theta_x = \sin^{-1}(\lambda f_x) \approx \lambda f_x$ and $\Theta_y = \sin^{-1}(\lambda f_y) \approx \lambda f_y$ ($\approx$: paraxial approximation). (b) Generation of the far field in the focal plane of a lens: The Fourier transformation ($d = f$) is changed by an additional phase term for $d \neq f$ with d: distance, f: focal length.

(3.1.40): *Fraunhofer's approximation*

– *Fresnel number*:

$$N_{\mathrm{F}} = a^2/\lambda z. \tag{3.1.46}$$

– Validity of Fraunhofer's approximation: $N_{\mathrm{F}} \ll 1$.
  $p \neq 1$ (parabolic phase): the *intensity* of diffracted light is the square of the *modulus* of the Fourier transform of $E(x, y, 0)$ only.
– Additional condition with second Fresnel number $N_{\mathrm{F}'} = b^2/\lambda z \ll 1$:
  $E(x, y, z)$ is the *Fourier transform* of $E(x, y, 0)$ in dependence on the spatial frequencies $f_x \approx (x/z)/\lambda \approx \Theta_x/\lambda$ and $f_y \approx (y/z)/\lambda \approx \Theta_y/\lambda$.
– *Different conventions* on the *spatial Fourier transform* $F(f_x)$ of a spatial distribution $f(x)$:

  – The convention of the plane-wave structure $\exp(\mathrm{i}\, kx - \mathrm{i}\,\omega t)$ is connected with the determination of $F(f_x)$ by

  $$F(f_x) = \int_{-\infty}^{\infty} \mathrm{d}x\, f(x)\, \mathrm{e}^{-\mathrm{i}\, 2\pi f_x x}$$

  [68Goo, 68Pap, 78Loh, 78Gas, 93Sto, 05Hod].
  – The plane-wave structure $\exp(\mathrm{i}\,\omega t - \mathrm{i}\, kx)$ can be combined with

  $$F(f_x) = \int_{-\infty}^{\infty} \mathrm{d}x\, f(x)\, \mathrm{e}^{\mathrm{i}\, 2\pi f_x x}$$

  [71Col, 73Men, 92Lug], but

  $$F(f_x) = \int_{-\infty}^{\infty} \mathrm{d}x\, f(x)\, \mathrm{e}^{-\mathrm{i}\, 2\pi f_x x}$$

  is defined also in [88Kle, 91Sal, 95Wil, 96Ped].

– *Different approximations* in (3.1.37) and (3.1.38):

$$r_{\rm SP} \approx r_0 + \frac{2x\xi - \xi^2 + 2y\eta - \eta^2}{2r_0}$$

[99Bor, 68Pap, 78Gra] with r_0 from Fig. 3.1.8 versus

$$r_{\rm SP} \approx z + \frac{2xx' - x'^2 + 2yy' - y'^2}{2z}$$

(references on lasers: [86Sie, 05Hod], optoelectronics: [68Goo, 72Mar, 91Sal]) for *grating diffraction*: The sine of the diffraction angle $\sin\Theta_x = x/r_0$ is derived from principle and not by a postpositive reasoning of the paraxial range $x/z = \tan\Theta_x \approx \sin\Theta_x$. x/z should be "translated" into $\sin\Theta_x$ for better approximation.

(3.1.41)–(3.1.43): *Plane-wave spectrum or angular-spectrum representation* (also Rayleigh-Sommerfeld-Debye diffraction theory) [78Loh, 99Pau] is the plane-wave formulation of (3.1.34) [78Loh, 97For]. Application: see *Fourier optics* [68Goo, 83Ste, 93Sto].

(3.1.44), (3.1.45): *Generation of the far field in the focal plane of a lens*: $d \neq f$ (object is outside the object-side focal plane) $\Rightarrow$ additional phase term p to the pure (inverse) Fourier transform ($d = f$), similarly to (3.1.40).

Applications: generation of the spectrum of a function, possibility of mathematical operations in the Fourier-space with complex filtering masks, correlation and convolution.

Another important diffraction theory

Diffraction theory after *Young, Maggi, Rubinowicz* [66Rub, 99Pau]: The light in point P of Fig. 3.1.8 results from the unperturbed light and local waves, which are emitted by the edge of the aperture A. Therefore, a line integral is to be calculated [99Pau]. There is an equivalence with Fresnel-Kirchhoff's theory.

3.1.4.3 Time-dependent diffraction theory

Two formulations of the time-dependent treatment of diffraction are possible:

1. A general Fresnel-Kirchhoff's integral formula exists for *time-dependent source functions* in the aperture A, see [99Bor, 99Pau].
2. Used more often now [96Die, 99Pau]: The time-dependent source functions are decomposed into a *superposition of monochromatic fields*. The diffracted field is calculated for every monochromatic component by the stationary diffraction given above. The superposition of all diffracted monochromatic components yields the time-dependent diffracted field.

3.1.4.4 Fraunhofer diffraction patterns

3.1.4.4.1 Rectangular aperture with dimensions $2a \times 2b$

In Fig. 3.1.10 the geometry of the diffraction from a rectangular aperture $2a \times 2b$ is shown. The x-part of the diffraction pattern in Fig. 3.1.10 is given in Fig. 3.1.11. In Table 3.1.5 the zeros and maxima of the intensity distribution are listed.

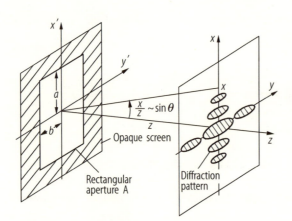
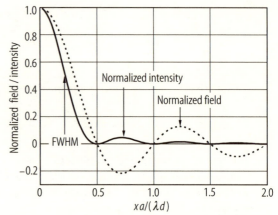

Fig. 3.1.10. Geometry of the diffraction from a rectangular aperture $2a \times 2b$.

Fig. 3.1.11. x-part of the diffraction pattern in Fig. 3.1.10. This is the diffraction pattern of a slit. For more exact electromagnetic solutions of a slit see [61Hoe, p. 266].

Table 3.1.5. Zeros and maxima of the intensity distribution.

Number n	$xa/\lambda z$	I_n/I_0
0	0	1
FWHM	2×0.221	0.5
1	0.5	0
1	0.715	0.0472
2	1	0
2	1.230	0.0168
3	1.5	0
3	1.735	0.0083
4	2	0
4	2.239	0.0050

Field distribution:

$$E(x,y,z) = \frac{4\,a\,b}{\mathrm{i}\,\lambda\,z}\, E_0 \exp\left\{-\mathrm{i}k\left(z + \frac{x^2+y^2}{2z}\right)\right\} \mathrm{sinc}\left\{\frac{2\pi a x}{\lambda z}\right\} \mathrm{sinc}\left\{\frac{2\pi b y}{\lambda z}\right\} \quad (3.1.47)$$

with $\mathrm{sinc}(x) = \dfrac{\sin x}{x}$ and E_0 the electric-field amplitude.

Intensity:

$$I(x,y,z) = I(0,0,z)\, \mathrm{sinc}^2\left\{\frac{2\pi a x}{\lambda z}\right\} \mathrm{sinc}^2\left\{\frac{2\pi b y}{\lambda z}\right\} . \quad (3.1.48)$$

If the Fraunhofer diffraction is observed in the focal plane, z has to be replaced by f.

3.1.4.4.2 Circular aperture with radius a

The circular aperture with radius a is discussed in [61Hoe, p. 453]. In Fig. 3.1.12 diffraction by a circular aperture is shown. In Fig. 3.1.13a the diffracted field and intensity and in Fig. 3.1.13b the encircled energy in the diffraction plane with a circular screen are given. The zeros and maxima of intensity for diffraction by a circular aperture are listed in Table 3.1.6.

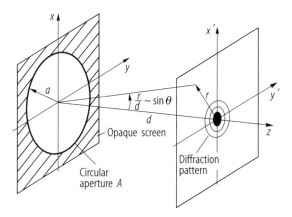

Fig. 3.1.12. Diffraction by a circular aperture.

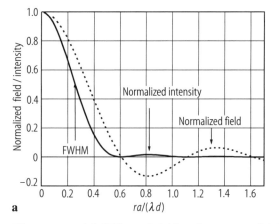

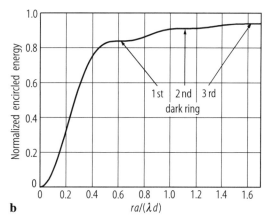

Fig. 3.1.13. (a) Diffracted field and intensity. (b) Encircled energy in the diffraction plane with a circular screen.

Table 3.1.6. Zeros and maxima of intensity for diffraction by a circular aperture.

Number n	$r_n a/(\lambda d)$	I_n/I_0
0	0	1
FWHM	2×0.257	0.5
1	0.610	0
1	0.817	0.0175
2	1.117	0
2	1.340	0.00415
3	1.619	0
3	1.849	0.00160
4	2.121	0
4	2.355	0.00078

Field distribution:

$$E(r,z) = \frac{\pi a^2}{\mathrm{i}\lambda z} E_0 \exp\left\{-\mathrm{i}k\left(z + \frac{kr^2}{2z}\right)\right\} \left\{2\,\frac{J_1[2\pi a r/(\lambda z)]}{2\pi a r/(\lambda z)}\right\} \tag{3.1.49}$$

with E_0 the electric-field amplitude and r the radius in the far-field plane.

Intensity:

$$I(r) = I(0,z) \left\{2\,\frac{J_1[2\pi a r/(\lambda z)]}{2\pi a r/(\lambda z)}\right\}^2 . \tag{3.1.50}$$

3.1.4.4.2.1 Applications

Airy's disc:

$$r_{1\,\text{Airy}} = 0.610\,\lambda/\sin\sigma \,, \tag{3.1.51}$$

1st-minimum radius of the intensity distribution in the focal plane of an aberration-free lens (Lommel 1885, Debye 1909, [86Sta, 99Bor]): Substitute in (3.1.50) $a/z \Rightarrow \sin\sigma$ (numerical aperture = sinus of the intersection angle σ with optical axis in the focal point, *generally: image point*) and $r = r_{1\,\text{Airy}}$ as above.

Annular aperture: obscuration of the central part in the circular aperture A of Fig. 3.1.12:

- Reduction of the central diffraction maximum width by $\approx 20\,\%$.
- Increase of secondary maximum by factor ≈ 7.
- See Bessel beams, Sect. 3.1.3.2.4, [05Hod].

3.1.4.4.3 Gratings

Grating equation:

$$\sin\alpha + \sin\beta = m\,\frac{\lambda}{g} \tag{3.1.52}$$

with

α: angle of incidence (see Fig. 3.1.14),
β: diffraction angle,
g: grating constant (grating period, groove distance),
m: order of diffraction. *Convention* [82Hut, p. 25] often used: If the diffraction order is on the same side with the zero order ($m = 0$) as the grating normal: $m > 0$, otherwise $m < 0$. In Fig. 3.1.14, the directions of the +1st transmitted order and the grating normal (dashed and dotted lines) are on the same side of the 0th transmitted order. Therefore $m = 1 > 0$.

Slit factor: represents the diffraction by a single slit of the grating. Its form regulates the energy distribution between the different orders m [82Hut, 99Bor]. For the real phase and reflection gratings, it is substituted by the *diffraction efficiency* curves in dependence on α or λ. There is an extreme diversity of cases. Catalogs of such curves: see [80Pet, 97Loe].

Theoretical spectral resolution of a grating:

$$R_{\text{theor}} = \lambda/(\Delta\lambda) = m\,N = W\,(\sin\alpha + \sin\beta)/\lambda \tag{3.1.53}$$

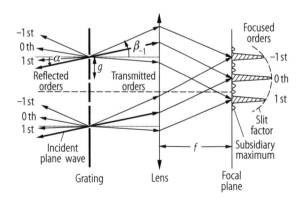

Fig. 3.1.14. Reflected and transmitted orders of a grating, here with $N = 4$ slits. The far-field distribution is *visualized* after focusing by an *ideal* lens. Between the main maxima occur $N - 2$ subsidiary maxima. The dashed envelope is the *slit factor*.

with

N : number of grooves of the grating,
W : width of the grating,
α, β : see (3.1.52).

Real resolution contains theoretical resolution and the aberrations of the optical elements for collimation and focusing of the grating-diffracted plane waves or by the aberrations of the *concave gratings* with imaging properties. [87Chr, 82Hut].

Holographical gratings [82Hut] show lower disturbations than mechanically produced gratings (application: external laser resonators).

Blazed gratings diffract light into an order m wanted with more than 60–90 % over one octave of wavelengths [80Pet, 82Hut, 97Loe].

Volume gratings: [81Sol, 81Rus].

Mountings of spectral devices: [82Hut].

3.1.4.5 Fresnel's diffraction figures

Fresnel's approximation is given in (3.1.39) in Table 3.1.4.

3.1.4.5.1 Fresnel's diffraction on a slit

In Fig. 3.1.15 Fresnel's diffraction pattern of a slit with width $2a$ is shown.

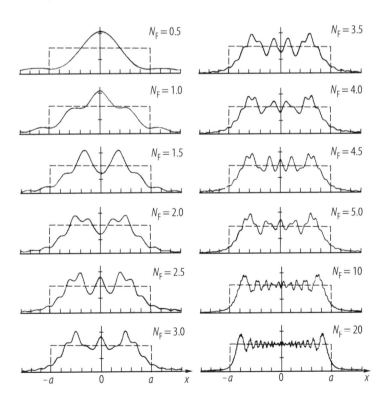

Fig. 3.1.15. Fresnel's diffraction pattern of a slit with width $2a$ (see Fig. 3.1.10 with $b \Rightarrow \infty$). Fresnel's number $N_F = a^2/(\lambda z)$ is the essential parameter to characterize the transition from farfield (Fraunhofer) approximation ($N_F < 0.2\ldots 0.5$) to near-field (Fresnel) approximation ($N_F > 0.5$). $N_F = 0.5$: one central maximum only, $N_F = 3$: three maxima, $N_F = N$: N maxima. *Hard-edge* diffraction results in a ripple in the near field, which can be avoided by *soft apertures*, for instance Gaussian-like [86Sie] (apodization in optics [99Bor]). Figure after [86Sie, p. 721].

3.1.4.5.2 Fresnel's diffraction through lens systems (paraxial diffraction)

Given: a system of lenses and the field distribution $E(x,y)$ to be propagated.

The *sequence of steps* easily taken is:

– Given: $E(x,y)$ in the plane $z = 0$. Required: the field in the plane $z = z$. Solution: (3.1.39).
– Given: $E(x,y)$ in the plane $z = 0$ and near to this plane a lens. Required: the field in the plane $z = z$. Solution: modification of (3.1.39) by an additional factor $L(x',y')$ to:

$$E_{\text{Fre}}(x,y,z) = \frac{\mathrm{i}\exp\{-\mathrm{i}kz\}}{\lambda d} \iint_A E(x',y',0)\, L(x',y')$$
$$\times \exp\left\{-\mathrm{i}\pi\frac{(x-x')^2 + (y-y')^2}{\lambda z}\right\} \mathrm{d}x'\,\mathrm{d}y' , \qquad (3.1.54)$$

$$L(x',y') = p(x',y')\, \exp\{-\mathrm{i}kn t_{\mathrm{L}}\}\, \exp\left\{\frac{\mathrm{i}k(x'^2 + y'^2)}{2f}\right\} \qquad (3.1.55)$$

with

n : refractive index of the lens,
t_{L} : thickness of the lens,
f : focal length of the lens,
$p(x',y')$: amplitude part, which can describe a marginal aperture or a Gaussian apodization.

A general complex function $L(x',y')$ can model diffractive optical elements.

Cases of integration:

– No transversal limitations (without stops) and quadratic arguments of the exponential functions due to analytical results. The *Collins integral* is the closed form of such a calculation (see Sect. 3.1.7.4).
– One stop (finite integration limits): The result includes the error function [70Abr].
– Two and more finite integration limits are not useful. Then, (commercial) numerical field propagation programs through systems should be consulted.

Examples: [68Goo, 91Sal, 71Col, 85Iiz, 92Lug, 68Pap].

The *Beam Propagation Method (BPM)* in integrated optics (many "infinitely thin lenses") is the generalization of this method [95Mae, 91Spl, 99Lau, 98Hec].

3.1.4.6 Fourier optics and diffractive optics

Fourier optics results from the transformation of the temporal frequency methods of electrical engineering to spatial frequency methods in optics, see Figs. 3.1.9, 3.1.10 and (3.1.41), (3.1.43), (3.1.44).

References: principles of Fourier optics: [68Goo, 78Loh, 83Ste, 85Iiz, 89Ars, 93Sto, 98Hec, 99Lau], filtering: [92Lug], filtering in connection with holography: [96Har, 71Col], noise suppression: [91Wyr].

Example 3.1.5. Spatial spectral filtering

In Fig. 3.1.16 low-pass filtering of a laser beam with a four-f-setup is shown.

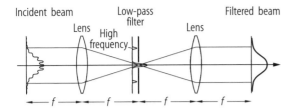

Fig. 3.1.16. Low-pass filtering of a laser beam with a four-f-setup [92Lug]. The mask is a low-pass filter, which transmits a zero mode only and suppresses the higher modes. The incident beam can also be modified by a transmission element which changes amplitude and phase.

Diffractive optical elements influence the propagation of light with help of amplitude- and/or phase-changing microstructures whose dimensions are of the order of the wavelength mostly. They extend the classical means of optical design. References: [67Loh, 84Sch, 97Tur, 00Tur, 00Mey, 01Jah].

Example 3.1.6.

– *Gratings* generated by mechanical or interference ruling [69Str, 67Rud] on either plane or concave substrates for the combination of dispersive properties with imaging [82Hut, 87Chr].
– *Fresnel's zone plates* acting as microoptic lenses of [97Her].
– *Mode transformation optics* ("modane") for transformation and filtering of modes of a laser [94Soi].
– Generation of theoretical ideal wavefronts for *optical testing* with interferometrical methods [95Bas, Vol. II, Chap. 31].
– Mode-discriminating and emission-forming elements in resonators [94Leg, 97Leg, 99Zei].

For pure imaging applications, refracting surfaces are still preferred, even in the micro-range [97Her]. Tasks with special dispersion requirements and special optical field transformations are the main application of the diffractive elements with increasing share.

The technology of dispersion compensation and weight reduction in large optical systems by special diffractive elements is partially solved, now.

3.1.5 Optical materials

Medium with absorption:

$$\hat{\varepsilon}_r = \hat{n}^2 \qquad (3.1.56)$$

with

$\hat{\varepsilon}_r$: complex relative dielectric constant (or tensor),
$\hat{n}$: complex refractive index,

weakly absorbing isotropic medium:

$$\alpha \ll k_0 : \quad \hat{n} = n - i\,k_e = n - i\,n\,\kappa = n - i\,\frac{\alpha}{2\,k_0}\,, \qquad (3.1.57)$$

damped plane wave (unity field amplitude):

$$\exp\{-i\,k z\} = \exp\{-i\,k_0(n - i\,\kappa)\,z\} \exp\left\{-i\,k_0\left(n - i\,\frac{\alpha}{2\,k_0}\right) z\right\}$$
$$= \exp\left\{-i\,k_0 z - \frac{\alpha}{2} z\right\}\,, \qquad (3.1.58)$$

intensity:

$$I(z) = I(0) \exp\{-\alpha z\} \quad (Lambert\text{-}Beer\text{-}Bouguer's\ law)\ , \tag{3.1.59}$$

amplification in pumped media:

$$I(z) = I(0) \exp\{g z\} \tag{3.1.60}$$

with

α [m^{-1}]: (linear) *absorption constant* (standard definition [95Bas, Vol. II, Chap. 35], [99Bor, 91Sal, 96Yar, 05Hod]) or extinction constant or attenuation coefficient,
g [m^{-1}]: *gain*,
k_e [m^{-1}]: [88Yeh, 95Bas] (or κ [m^{-1}]: [99Bor, 04Ber]) *extinction coefficient*, attenuation index.

Different *convention* after (3.1.6): α, g, k_e and κ are defined with other signs, for example $\hat{n} = n(1 + \mathrm{i}\kappa)$ if the other time separation (1$^\mathrm{st}$ convention) is used [99Bor, Chap. 13], [95Bas, Vol. I, Chap. 9].

Measurement of α: see [85Koh, 04Ber, 82Bru], [90Roe, p. 34], [95Bas, Vol. II, Chap. 35].

3.1.5.1 Dielectric media

In Fig. 3.1.17 the real- and imaginary part of the refractive index in the vicinity of a resonance in the UV are shown.

Single-resonance model for low-density media [99Bor, 96Ped]:

$$\begin{aligned}\hat{n} = n - \mathrm{i}\,k_\mathrm{e} &= 1 + \frac{N e^2}{2\varepsilon_0 m\,(\omega_0^2 - \omega^2 + \mathrm{i}\gamma\omega)} \\ &= \left\{1 + \frac{N e^2 \gamma\,(\omega_0^2 - \omega^2)}{2\varepsilon_0 m\,[(\omega_0^2 - \omega^2)^2 + \gamma^2\,\omega^2]}\right\} - \mathrm{i}\left\{\frac{N e^2 \gamma\,\omega}{2\varepsilon_0 m\,[(\omega_0^2 - \omega^2)^2 + \gamma^2\,\omega^2]}\right\}\end{aligned} \tag{3.1.61}$$

with

$e = -1.602 \times 10^{-19}$ C: elementary charge,
$m = 9.109 \times 10^{-31}$ kg: mass of the electron,
$\omega = 2\pi\nu$ [s^{-1}]: circular frequency of the light,
ω_0 [s^{-1}]: circular resonant frequency of the electron,

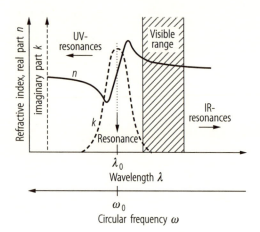

Fig. 3.1.17. Real- and imaginary part of the refractive index in the vicinity of a resonance in the UV. The principal shape is explained by the classical oscillator model after J.J. Thomson, P. Drude, and H.A. Lorentz [99Bor, 88Yeh].

γ [s^{-1}]: damping coefficient,
N [m^{-3}]: density of molecules,
$\varepsilon_0 = 8.8542 \times 10^{-12}$ As/Vm: electric permittivity of vacuum.

Examples see [96Ped, 88Kle], generalization to dense media see [96Ped, 88Kle, 99Bor].

The *Kramers-Kronig relation* connects $n(\omega)$ with $k(\omega)$ [88Yeh].

3.1.5.2 Optical glasses

Dispersion formula [95Bac]:

$$n^2(\lambda) = 1 + \frac{B_1 \lambda^2}{\lambda^2 - C_1} + \frac{B_2 \lambda^2}{\lambda^2 - C_2} + \frac{B_3 \lambda^2}{\lambda^2 - C_3} \quad (Sellmeier's\ formula)\ . \tag{3.1.62}$$

The dimensions of the constants are given in example 3.1.7. The available wavelength range is given by the transmission limits, usually.

Example 3.1.7. [96Sch]: Glass N-BK7: λ [µm], $B_1 = 1.03961212$, $B_2 = 2.31792344 \times 10^{-1}$, $B_3 = 1.01046945$, $C_1 = 6.00069867 \times 10^{-3}$ [µm^2], $C_2 = 2.00179144 \times 10^{-2}$ [µm^2], $C_3 = 1.03560653 \times 10^2$ [µm^2], $n(0.6328\ \text{µm}) = 1.51509$, $n(1.06\ \text{µm}) = 1.50669$.

Other interpolation formulae for $n(\lambda)$ are given in [95Bac], [95Bas, Vol. II, Chap. 32], [05Gro1, p. 121].

Further information is available from glass catalogs (see Sect. 3.1.5.10) and from subroutines in commercial optical design programs:

- *relative dispersive power* or *Abbe's number* $\nu_d = \dfrac{n_d - 1}{n_F - n_C}$ with $n_d(587.56\ \text{nm} = \text{yellow He-line})$, $n_F(486.13\ \text{nm} = \text{blue H-line})$, $n_C(656.27\ \text{nm} = \text{red H-line})$ [95Bac, 80Sch]; *application:* achromatic correction of systems [84Haf],
- spectral range of *transmission*,
- *temperature coefficients* of n and ν_d,
- *photoelastical coefficients*,
- *Faraday's effect* (Verdet's constant),
- chemical resistance, thermal conductivity, micro hardness etc.

Sellmeier-like formulae for crystals are available in [95Bas, Vol. II, Chap. 32]. Information in connection with laser irradiation damage is presented in [82Hac]. Specific values of *laser glasses* are given in tables in [01Iff].

3.1.5.3 Dispersion characteristics for short-pulse propagation

The parameters can be calculated from the dispersion interpolation (3.1.62) [91Sal, 96Die]:

$$\beta(\nu) = n(\nu) \frac{2\pi \nu}{c_0} \quad (propagation\ constant\ [\text{m}^{-1}])\ , \tag{3.1.63}$$

$$c_{\text{ph}} = \frac{c_0}{n(\nu)} \quad (phase\ velocity\ [\text{m s}^{-1}])\ , \tag{3.1.64}$$

$$v = \frac{2\pi}{\frac{d\beta}{d\nu}} = \frac{1}{\frac{d\beta}{d\omega}} \quad (group\ velocity\ [m\ s^{-1}]),\tag{3.1.65}$$

$$D_v = \frac{1}{2\pi}\frac{d^2\beta}{d\nu^2} = 2\pi\frac{d^2\beta}{d\omega^2} = \frac{d}{d\nu}\left(\frac{1}{v}\right) \quad (group\ velocity\ dispersion\ (GVD))\tag{3.1.66}$$

with

ν: frequency of light,
c_0: velocity of light in vacuum.

Application: Temporal pulse forming by the GVD of dispersive optical elements [96Die, 01Ben].

3.1.5.4 Optics of metals and semiconductors

The refractive index of *metals* is characterized by free-electron contributions ($\omega_0 = 0$ in (3.1.61)). One obtains from [67Sok, 72Woo], [95Bas, Vol. II, Chap. 35] with a plasma resonance (here collision-free: $\gamma = 0$):

$$n^2(\omega) = 1 - \left(\frac{\omega_p}{\omega}\right)^2\tag{3.1.67}$$

with

ω_p [s^{-1}]: plasma frequency, depending on free-electron density [88Kle].

From (3.1.67) follows

- $n(\omega) < 1$ for $\omega > \omega_p$, which means $\lambda < \lambda_p$ (example: $\lambda_p = 209$ nm for Na): transparency,
- pure imaginary $n(\omega)$ for $\omega < \omega_p$, $\lambda_p < \lambda$.

Other effects change the ideal case (3.1.67) [88Kle].

The complex refractive index of *semiconductors* is determined by transitions of electrons between or within the energy bands and by photon interaction with the crystal lattice (reststrahlen wavelength region). It depends strongly on the wavelength and is modified by heterostructures and dopands [71Pan, 95Kli], [95Bas, Vol. II, Chap. 36].

3.1.5.5 Fresnel's formulae

Fresnel's formulae describe the transmission and reflection of plane light waves at a plane interface between

- homogeneous isotropic media: [99Bor, 88Kle] and other textbooks on optics,
- homogeneous isotropic medium and anisotropic medium: special cases [99Bor, 86Haf] and other textbooks on optics,
- general case of anisotropic media: [58Fed],
- modification by photonic crystals: [95Joa, 01Sak].

Fresnel's formulae for the *amplitude* (field) *reflection and transmission coefficients* are listed in Table 3.1.7.

Plane of incidence: plane, containing the wave number vector $\boldsymbol{k}$ of the light and the normal vector $\boldsymbol{n}$ on the interface.

Table 3.1.7. Fresnel's formulae for the *amplitude* (field) *reflection and transmission coefficients*.

Case	The four values $\Theta, \Theta', \hat{n}$, and $\hat{n}'$ are considered	Using the angles Θ and Θ' only	$\sin \Theta'$ is eliminated $\bar{n} = \dfrac{\hat{n}'}{\hat{n}}$	
Reflection $r_s = \dfrac{E_s''}{E_s} =$	$\dfrac{\hat{n}\cos\Theta - \hat{n}'\cos\Theta'}{\hat{n}\cos\Theta + \hat{n}'\cos\Theta'}$	$-\dfrac{\sin(\Theta - \Theta')}{\sin(\Theta + \Theta')}$	$\dfrac{\cos\Theta - \sqrt{\bar{n}^2 - \sin^2\Theta}}{\cos\Theta + \sqrt{\bar{n}^2 - \sin^2\Theta}}$	(3.1.68)
Reflection $r_p = \dfrac{E_p''}{E_p} =$	$\dfrac{\hat{n}'\cos\Theta - \hat{n}\cos\Theta'}{\hat{n}'\cos\Theta + \hat{n}\cos\Theta'}$	$\dfrac{\tan(\Theta - \Theta')}{\tan(\Theta + \Theta')}$	$\dfrac{\bar{n}^2\cos\Theta - \sqrt{\bar{n}^2 - \sin^2\Theta}}{\bar{n}^2\cos\Theta + \sqrt{\bar{n}^2 - \sin^2\Theta}}$	(3.1.69)
Transmission $t_s = \dfrac{E_s'}{E_s} =$	$\dfrac{2\hat{n}\cos\Theta}{\hat{n}\cos\Theta + \hat{n}'\cos\Theta'}$	$\dfrac{2\sin\Theta'\cos\Theta}{\sin(\Theta + \Theta')}$	$\dfrac{2\cos\Theta}{\cos\Theta + \sqrt{\bar{n}^2 - \sin^2\Theta}}$	(3.1.70)
Transmission $t_p = \dfrac{E_p'}{E_p} =$	$\dfrac{2\hat{n}\cos\Theta}{\hat{n}'\cos\Theta + \hat{n}\cos\Theta'}$	$\dfrac{2\sin\Theta'\cos\Theta}{\sin(\Theta + \Theta')\cos(\Theta - \Theta')}$	$\dfrac{2\bar{n}\cos\Theta}{\cos\Theta + \bar{n}^2\sqrt{\bar{n}^2 - \sin^2\Theta}}$	(3.1.71)
Application of cases	Mostly used for pure dielectric media.	In a stack of films, the angles to the axis were calculated previously.	See remark in Sect. 3.1.5.5.	

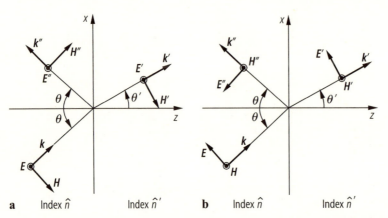

Fig. 3.1.18. Refraction at an interface, represented in the plane of incidence: (a) E_s-case, (b) E_p-case. The commonly used *convention* is shown for the orientation of the relevant vectors ($\mathbf{k}$: the wave number vector, $\mathbf{E}$: the electrical field, and $\mathbf{H}$: the magnetic field) ensuring that $\mathbf{k}$, $\mathbf{E}$, and $\mathbf{H}$ are a right-handed system in every case. The $\mathbf{E}$-field is important for the action on a nonmagnetic material.

Polarization:

- $\mathbf{E}$ *perpendicular* to the plane of incidence: s-polarization (TE-case or σ-case [88Kle]), the corresponding $\mathbf{E}$-component is called $E_\perp$ [99Bor] or E_s (s: "senkrecht" (German) which means "perpendicular") [88Yeh] or index E [97Hua] or index x [90Roe, 77Azz, 91Sal].
- $\mathbf{E}$ *parallel* to the plane of incidence: p-polarization (TM-case or π-case [88Kle]), the corresponding $\mathbf{E}$-component is called $E_\parallel$ [99Bor] or E_p [88Yeh] or index M [97Hua] or index y [90Roe, 77Azz, 91Sal].

Snell's law:

$$\hat{n}\,\sin\Theta = \hat{n}'\,\sin\Theta' \tag{3.1.72}$$

with

$\hat{n}, \hat{n}'$: refractive indices of both media, respectively,
Θ, Θ' : see Fig. 3.1.18.

Other convention than Fig. 3.1.18b [58Mac, 89Gha, 91Ish] (electrical engineering) on the orientation of the $\mathbf{E}$-vectors: $\mathbf{E}$ and $\mathbf{E}''$ point into the same direction for $\Theta \to 0$, $\mathbf{H}$ changes sign; application: $\mathbf{E}$-interferences.

Remark:

- $\hat{n}$ is real and $\hat{n}'$ is complex (*absorption* [76Fed, 77Azz] or *gain* [88Boi]).
- $\hat{n}$ and $\hat{n}'$ are real and $\bar{n} = \dfrac{\hat{n}'}{\hat{n}} < 1$ and $(\bar{n}^2 - \sin^2\Theta) < 0$ (*total reflection*). Then $\sqrt{\bar{n}^2 - \sin^2\Theta} = i\sqrt{\sin^2\Theta - \bar{n}^2}$ yields for (3.1.68) and (3.1.69) $r_s = \exp(i\,\delta_s)$ and $r_p = \exp(i\,\delta_p)$ (modulus = 1, all energy reflected) and $\tan\dfrac{\delta_s}{2} = -\dfrac{\sqrt{\sin^2\Theta - \bar{n}^2}}{\cos\Theta}$ and $\tan\dfrac{\delta_p}{2} = -\dfrac{\sqrt{\sin^2\Theta - \bar{n}^2}}{\bar{n}^2\cos\Theta}$.

The intensities in the media are calculated with help of the z-component of Poynting's vector [88Kle, 90Roe, 76Fed].

Reflectance (reflected part of intensity):

$$R_{s,p} = |r_{s,p}|^2 \,. \tag{3.1.73}$$

Transmittance (transmitted part of intensity):

$$T_{s,p} = \frac{\text{Re}(\hat{n}' \cos \Theta')}{\text{Re}(\hat{n} \cos \Theta)} |t_{s,p}|^2 \qquad (3.1.74)$$

with

Re: real part.

Energy conservation:

$$T_{s,p} + R_{s,p} = 1 .$$

3.1.5.6 Special cases of refraction

3.1.5.6.1 Two dielectric isotropic homogeneous media ($\hat{n}$ and $\hat{n}'$ are real)

$$r_s = \frac{n - n'}{n + n'} = -r_p . \qquad (3.1.75)$$

(The negative sign of r_p results from the convention of Fig. 3.1.18 that E_p is diffracted into $-E_p''$).

$$R_s = R_p = \left(\frac{n - n'}{n + n'}\right)^2 \quad \text{and} \quad T_s = T_p = 1 - R_s . \qquad (3.1.76)$$

Example 3.1.8. $n = 1$, $n' = 1.5$ (glass): $R_s = 0.04$.

3.1.5.6.2 Variation of the angle of incidence

3.1.5.6.2.1 External reflection ($n < n'$)

Brewster's angle (angle of polarization) Θ_B:

$$\Theta_B + \Theta_B' = 90°, \quad R_p = 0, \quad \tan \Theta_B = \frac{n'}{n} . \qquad (3.1.77)$$

Example 3.1.9. $n = 1$, $n' = 1.5$, $\Theta_B = 56.3°$. See Fig. 3.1.19.

3.1.5.6.2.2 Internal reflection ($n > n'$)

Critical angle of total reflection:

$$\sin \Theta_C = \frac{n'}{n} . \qquad (3.1.78)$$

Total reflection: $\Theta > \Theta_C$ with $|r_s| = |r_p| = 1$ and the phases of the reflected waves: $r_s = \exp(i\Phi_s)$ and $r_p = \exp(i\Phi_p)$.

Brewster's angle:

$$\tan \Theta_B = \frac{n'}{n} . \qquad (3.1.79)$$

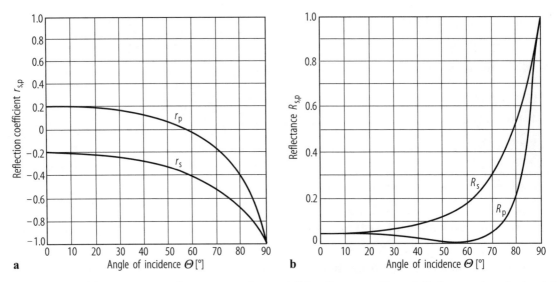

Fig. 3.1.19. (a) Reflection coefficients r_p and r_s and (b) reflectances R_p and R_s for $n = 1$ and $n' = 1.5$.

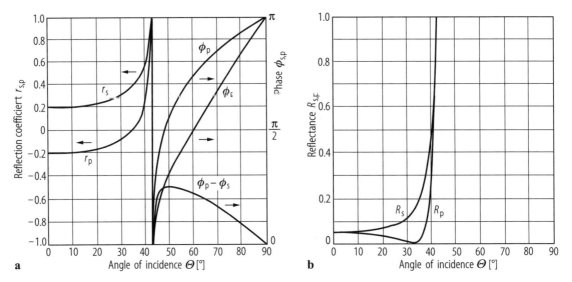

Fig. 3.1.20. Internal reflection ($n = 1.5$, $n' = 1$). (a) Reflection coefficients r_p and r_s for $\Theta < \Theta_C$ and phases Φ_p and Φ_s for $\Theta > \Theta_C$. (b) Reflectances R_p and R_s (= 1 for $\Theta > \Theta_C$).

Example 3.1.10. $n = 1.5$, $n' = 1$, $\Theta_C = 41.8°$, $\Theta_B = 33.7°$. See Fig. 3.1.20.

Penetration depth in Fig. 3.1.21 [88Kle, p. 67]:

$$d_{\mathrm{pen}} = \frac{\lambda_0}{2\pi\sqrt{n^2 \sin^2\Theta - n'^2}} \,. \tag{3.1.80}$$

Goos–Hänchen shift [88Yeh, p. 74], see Fig. 3.1.22:

$$d_{\mathrm{G.-H.},s,p} = \frac{\mathrm{d}\Phi_{s,p}}{\mathrm{d}\Theta} \tag{3.1.81}$$

with Φ_p and Φ_s from Fig. 3.1.20. For a more precise treatment of the Goos–Hänchen shift for Gaussian beams see [05Gro1, p. 100].

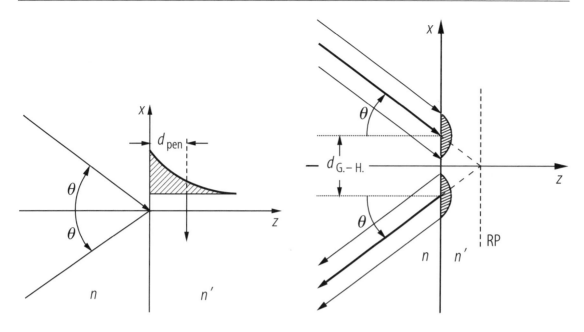

Fig. 3.1.21. Total reflection of plane waves with an inhomogeneous wave in the medium with the refractive index n' (d_{pen}: amplitude $\Rightarrow 1/e$).

Fig. 3.1.22. Goos–Hänchen shift of a total reflected beam with finite (exaggerated small) cross section (RP: effective reflection plane).

3.1.5.6.3 Reflection at media with complex refractive index (Case $\hat{n} = 1$ and $\hat{n}' = n' + \mathrm{i}k'$)

In Fig. 3.1.23 the refractive index n and the attenuation index k of gold (Au) is shown, in Fig. 3.1.24 the reflectance for both polarization cases of gold is given.

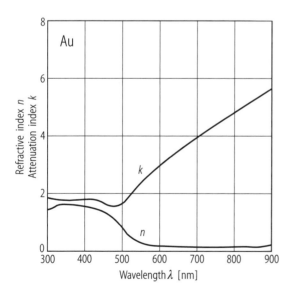

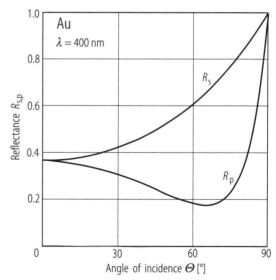

Fig. 3.1.23. Refractive index n and attenuation index k of gold (Au).

Fig. 3.1.24. Reflectance for both polarization cases of gold (Au). There is a minimum of R_p which is connected with a *pseudo Brewster angle*.

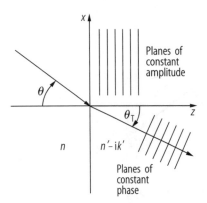

Fig. 3.1.25. Refraction at a medium with absorption: generation of an inhomogeneous wave.

Inhomogeneous wave (Fig. 3.1.25): *Snell's refraction law* is modified:

$$\sin\Theta_T = \frac{n}{n_T}\sin\Theta \tag{3.1.82}$$

with

$$2n_T^2 = n'^2 - k'^2 + n^2\sin^2\Theta + \sqrt{\left(n'^2 - k'^2 - n^2\sin^2\Theta\right)^2 + 4n'^2k'^2} \quad \text{(Ketteler's formula)}.$$

The effective refractive index n_T determines the direction angle Θ_T of planes of constant phase in Fig. 3.1.25 via (3.1.82) [88Kle, p. 78], [41Str, p. 503], [99Bor, p. 740]. The full inhomogeneous wave can be calculated using [99Bor, p. 740].

Example 3.1.11. $\Theta = 45°$, Au: $\lambda = 800$ nm, $n' = 0.19$, $k' = 4.9$, $n_T = 0.73$, $\Theta_T = 75.1°$ (see [28Koe, p. 209]).

Intensity attenuation in the case $\Theta = 0°$:

$$I = I_0 \exp\left\{-2\left(\omega/c\right)k'z\right\}. \tag{3.1.83}$$

Example 3.1.12. $\Theta = 0°$, Au: $\lambda = 800$ nm, $n' = 0.19$, $k' = 4.9$, $I = I_0 \exp\left(-7.7 \times 10^4 z[\text{mm}]\right)$, $1/e-\text{depth} = 13$ nm.

Ellipsometry: $\delta_p - \delta_s$ and moduli $\frac{|r_p|}{|r_s|}$ of the reflected light $r_p = |r_p|\exp(i\delta_p)$ and $r_s = |r_s|\exp(i\delta_s)$ can be measured. The complex refractive index of a material results [77Azz, 90Roe]. Application: Measurements for the optical constants of metals, semiconductors, and thin-film systems.

3.1.5.7 Crystal optics

3.1.5.7.1 Classification

The dielectric tensor $\varepsilon_r = \varepsilon_{ij}$ in (1.1.8) is symmetrical and real in the case of a nonabsorbing medium.

In Fig. 3.1.26 vectors connected with wave propagation in crystal optics are depicted. In Table 3.1.8 optical crystals are listed. In Table 3.1.9 three of the eight surfaces for visualization of wave propagation in crystals are presented.

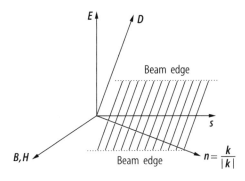

Fig. 3.1.26. Vectors connected with wave propagation in crystal optics [99Bor]: s: ray direction unit vector $\parallel$ Poynting vector $\boldsymbol{E} \times \boldsymbol{H}$, $\boldsymbol{n}$: unit vector in the normal direction $\parallel \boldsymbol{k}$ and $\perp$ phase planes, orthogonalities: $\boldsymbol{B}, \boldsymbol{H} \perp \boldsymbol{E}, \boldsymbol{D}, \boldsymbol{n}, \boldsymbol{s}$; $\boldsymbol{E} \perp \boldsymbol{s}$; $\boldsymbol{D} \perp \boldsymbol{n}$.

Table 3.1.8. Optical crystals.

Classification: system (syngony) of crystal	Refractive index in the main axis system	Optical type of crystal	Example	Values of the refractive index for $\lambda = 589.3$ nm
triclinic, monoclinic, orthorhombic	$n_x \neq n_y \neq n_z \neq n_x$	biaxial crystal, no ordinary waves	NaNO$_3$	$n_x = 1.344$, $n_y = 1.411$, $n_z = 1.651$
trigonal, tetragonal, hexagonal	$n_x = n_y = n_o$ (ordinary index)	positive uniaxial crystal: $n_o < n_e$	SiO$_2$ (quartz)	$n_o = 1.544$, $n_e = 1.553$
	$n_x \neq n_z = n_e$ (extraordinary index)	negative uniaxial crystal: $n_o > n_e$	CaCO$_3$ (calcite)	$n_o = 1.658$, $n_e = 1.486$
cubic	$n_x = n_y = n_z = n$	isotropic crystal	NaCl	$n = 1.544$

Table 3.1.9. Three of the eight surfaces for visualization of wave propagation in crystals.

Surface	Given	Found by construction are the
Index ellipsoid (indicatrix) (one-shell surface)	normal direction $\boldsymbol{n}$	$\boldsymbol{D}$-vectors for the two polarization cases and the two refractive indices for phase propagation
Index surface, wave vector surface (two-shell surface)	normal direction $\boldsymbol{n}$	ray directions $\boldsymbol{s}$, which are perpendicular to the surface for both polarization cases
Ray surface, wave surface, representing Huygens' elementary wave for both polarization cases (two-shell surface)	ray direction $\boldsymbol{s}$	normal direction $\boldsymbol{n}$, which is perpendicular to the surface

Main feature of *crystal optics*: $\boldsymbol{s}$ *is not parallel with* $\boldsymbol{n}$ for wave propagation, mostly.

– $\boldsymbol{s}$ is essential for description of the energy propagation (edges of bundles, rays),
– $\boldsymbol{n}$ is essential for description of the interferences of infinite broad waves.

References: [28Szi, 54Bel, 58Shu, 61Ram, 76Fed, 79Wah, 84Yar, 04Ber, 99Pau, 99Bor]. A detailed comparison between that surfaces is given in [28Szi].

3.1.5.7.2 Birefringence (example: uniaxial crystals)

Uniaxial crystals in the plane of incidence:

- *Refraction of the normal direction* $\mathbf{n}$ *of wavefronts*: The wavevector surface is shown in Fig. 3.1.27.

$$\sin\Theta_o = \frac{n}{n_o}\sin\Theta \quad (\text{ordinary wave } (\mathbf{k}_o)) \tag{3.1.84}$$

(n_o does not depend on the angle of incidence),

$$\sin\Theta_e = \frac{n}{n_{\theta\,e}(\Theta_e(\Theta))}\sin\Theta \quad (\text{extraordinary wave } (\mathbf{k}_e)) \tag{3.1.85}$$

(n_e depends on the angle of incidence).

- *Refraction of rays* (Poynting vector): $\mathbf{s}_e$ and $\mathbf{s}_o$ are given by tangent construction in Fig. 3.1.28.
- *Algorithm* for the calculation of $\mathbf{k}_o$ ($\|\mathbf{s}_o$), $\mathbf{k}_e$, $\mathbf{s}_e$ of Fig. 3.1.28 with n, n_o, n_e, η, θ of Fig. 3.1.29 [86Haf]:

$$n_{\theta\,e}^2 = \frac{A}{B} + \frac{n^2(n_o^2 - n_e^2)^2 \sin^2\Theta \, \sin^2(2\eta)}{2B^2}$$
$$\pm \frac{n(n_o^2 - n_e^2)^2 \sin\Theta \sin(2\eta)}{B} \times \sqrt{n^2 \sin^2\Theta \left[\frac{(n_o^2 - n_e^2)^2 \sin^2(2\eta)}{4B^2} - 1\right] + \frac{A}{B}} \tag{3.1.86}$$

(*refractive index for the extraordinary wave*)

with

$$A = (n_e^2 - n_o^2)\,n^2 \sin^2\Theta \, \cos(2\eta) - n_o^2\,n_e^2 \,,$$
$$B = n_o^2 + (n_e^2 - n_o^2)\sin^2\eta \,,$$

where the decision on the $\pm$ sign in (3.1.86) can be made by controlling the satisfaction of

$$n_{\theta\,e}^2 \left[n_o^2 + (n_e^2 - n_o^2)\sin^2(\eta + \Theta_e)\right] = n_e^2\,n_o^2 \,.$$

The resulting angles are:

$$\Theta_o = \arcsin(n \sin\Theta/n_o) \,, \tag{3.1.87}$$

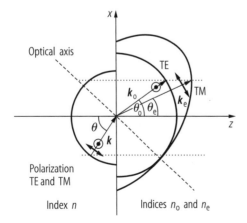

Fig. 3.1.27. Construction of wavefront birefringence with the wavevector surface: The wavefronts show no transversal limitation.

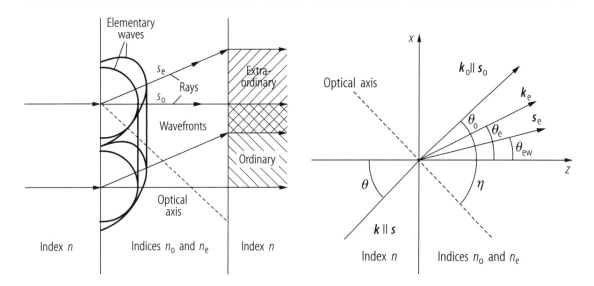

Fig. 3.1.28. Huygens' tangent construction of birefringence in a crystal slab for transversal-limited beams.

Fig. 3.1.29. Refraction for normal and ray directions. η: angle between z-axis and optical axis.

$$\Theta_e = \arcsin(n \sin\Theta / n_{\Theta e}) ,\tag{3.1.88}$$

$$\Theta_{ew} = \arctan \frac{\tan\eta - C}{1 + C \tan\eta} \tag{3.1.89}$$

with

$$C = \frac{n_o^2}{n_e^2} \times \frac{\sqrt{n_{\Theta e}^2 - n^2 \sin^2\Theta}\ \tan\eta + n \sin\Theta}{\sqrt{n_{\Theta e}^2 - n^2 \sin^2\Theta}\ - n \sin\Theta \tan\eta} .$$

Application: Θ_e, Θ_o, n_o, and $n_{\Theta e}$ ⇒ phase differences (interferences) and reflection coefficient, Θ_o and Θ_{ew} ⇒ ray separation in a crystal.

Example 3.1.13. Calcite: $n_o = 1.658$, $n_e = 1.486$, $\eta = 45°$: #1: $\Theta = 0°$: $C = 1.244822$, $n_{\Theta e} = 1.565$, $\Theta_o = \Theta_e = 0°$, $\Theta_{ew} = -6.224°$; #2: $\Theta = 45°$: $n_{\Theta e} = 1.636$, $C = 0.438329$, $\Theta_o = 25.23°$, $\Theta_e = 25.6°$, $\Theta_{ew} = 21.33°$.

General formulation of (3.1.85)–(3.1.89): see [76Fed, Table 9.1] for more detailed discussions.

3.1.5.8 Photonic crystals

Starting with the forbidden (stop) bands in case of multi-layer Bragg reflection [88Yeh, p. 123] a material class is under development which stops light propagation along as many directions and for as many wavelengths as possible. This suppresses the spontaneous emission for laser applications and opens new possibilities in the micro- and nano-optics [95Joa, 01Sak, 04Bus]. Photonic crystal fibers [04Bus] can be designed for special light propagation properties and high-power fiber lasers [03Wad].

3.1.5.9 Negative-refractive-index materials

The common excitation of electrical dipoles and magnetical dipoles by light in a medium can result in a negative dielectric permittivity $\text{Re}(\varepsilon) < 0$ in combination with a negative magnetic permeability $\text{Re}(\mu) < 0$. Then, in Snell's law (3.1.72) an effective index $\hat{n}' < 0$ is possible [68Ves] which results in imaging by a slab of this material without curved surfaces [00Pen] and other interesting effects [05Ram]. Such metamaterials can be generated by microtechnology, now for mm- and terahertz-waves, but with the trend towards visible radiation [05Ele].

3.1.5.10 References to data of linear optics

[62Lan] contains optical constants, only. In later editions, the optical constants are listed together with other properties of substances. An overview is given in the content volume [96Lan].

Optical glass:	[62Lan, Chap. 283], [97Nik], [95Bas, Vol. 2, Chap. 33], catalogs of producers: [96Sch, 98Hoy, 96Oha, 92Cor], and commercial optical design programs.
Infrared materials:	[98Pal, 91Klo], [96Sch, infrared glasses], commercial optical design programs.
Crystals:	[62Lan, Chap. 282], [95Bas, Vol. 2, Chap. 33], [97Nik, 91Dmi, 81Kam].
Photonic crystals:	[95Joa, 01Sak, 04Bus].
Negative-refractive-index materials:	[05Ram].
Polymeric materials:	[62Lan, Chap. 283], [95Bas, Vol. 2, Chap. 34], [97Nik].
Metals:	[62Lan, Chap. 281], [98Pal], [95Bas, Vol. 2, Chap. 35].
Semiconductors:	[96Lan, 98Pal, 87EMI], [95Bas, Vol. 2, Chap. 36].
Solid state laser materials:	[01Iff, 97Nik, 81Kam].
Liquids:	[62Lan, Chaps. 284, 285], [97Nik].
Gases:	[62Lan, Chap. 286].

3.1.6 Geometrical optics

Geometrical optics represents the limit of the wave optics for $\lambda \Rightarrow 0$.

The development $\sin\sigma = \sigma - \frac{1}{3!}\sigma^3 + \frac{1}{5!}\sigma^5 - \ldots$ with σ the angle in Snell's law characterizes the different approaches of geometrical optics. Table 3.1.10 gives an overview of different approximations of geometrical optics.

3.1.6.1 Gaussian imaging (paraxial range)

The signs of the parameters determined in [03DIN, 96Ped] are applied in Sect. 3.1.6.1.1, later on $f = f'$ is used.

Table 3.1.10. Different approximations of geometrical optics.

Problem to be treated	Algorithm for solving
Given: object point O in the paraxial range, *asked*: image point O' in the *paraxial range* approximation: $\sin \sigma \approx \sigma$.	– Gaussian collineation and Listing's construction: see Sect. 3.1.6.1. – Gaussian matrix formalism (**ABCD**-matrix): see Sect. 3.1.6.2, ref.: [04Ber, 99Bor].
Imaging in *Seidel's range*, *asked*: imaging quality approximation: $\sin \sigma \approx \sigma - \frac{1}{3!}\sigma^3$.	Formulae for Seidels aberrations: see Sect. 3.1.6.3, ref.: [70Ber, 80Hof, 84Haf, 84Rus, 86Haf, 91Mah].
General image formation.	(Commercial) raytracing programs with geometric and wave optical merit functions and tolerancing, ref.: [84Haf, 86Haf].

3.1.6.1.1 Single spherical interface

Figure 3.1.30 shows the imaging by a spherical interface in the paraxial range (small x, x', h).

Gaussian imaging equation:

$$n\left(\frac{1}{r} - \frac{1}{s}\right) = n'\left(\frac{1}{r} - \frac{1}{s'}\right) \quad \text{or} \quad \frac{n'}{s'} = \frac{n}{s} + \frac{n'-n}{r}. \tag{3.1.90}$$

Abbe's invariant $n\left(\dfrac{1}{r} - \dfrac{1}{s}\right)$ is a constant on both sides of the interface.

Object-space focal length:

$$f = -\frac{nr}{n'-n}. \tag{3.1.91}$$

Image-side focal length:

$$f' = \frac{n'r}{n'-n}. \tag{3.1.92}$$

Remark: The symbol f means outside this section, Sect. 3.1.6.1, the positive focal length for a positive (converging) lens.

Newton's imaging equation:

$$z\,z' = f\,f'. \tag{3.1.93}$$

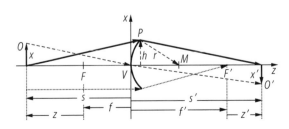

Fig. 3.1.30. Imaging by a spherical interface in the paraxial range (small x [object height], x' [image height], h [zonal height]). Full line: axial imaging, dashed line: off-axis imaging, dotted line: focusing to image side F'. *Sign conventions*: $s, s' > 0$, if they point to the right-hand side of the vertex V, $r > 0$, if the center of curvature of the interface is on the right-hand side in comparison with V. Here: $s < 0$, $s' > 0$, $r > 0$. M: center of curvature of the sphere. The left-hand-side-directed arrows symbolize negative values for the corresponding parameters here.

Lagrange's invariant:

$$x' n' s' = x n s \tag{3.1.94}$$

with

- n: object-space refractive index,
- n': image-space refractive index,
- s: object distance,
- s': image distance,
- r: radius of curvature of the interface,
- x: height of the object point,
- x': height of the image point,
- z: focus-related object distance,
- z': focus-related image distance.

Imaging through an optical system: concatenation of the imaging of the spherical surfaces in succession via (3.1.90) by using $s_{\text{following surface}} = s'_{\text{prior surface}} - d$, d: the distance between the surfaces, and (3.1.94) for an object height $x \neq 0$.

3.1.6.1.2 Imaging with a thick lens

Figure 3.1.31 shows the axial imaging with a thick lens, Fig. 3.1.32 depicts *Listing's construction* for thick-lens imaging of a finite-height object point O to image point O'.

Thick-lens imaging equation:

$$-\frac{1}{a} + \frac{1}{a'} = \frac{1}{f'} = (n'-1)\left(\frac{1}{r_1} - \frac{1}{r_2}\right) + \frac{t(n'-1)^2}{n' r_1 r_2}. \tag{3.1.95}$$

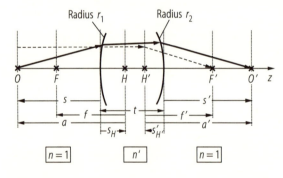

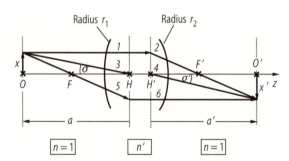

Fig. 3.1.31. Axial imaging with a thick lens. *Cardinal planes and points* are: *object-space principal plane* with object principal point H on axis, *image-space principal plane* with image principal point H' on axis, *object-space focal point* F, *image-space focal point* F'. *Nodal points* [98Mah, 96Ped] are equal to the principal points if O and O' are embedded in media with equal refractive index as here. Then $f = -f'$. The *sign convention* used here means: Parameters characterized by an arrow pointing to the left (right) hand side show a negative (positive) sign [80Hof, 86Haf]. The dashed line shows the use of H' for simplifying the plot for a ray focusing.

Fig. 3.1.32. *Listing's construction* for thick-lens imaging of a finite-height object point O to image point O'. Scheme of construction: Ray 1 (parallel with axis) is sharply bent at plane H' towards F'. Ray 3 towards H is continued at H' with the angle $\sigma' = \sigma$. Ray 5 through F is bent sharply parallel with axis at H-plane. The magnification $x'/x = a'/a$ can be calculated by elimination of a' from (3.1.95) $\Rightarrow x'$.

Position of the principal point H:

$$s_H = -\frac{n'-1}{n'r_2} f't \,. \tag{3.1.96}$$

Position of the principal point H':

$$s_{H'} = -\frac{n'-1}{n'r_1} f't \,. \tag{3.1.97}$$

Distance between the principal planes:

$$\overline{HH'} = t\left(1 - f'\frac{n'-1}{n'}\left(\frac{1}{r_1} - \frac{1}{r_2}\right)\right) \,. \tag{3.1.98}$$

Thin lens: $t \Rightarrow 0$: (3.1.95) $\Rightarrow$ "Lens maker's formula".

3.1.6.2 Gaussian matrix (ABCD-matrix, ray-transfer matrix) formalism for paraxial optics

Three tasks can be treated with the help of the ray-transfer matrix:

1. full description of *paraxial optics* (this section, Sect. 3.1.6.2),
2. *Gaussian beam propagation (coherent radiation)* by combination with a special beam calculation algorithm (see Sect. 3.1.7 on beam propagation),
3. *propagation of the second-order moments of the radiation field (inclusion of partial coherent radiation)* (see Chap. 2.2 on beam characterization).

The optical system can be the separating distance in an optical medium, a single spherical optical surface or a true, more complicated optical system.

There are *different definitions* for the ABCD-matrices:

Here: The slope components of the input and output rays are the real angles without any relation to the refractive indices at input and output spaces of Fig. 3.1.33 [66Kog1, 66Kog2, 84Hau, 91Sal, 95Bas, 96Ped, 96Yar, 98Hec, 98Sve, 01Iff, 05Gro1, 05Hod]. Then, the determinant of the matrix M: $\|M\| = n'/n$ with n the index of the medium of the input plane and n' the index of the medium of the output plane.

Other authors [75Ger, 86Sie, 88Kle, 04Ber] use:

slope parameter = (angle) × (related refractive index). Then the equation $\|M\| = 1$ applies.

In Fig. 3.1.34 the concatenation of different ray-transfer matrices for different types of subsystems is shown.

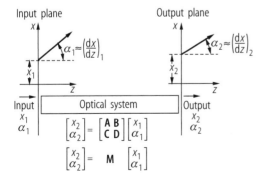

Fig. 3.1.33. Transfer of the input height x_1 and slope α_1 into the output height x_2 and slope α_2 with the ray-transfer matrix M. The *sign* of slope α_1 is positive in this figure. The German standard DIN 1335 uses a different sign with *change of some signs in the* ABCD *matrices* [96Ped].

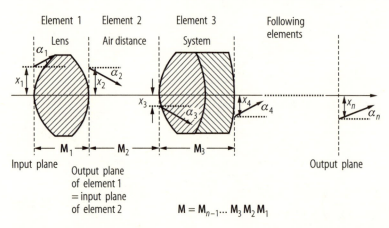

Fig. 3.1.34. Concatenation of different ray-transfer matrices for different types of sub-systems. Matrices known for systems before can be used to construct the matrix for a larger system containing the known systems. The sequence of the matrices is shown at the bottom of the figure.

3.1.6.2.1 Simple interfaces and optical elements with rotational symmetry

In Table 3.1.11 ABCD-matrices for simple interfaces and optical elements with rotational symmetry are listed.

3.1.6.2.2 Non-symmetrical optical systems

Rotational symmetry lacks and the axis is tilted due to the *non-symmetrical optical system*. In such a system, the central ray of imaging is called the basic ray. The optics in a narrow region around the basic ray is called *parabasal optics* [95Bas, Vol. 1, p. 1.47] as analogon to paraxial optics. For treatment of astigmatic pencils see [72Sta].

A special case of the non-symmetrical optical system is a *system without torsion*: Two orthogonal cases do not mix during propagation. Examples are different setups of spectroscopy and laser physics (ring resonators).

In Table 3.1.12 ABCD-matrices for non-symmetrical optical elements without torsion are listed.

3.1.6.2.3 Properties of a system

Properties of a system included in its ABCD-matrix are discussed in [75Ger, 96Ped, 05Hod, 05Gro1].

In Table 3.1.13 distances between *cardinal elements* of an optical system are listed, in Table 3.1.14 the meaning of the vanishing of different elements of the ABCD-matrix is depicted.

3.1.6.2.4 General parabolic systems without rotational symmetry

The generalization of the two-dimensional ray transfer after Fig. 3.1.33 to three dimensions [69Arn] is shown in Fig. 3.1.35. The ray in the input plane is characterized by two coordinates x_1 and y_1 of the piercing point P and two small (paraxial range) angles α_1 and β_1.

The matrix S relates these parameters to the corresponding parameters in the output plane like in Fig. 3.1.33:

Table 3.1.11. ABCD-matrices for simple interfaces and optical elements with rotational symmetry.

Effect	Figure	ABCD-matrix	Remark
Propagation		$\begin{bmatrix} 1 & d \\ 0 & 1 \end{bmatrix}$	The rays propagate from I to O within the same medium.
Spherical surface		$\begin{bmatrix} 1 & 0 \\ \dfrac{n_1 - n_2}{n_2 r} & \dfrac{n_1}{n_2} \end{bmatrix}$	Sign: $r > 0$ for convex surface seen by the propagating light.
Plane		$\begin{bmatrix} 1 & 0 \\ 0 & \dfrac{n_1}{n_2} \end{bmatrix}$	Corresponds to a spherical surface with $r \Rightarrow \infty$.
Planar plate		$\begin{bmatrix} 1 & \dfrac{n_1}{n_2} d \\ 0 & 1 \end{bmatrix}$	Contains two refractions.
Thin lens		$\begin{bmatrix} 1 & 0 \\ -\dfrac{1}{f} & 1 \end{bmatrix}$	$\dfrac{1}{f} = \dfrac{n_2 - n_1}{n_1} \left[\dfrac{1}{r_1} - \dfrac{1}{r_2} \right]$, air: $n_1 = 1$.
Thick lens in air		$\begin{bmatrix} 1 - \dfrac{s_{H'}}{f} & \dfrac{d}{n} \\ -\dfrac{1}{f} & 1 + \dfrac{s_H}{f} \end{bmatrix}$	$\dfrac{1}{f} = (n-1)\left[\dfrac{1}{r_1} - \dfrac{1}{r_2}\right] + \dfrac{(n-1)^2 t}{n r_1 r_2}$, $s_H = -\dfrac{(n-1) f t}{n r_2}$, see (3.1.96), $s_{H'} = -\dfrac{(n-1) f t}{n r_1}$, see (3.1.97), H, H': principal planes.
Spherical mirror		$\begin{bmatrix} 1 & 0 \\ -\dfrac{2}{r} & 1 \end{bmatrix}$	Unfolding of the mirror; sign$(r) > 0$, if the incident light sees a concave mirror surface.
Gradient-index lens or thermal lens		$\begin{bmatrix} A & B \\ C & D \end{bmatrix}$ $n = n_0 (1 - \gamma x^2)$; $\gamma > 0$: higher index on axis	$A = \cos\left(\sqrt{2\gamma}\, t\right)$; $B = n_1 \sin\left(\sqrt{2\gamma}\, t\right) / \left(n_0 \sqrt{2\gamma}\right)$; $C = -\left(\sqrt{2\gamma}\, n_0/n_1\right) \sin\left(\sqrt{2\gamma}\, t\right)$; $D = \cos\left(\sqrt{2\gamma}\, t\right)$; development of the trigonometric functions for $\sqrt{2\gamma}\, t \ll 1 \Rightarrow$ simplifications. Gradient optics: see [02Gom, 05Gro1].

(continued)

Table 3.1.11 continued.

Effect	Figure	ABCD-matrix	Remark
Gaussian apodization, usable for q-parameter transfer (Table 3.1.18)		$\begin{bmatrix} 1 & 0 \\ -\dfrac{\mathrm{i}\lambda a}{2\pi} & 1 \end{bmatrix}$ λ: wavelength of light	The amplitude transmission function between I and O is $\exp\left(-a\,x^2/2\right)$, x: transverse coordinate [86Sie, p. 787]

Remark: Other treatments of the mirror see [86Sie, 98Sve, 75Ger].

Table 3.1.12. ABCD-matrices for non-symmetrical optical elements without torsion.

Effect	Figure	ABCD-matrix	Remark
Refraction at a sphere Tangential (meridional) plane		$\begin{bmatrix} \dfrac{\cos(\theta_1)}{\cos(\theta_2)} & 0 \\ \dfrac{\Delta n_\mathrm{t}}{r\,n_2} & \dfrac{n_1\cos(\theta_2)}{n_2\cos(\theta_1)} \end{bmatrix}$	$n_1\sin(\theta_1) = n_2\sin(\theta_2)$ (Snell's law) $\Delta n_\mathrm{t} = \dfrac{n_2\cos(\theta_2) - n_1\cos(\theta_1)}{\cos(\theta_1)\cos(\theta_2)}$
Sagittal plane		$\begin{bmatrix} 1 & 0 \\ \dfrac{\Delta n_\mathrm{s}}{r\,n_2} & \dfrac{n_1}{n_2} \end{bmatrix}$	$\Delta n_\mathrm{s} = n_2\cos(\theta_2) - n_1\cos(\theta_1)$
Rowland concave grating (unfolded) Tangential (meridional) plane Radius of curvature r		$\begin{bmatrix} A & B \\ C & D \end{bmatrix}$, $A = \dfrac{\cos(\theta_1)}{\cos(\theta_2)}$; $B = 0$; $C = -\dfrac{2\cos(\theta_2)}{r_\mathrm{t}\cos(\theta_1)}$; $D = A$.	Grating equation (3.1.52): $\sin(\theta_1) + \sin(\theta_2) = m\,\dfrac{\lambda}{g}$, $r_\mathrm{t} = \dfrac{2\,r\cos^2(\theta_2)}{\cos(\theta_1) + \cos(\theta_2)}$
Sagittal plane		$\begin{bmatrix} 1 & 0 \\ -\dfrac{2}{r_\mathrm{s}} & 1 \end{bmatrix}$	$r_\mathrm{s} = \dfrac{2\,r}{\cos(\theta_1) + \cos(\theta_2)}$, general corrected holographical gratings: see [81Gue]
Spherical concave mirror		Specialization of the Rowland grating to $g \Rightarrow \infty$, $\theta_1 = \theta_2$.	

Table 3.1.13. Distances between *cardinal elements* of an optical system: F, F': object- and image-space focal points, respectively; H, H': object- and image-space principal points, respectively; I, O: input and output plane, respectively. The order of points determines the signs.

Distance between two points	A, B, C, and D for $n_1 = n_2$
$\overline{IF}$	$\dfrac{D}{C}$
$\overline{FH}$	$-\dfrac{1}{C}$
$\overline{OF'}$	$-\dfrac{A}{C}$
$\overline{HF'}$	$-\dfrac{1}{C}$

Table 3.1.14. The meaning of the vanishing of different elements of the ABCD-matrix.

Element	Figure	Remark
$A = 0$	$\begin{bmatrix} 0 & B \\ C & D \end{bmatrix}$	$x_2 = B\,\alpha_1$ *Focusing* of collimated light into the image-side focal plane.
$B = 0$	$\begin{bmatrix} A & 0 \\ C & D \end{bmatrix}$	$x_2 = A\,x_1$ The input plane is *imaged* to the output plane (conjugated planes). A: magnification of imaging; appl.: calculation of image plane.
$C = 0$	$\begin{bmatrix} A & B \\ 0 & D \end{bmatrix}$	$\alpha_2 = D\,\alpha_1$ Transformation of collimated light into collimated light. D: angular magnification; telescope (*afocal* system).
$D = 0$	$\begin{bmatrix} A & B \\ C & 0 \end{bmatrix}$	$\alpha_2 = C\,x_1$ *Collimation* of divergent pencil of rays. C: *power* of the element or system.

$$\begin{bmatrix} x_2 \\ y_2 \\ \alpha_2 \\ \beta_2 \end{bmatrix} = \begin{bmatrix} A_{xx} & A_{xy} & B_{xx} & B_{xy} \\ A_{yx} & A_{yy} & B_{yx} & B_{yy} \\ C_{xx} & C_{xy} & D_{xx} & D_{xy} \\ C_{yx} & C_{yy} & D_{yx} & D_{yy} \end{bmatrix} \begin{bmatrix} x_1 \\ y_1 \\ \alpha_1 \\ \beta_1 \end{bmatrix} \quad \text{or} \quad \begin{bmatrix} \mathbf{r}_2 \\ \boldsymbol{\gamma}_2 \end{bmatrix} = \begin{bmatrix} \mathsf{A} & \mathsf{B} \\ \mathsf{C} & \mathsf{D} \end{bmatrix} \begin{bmatrix} \mathbf{r}_1 \\ \boldsymbol{\gamma}_1 \end{bmatrix} = \mathsf{S} \begin{bmatrix} \mathbf{r}_1 \\ \boldsymbol{\gamma}_1 \end{bmatrix} \qquad (3.1.99)$$

with the matrices A, B, C, D, and S given by comparison with the more detailed representations. Identities between the matrices, characteristic for the symplectic geometry (see Sect. 3.1.6.2.6), are: $\mathsf{A}\mathsf{D}^\mathsf{T} - \mathsf{B}\mathsf{C}^\mathsf{T} = \mathsf{I}$; $\mathsf{A}\mathsf{B}^\mathsf{T} = \mathsf{B}\mathsf{A}^\mathsf{T}$; $\mathsf{C}\mathsf{D}^\mathsf{T} = \mathsf{D}\mathsf{C}^\mathsf{T}$, and $\det \begin{vmatrix} \mathsf{A} & \mathsf{B} \\ \mathsf{C} & \mathsf{D} \end{vmatrix} = \dfrac{n'}{n}$, where T means the

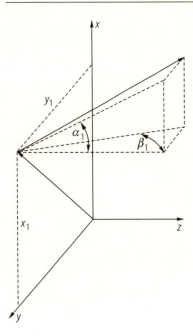

Fig. 3.1.35. Three-dimensional ray in the input plane I.

transposition of the matrix and I the identity matrix [86Sie, 05Hod]. The matrix S contains at most 10 independent parameters [76Arn, 86Sie, 05Hod].

In Table 3.1.15 general ray-transfer matrices are given.

3.1.6.2.5 General astigmatic system

A *general astigmatic system* can be generated by two cylindrical lenses with their axes non-parallel and non-orthogonal, separated by a distance L: $S_{GA} = R^{-1} S_{cyl,1} R S_L S_{cyl,2}$.

3.1.6.2.6 Symplectic optical system

Symplectic optical systems in the paraxial range can be described by the formalism of the symplectic geometry [03Wal]. They can be generated by a finite number of cylindrical and spherical lenses separated by free spaces. The mathematical formulation is connected with the matrix properties given in Sect. 3.1.6.2.4. For theoretical foundation and practical calculations see [64Lun, p. 216], [83Mac, 85Sud, 86Sie, 99Gao, 05Gro1, 05Hod].

3.1.6.2.7 Misalignments

The geometric optical calculations of *misalignments* with matrix techniques require, generally, higher dimensional matrices [05Gro1, p. 51], for example 3×3-matrices [86Sie] or 4×4-matrices [85Wan] for two-dimensional problems or 6×6-matrices for three-dimensional problems [76Arn].

Table 3.1.15. General ray-transfer matrices [99Gao, 05Hod].

Effect of the matrix	Matrix
Free propagation, index n_0, length z	$S_L = \begin{bmatrix} 1 & 0 & \dfrac{z}{n_0} & 0 \\ 0 & 1 & 0 & \dfrac{z}{n_0} \\ 0 & 0 & 1 & 0 \\ 0 & 0 & 0 & 0 \end{bmatrix}$
Aligned spherical thin lens, focal length f	$S_{sph} = \begin{bmatrix} 1 & 0 & 0 & 0 \\ 0 & 1 & 0 & 0 \\ \dfrac{-1}{f} & 0 & 1 & 0 \\ 0 & \dfrac{-1}{f} & 0 & 1 \end{bmatrix}$
Aligned cylindrical thin lens	$S_{cyl} = \begin{bmatrix} 1 & 0 & 0 & 0 \\ 0 & 1 & 0 & 0 \\ \dfrac{-1}{f_x} & 0 & 1 & 0 \\ 0 & 0 & 0 & 1 \end{bmatrix}$
Cylindrical telescope, m and n are the magnifications along x- and y-axis, respectively	$S_M = \begin{bmatrix} m & 0 & 0 & 0 \\ 0 & n & 0 & 0 \\ 0 & 0 & m^{-1} & 0 \\ 0 & 0 & 0 & n^{-1} \end{bmatrix}$
Rotation of the x-y-plane by the angle θ: given a system matrix S, then the rotated system matrix $S_{rot} = R^{-1}(\theta)\, S(\theta=0)\, R(\theta)$ with $R^{-1}(\theta) = R(-\theta) = R^T(\theta)$	$R = \begin{bmatrix} \cos\theta & \sin\theta & 0 & 0 \\ -\sin\theta & \cos\theta & 0 & 0 \\ 0 & 0 & \cos\theta & \sin\theta \\ 0 & 0 & -\sin\theta & \cos\theta \end{bmatrix}$

3.1.6.3 Lens aberrations

Corrections beyond the paraxial range are required by large object-space aperture light sources like *semiconductor lasers* (large vertical far-field angles) or large image-space aperture laser focusing optics like *CD-optics*.

Shape factor of a lens:

$$q = \frac{r_2 + r_1}{r_2 - r_1}\,. \tag{3.1.100}$$

Shape factor and spherical aberration for focusing of light:

– *Minimum of spherical aberration*:

$$\frac{r_1}{r_2} = \frac{n(2n-1)-4}{n(2n+1)}\,.$$

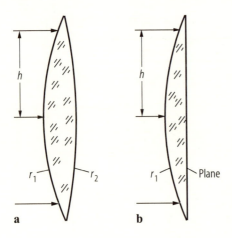

Fig. 3.1.36. Focusing of incident collimated light by (**a**) a general lens with curvature radii r_1 and r_2, (**b**) a plano-convex lens with shape factor $q = 1$.

- Refractive index $n = 1.5 \Rightarrow \dfrac{r_1}{r_2} = -\dfrac{1}{6} \Rightarrow q = 0.7$.

- $\left|\dfrac{r_1}{r_2}\right| \Rightarrow \dfrac{1}{\infty} \Rightarrow q = 1$ (plano-convex lens), spherical aberration near to minimum.

In Fig. 3.1.36 the focusing of incident collimated light by (**a**) a general lens with curvature radii r_1 and r_2 and (**b**) a plano-convex lens with shape factor $q = 1$ is shown.

In Table 3.1.16 the third-order spherical aberration and coma for a thin plano-convex lens is given in comparison with the diffraction-limited resolution for a plane wave or Gaussian illumination.

Remark 1: Third-order formulae for finite object distance: see [88Kle, 76Jen], more general: [80Hof, 86Haf, 96Ped, 99Bor].

Remark 2: About further third-order aberrations as astigmatism, field curvature, image distortion: see [76Jen, 78Dri, 80Hof, 86Haf, 88Kle, 96Ped, 99Bor].

Remark 3: The third-order aberrations are not exactly valid for higher apertures. Example: The third-order spherical aberration deviates for $2h/f = 1/5$ by $\approx 2\,\%$ from the ray-tracing values (the limit, recommended in [74Sle] for estimations), $h/f = 3/10$: $\approx 15\,\%$ deviation [76Jen]. Therefore, the *ray tracing* should be preferred for larger deviations from the paraxial case. It is the base of modern commercial optical design programs.

Example 3.1.14. Given: a plano-convex lens after Fig. 3.1.36b with the radius of the spherical surface $r_1 = 5$ mm, $n = 1.5$, collimated light with wavelength $\lambda = 1$ μm, stop with a height $h = 1.5$ mm, and a fiber with core diameter $2r = 100$ μm and numerical aperture $N.A. = 0.2$. Required: a geometric-optical estimation on the hits of the core of the fiber by the rays in the paraxial focal point and in the point of least confusion (Fig. 3.1.37). From (3.1.101)–(3.1.105): $f = 10$ mm, $\Delta s'_l = -262$ μm, $|\Delta s'_t| = 39$ μm, $\Delta s'_{lc} = -210$ μm, $|\Delta s'_{tc}| = 16$ μm, $\Delta s'_{tb} = 4$ μm, and $\Delta s'_{tg} = 2.1$ μm. In the paraxial focal plane as well as in the plane of least confusion, the hits of the fiber core by rays are closer than 50 μm to the optical axis and the angles of the rays with the optical axis are ≤ 0.15 within the fiber aperture. Therefore, all rays are accepted by a step-index fiber. About the analog task for Gaussian beams see references in Sect. 3.1.7.5.4 and commercial optical design programs, which show in this case, that a large part of radiation is coupled in higher-order modes.

Table 3.1.16. Third-order spherical aberration and coma for a thin plano-convex lens [76Jen, p. 152], [88Kle, p. 185], [87Nau, p. 109] in comparison with the diffraction-limited resolution for a plane wave or Gaussian illumination.

Figures	Formulae
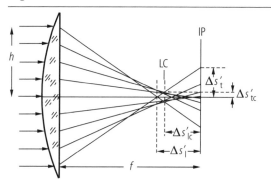 **Fig. 3.1.37.** Spherical aberration at a plano-convex lens. IP: paraxial image plane, LC: least confusion.	Lens equation (3.1.95) with $t = 0$, $a \Rightarrow -\infty$, $r_2 \Rightarrow -\infty$, $f = f'$, which is modified outside Sect. 3.1.6.1: $$\frac{1}{f} = (n-1)\frac{1}{r_1}, \quad (3.1.101)$$ $$\frac{\Delta s_l'}{f} = -\frac{n^3 - 2n^2 + 2}{2n(n-1)^2}\left(\frac{h}{f}\right)^2, \quad (3.1.102)$$ $$\Delta s_t' = \Delta s_l' \frac{h}{f}, \quad (3.1.103)$$ plane of least confusion [87Nau, 99Pau, 99Bor], [01Iff, p. 214]: $$\Delta s_{lc}' \approx 0.8 \Delta s_l', \quad (3.1.104)$$ $$\Delta s_{tc}' \approx 0.4 \Delta s_t'. \quad (3.1.105)$$ Gaussian weights of the illumination change the geometric optical position of least confusion [01Mah].
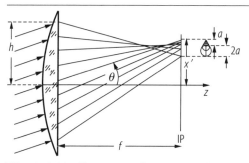 **Fig. 3.1.38.** Coma at a plano-convex lens.	$$x' = \theta f, \quad (3.1.106)$$ $$\frac{a}{f} = -\frac{n^2 - n - 1}{2n(n-1)}\left(\frac{h}{f}\right)^2 \theta \quad (3.1.107)$$ with θ: angle of incidence.
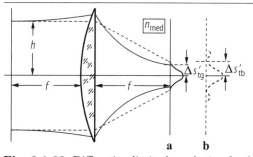 **Fig. 3.1.39.** Diffraction-limited resolution for (**a**) a Gaussian beam with waist h ($1/e^2$-intensity level) in the object-side focal plane, (**b**) a plane wave at circular stop with radius h.	$$\Delta s_{tg}' = \frac{\lambda}{\pi n_{\mathrm{med}}(h/f)}, \quad (3.1.108)$$ $$\Delta s_{tb}' = 0.61\frac{\lambda}{n_{\mathrm{med}}\sin\sigma'}$$ $$\approx 0.61\frac{\lambda}{n_{\mathrm{med}}(h/f)} \quad (3.1.109)$$ with λ: wavelength [m], h: zonal height [m], f: focal length [m], n_{med}: refractive index of the image space.

3.1.7 Beam propagation in optical systems

Paraxial propagation of light in a system given by its ABCD-matrix can be calculated
- for (*coherent*) *Gaussian beams* by q-parameter propagation (Sect. 3.1.7.2),
- for *general field distributions* by Collins integral (Sect. 3.1.7.4),
- for *second-order moments of the electric field* by propagation of the Wigner distribution in Chap. 2.2 (beam characterization).

3.1.7.1 Beam classification

In Table 3.1.17 various types of beams are listed.

3.1.7.2 Gaussian beam: complex q-parameter and its ABCD-transformation

3.1.7.2.1 Stigmatic and simple astigmatic beams

3.1.7.2.1.1 Fundamental Mode

- *Stigmatic beam* and rotational-symmetric system:
  ⇒ both longitudinal cross sections are treated equally,
- *Simple astigmatic beam* and elements with a symmetry plane:
  ⇒ two different sets of ABCD-matrices for the tangential and sagittal cut (see Table 3.1.12).

The *introduction* of the *complex q-parameter* [66Kog1, 66Kog2]

$$\frac{1}{q_x(z)} = \frac{1}{R_x(z)} - \frac{i\lambda}{\pi w_x(z)^2} \qquad (3.1.110)$$

formalizes the x-part of the fundamental-mode equation (3.1.31)

$$U_0(x,z) = \sqrt{\frac{w_{0x}}{w_x(z)}} \exp\left\{-\frac{x^2}{w_x(z)^2} - i\frac{kx^2}{2R_x(z)}\right\} \qquad (3.1.111)$$

to the simple complex shape

$$U_0(x,z) = \frac{1}{\sqrt{1+i\frac{z}{z_0}}} \exp\left\{-i\frac{kx^2}{q_x(z)}\right\} . \qquad (3.1.112)$$

In Fig. 3.1.40 the transfer of a field distribution by an optical system given by its ABCD-matrix is shown. In Table 3.1.18 the q-parameter transfer for stigmatic and simple astigmatic beams is given.

Table 3.1.17. Types of beams.

Beam [69Arn, 05Hod]	Generated by	Beam type is characterized by the shape of the matrix S (3.1.99)	Examples	References with practical example calculations
Stigmatic	Fundamental-mode laser	$A_{xx} = A_{yy}$; $A_{xy} = A_{yx} = 0$, and the same for B, C, D	TE00-mode handling in laser applications	[75Ger, 86Sie, 91Sal], [96Yar, 01Iff, 05Hod], see Sect. 3.1.7.2.1
Simple astigmatic	Semiconductor lasers or: Anamorphic optical system (f.e. cylindrical lens) in combination with a stigmatic beam	$A_{xx} \neq A_{yy}$; $A_{yx} = A_{xy} = 0$ (no mixing of both orthogonal planes), and the analog for B, C, D	– ring lasers, – lasers, including dispersive elements (dye-lasers), – tolerance calculations for resonators and beam-guiding optics	[05Hod, 99Gao, 86Sie], see Sect. 3.1.7.2.1
General astigmatic	General rotation of an anamorphic optics in relation with a simple astigmatic beam	General case	Transformation of higher-order radiation modes	[05Hod, 99Gao], see Sect. 3.1.7.2.2

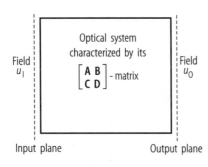

Fig. 3.1.40. Transfer of a field distribution by an optical system given by its ABCD-matrix.

Table 3.1.18. q-parameter transfer for stigmatic and simple astigmatic beams.

Given	Propagated field
– Gaussian beam in the *input* plane: $$u_I(x,z) = \exp\left\{-\mathrm{i}\frac{kx^2}{2q_{Ix}}\right\}. \quad (3.1.113)$$ – ABCD-matrix of the optical system (see Tables 3.1.11 and 3.1.12). – Starting point: $R_I = 1/\mathrm{Re}\,(1/q_I)$, $w_I = 1/\sqrt{-\pi\,\mathrm{Im}\,(1/q_I)/\lambda}$.	– Transformation of the q-parameter: $$q_{Ox} = \frac{Aq_{Ix}+B}{Cq_{Ix}+D}. \quad (3.1.114)$$ – Field in the output plane: $$u_O(x,z) = \exp\left\{-\mathrm{i}\frac{kx^2}{2q_{Ox}}\right\} \quad (3.1.115)$$ with the real parameters of the *output* beam [96Gro]: – beam radius: $$w_{Ox} = w_{Ix}\sqrt{\left(\frac{\lambda B}{\pi w_{Ix}^2}\right)^2 + \left(A + \frac{B}{R_{Ix}}\right)^2}, \quad (3.1.116)$$ – curvature radius of the wavefront: $$R_{Ox} = \frac{\left(A+\frac{B}{R_{Ix}}\right)^2 + \left(\frac{\lambda B}{\pi w_{Ix}^2}\right)^2}{\left(A+\frac{B}{R_{Ix}}\right)\left(C+\frac{D}{R_{Ix}}\right) + D\left(\frac{\lambda B}{\pi w_{Ix}^2}\right)^2}. \quad (3.1.117)$$

Example 3.1.15. Given: the waist of a Gaussian beam

$$u_I = \exp\left\{-\frac{x^2}{w_0^2}\right\} = \exp\left\{-\mathrm{i}\frac{kx^2}{2q_1}\right\}$$

with $q_1 = \mathrm{i}\,z_0$ in comparison with (3.1.110).

Asked: free-space propagation along the distance z with the ABCD-matrix $\begin{bmatrix} 1 & z \\ 0 & 1 \end{bmatrix}$.

Solution:

$$q_2 = \frac{Aq_1 + B}{Cq_1 + D} = z + \mathrm{i}\,z_0$$

and

$$U_O = \left(A + \frac{B}{q_1}\right)^{-1/2} \exp\left\{-\mathrm{i}\frac{kx^2}{2q_2}\right\} = \left(1 + \frac{z}{\mathrm{i}\,z_0}\right)^{-1/2} \exp\left\{-\mathrm{i}\frac{kx^2}{2(z+\mathrm{i}\,z_0)}\right\}.$$

3.1.7.2.1.2 Higher-order Hermite-Gaussian beams in simple astigmatic beams

Treatment of the x- or y-component of Hermite-Gaussian-beams after (3.1.27): The complex q-parameter transformation is treated as above, the fundamental mode part is given as above, the new beam radius for the Hermite polynom of order m, $H_m(\sqrt{2}\,x/w_{\mathrm{I}x})$ is calculated from the new q-parameter and the phase is derived from it, too [70Col].

For complex Hermite-Gaussian beams: see [86Sie].

3.1.7.2.2 General astigmatic beam

In Table 3.1.19 the Q^{-1}-matrix transfer for general astigmatic beams is given. The matrix Q^{-1} is the matrix scheme of inverses of q-parameters and no inverted matrix [96Gro].

Table 3.1.19. Q^{-1}-matrix transfer for general astigmatic beams.

Given	Propagated field
– General Gaussian beam in the input plane: $$U_{\mathrm{I}}(\boldsymbol{r}) = \exp\left\{-\mathrm{i}\frac{k}{2}\boldsymbol{r}\,\mathsf{Q}_{\mathrm{I}}^{-1}\,\boldsymbol{r}\right\},\quad (3.1.118)$$ $\boldsymbol{r} \sim (x,y)$ the transverse position vector perpendicular to the propagation axis z. – $\mathsf{Q}_{\mathrm{I}}^{-1}$-matrix: $$\mathsf{Q}_{\mathrm{I}}^{-1} = \begin{bmatrix} \frac{1}{q_{xx}} & \frac{1}{q_{xy}} \\ \frac{1}{q_{xy}} & \frac{1}{q_{yy}} \end{bmatrix}\quad (3.1.119)$$ with q_{xx}, q_{xy}, q_{yy} complex terms describing the general amplitude- and phase-distribution of U_{I}, and $$\boldsymbol{r}\,\mathsf{Q}_{\mathrm{I}}^{-1}\,\boldsymbol{r} = \frac{x^2}{q_{xx}} + 2\frac{xy}{q_{xy}} + \frac{y^2}{q_{yy}}.\quad (3.1.120)$$ – S-matrix of the optical system (see Table 3.1.15) with $$\mathsf{S} = \begin{bmatrix} \mathsf{A} & \mathsf{B} \\ \mathsf{C} & \mathsf{D} \end{bmatrix}$$ after (3.1.99).	– Transformation of the $\mathsf{Q}_{\mathrm{I}}^{-1}$-matrix to its output value: $$\mathsf{Q}_{\mathrm{O}}^{-1} = (\mathsf{C} + \mathsf{D}\,\mathsf{Q}_{\mathrm{I}}^{-1})(\mathsf{A} + \mathsf{B}\,\mathsf{Q}_{\mathrm{I}}^{-1})^{-1},\quad (3.1.121)$$ see [88Sim, 96Gro, 05Hod]. – Field in the output plane: $$U_{\mathrm{O}}(\boldsymbol{r}) = \exp\left\{-\mathrm{i}\frac{k}{2}\boldsymbol{r}\,\mathsf{Q}_{\mathrm{O}}^{-1}\,\boldsymbol{r}\right\}.\quad (3.1.122)$$

Example 3.1.16. Transformation of a simple astigmatic Gaussian beam (no mixing between x and y) with a θ-rotated cylindrical lens to a general astigmatic beam: We start with $\mathsf{Q}_{\mathrm{I}}^{-1} = \begin{bmatrix} \frac{1}{q_{xx}} & 0 \\ 0 & \frac{1}{q_{yy}} \end{bmatrix}$.
The rotation of an x-aligned cylindrical lens, given as $\mathsf{S}_{\mathrm{cyl}}$ in Table 3.1.15, is performed by multiplying first $\mathsf{S}_{\mathrm{cyl}}$ with the rotation matrix R of Table 3.1.15, and then the product with the inverse of R is:

$$S_{\text{rotated cyl.}} = R^{-1} S_{\text{cyl}} R = \begin{bmatrix} A & B \\ C & D \end{bmatrix}$$

with

$$A = \begin{bmatrix} 1 & 0 \\ 0 & 1 \end{bmatrix}, \quad B = \begin{bmatrix} 0 & 0 \\ 0 & 0 \end{bmatrix}, \quad C = \begin{bmatrix} -\cos^2\theta/f_x & -\sin\theta\cos\theta/f_x \\ -\sin\theta\cos\theta/f_x & -\sin^2\theta/f_x \end{bmatrix}, \quad D = \begin{bmatrix} 1 & 0 \\ 0 & 1 \end{bmatrix}$$

and

$$Q_O^{-1} = \begin{bmatrix} -\cos^2\theta/f_x + 1/q_{xx} & -\sin\theta\cos\theta/f_x \\ -\sin\theta\cos\theta/f_x & -\sin^2\theta/f_x + 1/q_{yy} \end{bmatrix} .$$

Therefore, the output field

$$u_O(r) = \exp\left\{-\mathrm{i}\frac{k}{2}\left[\left(-\frac{\cos^2\theta}{f_x} + \frac{1}{q_{xx}}\right)x^2 - 2\frac{\sin\theta\cos\theta}{f_x}xy + \left(-\frac{\sin^2\theta}{f_x} + \frac{1}{q_{yy}}\right)y^2\right]\right\}$$

is a general astigmatic Gaussian beam with a mixing term between the coordinates x and y.

3.1.7.3 Waist transformation

Often, the transfer of the beam waist is required for instance for focusing of laser light. Then, the following algorithms are *much more simple* than the q parameter algorithm.

3.1.7.3.1 General system (fundamental mode)

In Table 3.1.20 the waist transformation for a general system is given.

3.1.7.3.2 Thin lens (fundamental mode)

The formulae (3.1.123)–(3.1.126) are further simplified using the focal length f for the thin lens only, see Table 3.1.21.

Remark: Discussion of equation (3.1.127):

The right-hand-side term of (3.1.127) containing z_0 represents the *modification* introduced by the Gaussian beam optics to the thin-lens equation ((3.1.95), $t \Rightarrow 0$) shown in Fig. 3.1.42.
    In Fig. 3.1.43 the relation of the Gaussian waist transfer to the thin-lens equation of geometrical optics for different influences of diffraction is shown.

Main modifications of the geometrical optics:

- No "image distance" is at infinity.
- For $z = f$ (point P) the image is at $z' = f$ (transfer of the object-side focal plane to the image-side focal plane after (3.1.130), not $\Rightarrow \infty$).
- If a target z'-position is given, then two starting z-positions are possible.

Example 3.1.17. Given for Fig. 3.1.42: $z = 1179$ mm, $w_0 = 0.22$ mm, $\lambda = 1.06$ µm; it follows $z' = 109$ mm, $w_0' = 0.02$ mm, $\theta' = 0.96°$, and $z_0' = 1.21$ mm. The second right-hand term of (3.1.127) translates the Gaussian waist image by 0.16 mm in comparison with the geometrical optical image towards the lens.

Table 3.1.20. Waist transformation for a general system.

Given	Solution
– ABCD-matrix of the system, – waist w_0, – wavelength λ, including $z_0 = \pi w_0^2/\lambda$, – distance z to the input plane of the system.	$z' = \begin{cases} \dfrac{-(Az+B)(Cz+D) - ACz_0^2}{C^2 z_0^2 + (Cz+D)^2} & \text{for } C \neq 0, \\ -\dfrac{Az+B}{D} & \text{for } C = 0, \end{cases}$ (3.1.123)

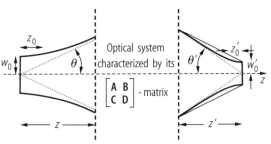

Fig. 3.1.41. Waist transformation by an optical system.

Asked: Waist w_0' and distance z' to the output plane of the system including z_0'.

$$z_0' = z_0 \left[\frac{Cz'+A}{Cz+D}\right] = \frac{z_0}{C^2 z_0^2 + (Cz+D)^2},$$ (3.1.124)

$$w_0' = \sqrt{\frac{\lambda z_0'}{\pi}},$$ (3.1.125)

$$\Theta_0' = \sqrt{\frac{\lambda}{\pi z_0'}}.$$ (3.1.126)

The *beam parameter product* is invariant:

$$w_0' \Theta_0' = w_0 \Theta_0 = \lambda/\pi.$$

Table 3.1.21. Waist transformation by a thin lens.

Given	Solution
– Focal length f of the lens, – wavelength λ, – waist w_0, including $z_0 = \pi w_0^2/\lambda$, – distance z to the input plane of the system.	$\dfrac{1}{z} + \dfrac{1}{z'} = \dfrac{1}{f} + \dfrac{z_0^2}{z\left[z^2 + z_0^2 - zf\right]},$ (3.1.127) see Fig. 3.1.42,

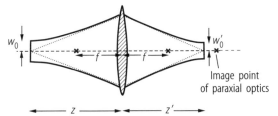

Fig. 3.1.42. Waist transformation by a thin lens.

Asked: Waist w_0' and distance z' to the output plane of the system and z_0'.

$$w_0' = w_0 \frac{f}{\sqrt{z_0^2 + (z-f)^2}},$$ (3.1.128)

$$z_0' = \frac{\pi w_0'^2}{\lambda}.$$ (3.1.129)

If $z = f$, then

$$z' = f \quad \text{and} \quad w_0' = \frac{w_0 f}{z_0}.$$ (3.1.130)

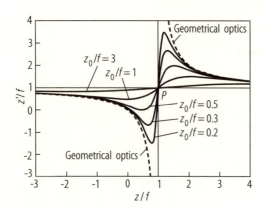

Fig. 3.1.43. Relation of the Gaussian waist transfer (full lines) to the thin-lens equation (dashed) of geometrical optics for different influences of diffraction (wavelength λ respectively z_0).

3.1.7.4 Collins integral

For Fresnel's approximation of diffraction in paraxial systems see [68Goo, 71Col, 78Loh, 94Roe]. It was generalized to the propagation of field distributions in ABCD-described systems by [70Col, 76Arn, 05Gro2, 05Hod].

3.1.7.4.1 Two-dimensional propagation

In Table 3.1.22 the propagation in rotational symmetric systems and simple astigmatic systems is given.

Table 3.1.22. Propagation in rotational symmetric systems and simple astigmatic systems.

Given	Solution
– ABCD-matrix of the optical system (see Tables 3.1.11 and 3.1.12), – field distribution in the input plane $U_\mathrm{I}(x)$, – path length along the optical axis L.	Field $U_\mathrm{O}(x_2)$ in the output plane (*Collins integral*): $$U_\mathrm{O}(x_2) = \sqrt{\frac{\mathrm{i}}{\lambda B}}\ \mathrm{e}^{-\mathrm{i}kL} \\ \times \int_{-\infty}^{\infty} \mathrm{d}x_1\, U_\mathrm{I}(x_1) \exp\left\{-\mathrm{i}\frac{k}{2B}\left[Ax_1^2 - 2x_1 x_2 + Dx_2^2\right]\right\}. \quad (3.1.131)$$

Example 3.1.18. The waist of a Gaussian beam is given with $U_\mathrm{I}(x_1) = \exp\left(-x_1^2/w_\mathrm{I}^2\right)$ in the input plane. The system consists of a thin lens with the focal length f followed by a free-space propagation by distance z. The ABCD-matrix is calculated from Fig. 3.1.34 and Table 3.1.11:

$$\begin{bmatrix} A & B \\ C & D \end{bmatrix} = \begin{bmatrix} 1 - z/f & z \\ -1/f & 1 \end{bmatrix}.$$

$$U_\mathrm{O}(x_2) = \sqrt{\frac{\mathrm{i}}{\lambda z}}\ \mathrm{e}^{-\mathrm{i}kL} \int_{-\infty}^{\infty} \mathrm{d}x_1 \exp\left(-\frac{x_1^2}{w_\mathrm{I}^2}\right) \exp\left\{-\mathrm{i}\frac{k}{2z}\left[\left(1 - \frac{z}{f}\right)x_1^2 - 2x_1 x_2 + x_2^2\right]\right\}.$$

The result is an output Gaussian intensity distribution with the waist radius

$$w_O = \frac{w_I}{\sqrt{1 + \left(\frac{\pi w_I^2}{\lambda f}\right)^2}},$$

the waist position $z = z_{\text{waist}} = f \left(\frac{\pi w_O w_I}{\lambda f}\right)^2$, and $z_O = \pi w_O^2/\lambda$.

For inclusion of *displacements and misalignments* in Collins Integral see [96Tov].

3.1.7.4.2 Three-dimensional propagation

In Table 3.1.23 the propagation in in general astigmatic systems is given.

Table 3.1.23. Propagation in general astigmatic systems.

Given	Solution
– S: matrix of the optical system, see Table 3.1.15 and (3.1.99) with $$S = \begin{bmatrix} A & B \\ C & D \end{bmatrix},$$ – field distribution in the input plane: $U_I(r_1)$, where r_1 is the position vector in the input plane.	Field $U_O(r_2)$ in the output plane (*Collins integral*): $$U_O(r_2) = \frac{-i \exp(-ikL)}{\lambda \sqrt{\det B}} \int\int d r_1 U_I(r_1)$$ $$\times \exp\left\{-i\frac{k}{2}\left[r_1 B^{-1} A r_1 - 2 r_1 B^{-1} r_2 + r_2 D B^{-1} r_2\right]\right\} \quad (3.1.132)$$ with det B the determinant and B^{-1} the inverse of the matrix B. Examples in [70Col, 05Gro2, 05Hod].

3.1.7.5 Gaussian beams in optical systems with stops, aberrations, and waveguide coupling

3.1.7.5.1 Field distributions in the waist region of Gaussian beams including stops and wave aberrations by optical system

Classical cases of *optical system design* are given in [99Bor, 80Hof, 86Haf]. [82Wag, 95Gae] use the calculation of the field distribution in the image by a stop and wave aberrations in the exit pupil.
The analog is modeled for Gaussian beams on the exit pupil in the following references:

– focused Gaussian beams with aberrations and stops: see [69Cam, 71Sch],
– obscuration of a rotationally symmetrical Gaussian beam including longitudinal focal shift: see [82Car, 86Sta],
– extended systematic discussion of diffraction with stops, obscuration, and aberrations: see [86Mah, 01Mah],
– spherical aberration: see [98Pu].

3.1.7.5.2 Mode matching for beam coupling into waveguides

The calculation of the excitation coefficient of an eigenmode in a waveguide (output mode) by the incident mode (input mode) at the surface of the waveguide is described in Table 3.1.24.
This task occurs

- if a laser beam is formed by an optical system and coupled afterwards into an optical fiber,
- if a laser beam of a master oscillator is to be coupled into a power amplifier,
- in the case of waveguide-waveguide coupling especially fiber-fiber coupling or coupling between semiconductor lasers.

Solutions are available in commercial optical design programs.

Table 3.1.24. Definitions for waveguide coupling.

Given	Solution
– Incident beam (emitted by a laser (and) transformed by an optical system): $E_{\text{input}}(x,y)$. – Waveguide with an eigenmode field the coupling to which is asked: $E_{\text{output}}(x,y)$.	*Coupling coefficient* (power relation): $$\eta = \frac{O_{\text{IO}} O_{\text{IO}}^*}{N_{\text{I}} N_{\text{O}}} \ . \quad (3.1.133)$$ *Overlap integral*: $$O_{\text{IO}} = \int_{-\infty}^{\infty} \mathrm{d}x \int_{-\infty}^{\infty} \mathrm{d}y \, E_{\text{I}}(x,y) \, E_{\text{O}}^*(x,y) \ . \quad (3.1.134)$$ *Normalization*: $$N_{\text{I}} = \int_{-\infty}^{\infty} \mathrm{d}x \int_{-\infty}^{\infty} \mathrm{d}y \, E_{\text{I}}(x,y) \, E_{\text{I}}^*(x,y) \ . \quad (3.1.135)$$ *Normalization*: $$N_{\text{O}} = \int_{-\infty}^{\infty} \mathrm{d}x \int_{-\infty}^{\infty} \mathrm{d}y \, E_{\text{O}}(x,y) \, E_{\text{O}}^*(x,y) \ . \quad (3.1.136)$$ Effective antireflection layers are assumed to be on the waveguide.
Fig. 3.1.44. Mode matching.	
Asked: Part of power transmitted into the waveguide (fiber, laser, integrated optical waveguide).	

3.1.7.5.3 Free-space coupling of Gaussian modes

For the case that a Gaussian output waist of a source waveguide and a Gaussian input waist of a receiver waveguide are separated by air, the coupling of both waveguides is generally treated in [64Kog]. Higher-order modes are also included. The approximation of small misalignments (offset and tilt) is given in Table 3.1.25, large offsets and tilts are treated in [64Kog, 91Wu].

Table 3.1.25. Coupling of waveguides.

Given	Solution
– Source WG1 (laser, waveguide) which emits a Hermite-Gaussian beam, – receiver WG2 (laser, waveguide) which can accept Hermite-Gaussian eigenmodes:	$$\eta_{00-00} = \frac{4}{\left(\dfrac{w_I}{w_O}+\dfrac{w_O}{w_I}\right)^2 + \left(\dfrac{\pi w_I w_O}{\lambda}\right)^2 \left(\dfrac{1}{R_I}-\dfrac{1}{R_O}\right)^2}$$ $$- \frac{8\,(w_{0I}\,w_{0O}\,\Delta x)^2}{(w_{0I}^2+w_{0O}^2)^3} - \frac{k^2\psi^2}{2}(w_I^2+w_O^2)\,.$$ (3.1.137)

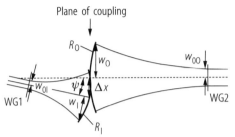

Fig. 3.1.45. Coupling of Gaussian beams. w_{0I} and w_{0O}: beam waist radii for WG1 and WG2, respectively; w_I and w_O: beam radii in the coupling plane; R_I and R_O: curvature radii of the beam wavefronts in the coupling plane; $k = 2\pi/\lambda$; λ: the wavelength of light; Δx: the lateral offset between the waveguides; ψ: the tilt of the axis.

Asked: The efficiency of the excitation of the modes in WG2, here the fundamental mode 00.

$\eta_{00-00} = 1$ for $\Delta x = \psi = 0$ and the exact beam radii and curvature fitting $w_I = w_O$ and $R_I = R_O$, otherwise $\eta_{00-00} < 1$.

Equation (3.1.137) contains the approximations:

– Gaussian beams (paraxial optics).
– Right-hand side of (3.1.137): 2nd and 3rd term $\ll 1^{st}$ term.

About coupling coefficients for higher-order modes and without the approximation: see [64Kog]; on couplings with Hermite-Gaussian modes and Laguerre-Gaussian modes: see [94Kri, 80Gra].

3.1.7.5.4 Laser fiber coupling

Methods of treatment:

– Launching of *fundamental-mode laser radiation* into the *fundamental mode* of a single-mode fiber:

  – Calculation of the overlap integral (3.1.134) for a Gaussian mode and the mode field for different fiber cross sections: see [88Neu, p. 179], [80Gra].
  – Approximation of the exact fiber fundamental modes by a Gaussian field distribution (see [88Neu, pp. 68]) and the application of the waist transformation from laser via an optical system with the methods of Sects. 3.1.7.2–3.1.7.4 and calculation of the overlap integral equation (3.1.134) or mode-coupling equation (3.1.137) [91Wu].

– Launching of fundamental-mode laser radiation or *multimode laser radiation* or incoherent light sources into *multimode fibers*:

  – Overlap integral techniques in the framework of partial coherence theory: see [87Hil].
  – Geometric optical methods (ray tracing and phase space techniques): see [90Gec, 95Sny, 91Gra, 91Wu, 01Iff].

- Inclusion of the *aberrations and stops* of the optical system used:
    - Monomodal and partial coherent case: calculation of the wave aberrations of the optical system by ray-tracing methods and inclusion of these aberrations into the overlap integral: see [82Wag, 95Gae, 89Hil, 99Gue].
    - Ray-tracing methods are adequate for stops and aberrations, but not reliable for a few mode waveguides: rough design [01Iff]: the spot diagram of the ray tracing in the fiber facet should be within the core area and the angles of incidence should be smaller then the aperture angle [88Neu] of the fiber.

References for 3.1

23Kot   Kottler, F.: Ann. Phys. (Leipzig) **71** (1923) 457; **72** (1923) 320.

28Koe   König, W.: Elektromagnetische Lichttheorie, in: Geiger, H., Scheel, K. (eds.): Handbuch der Physik, Vol. 20, Berlin: Springer-Verlag, 1928, p. 141–262.
28Szi   Szivessy, G.: Kristalloptik, in: Geiger, H., Scheel, K. (eds.): Handbuch der Physik, Vol. 20, Berlin: Springer-Verlag, 1928, p. 635–904.

41Jon   Jones, R.C.: J.Opt. Soc. Am. **31** (1941) 488.
41Str   Stratton, J.A.: Electromagnetic theory, New York: McGraw-Hill, 1941.

54Bel   Beljankin, D.S., Petrow, W.P.: Kristalloptik, Berlin: Verlag der Technik, 1954.

57Mue   Müller, G.: Grundprobleme der mathematischen Theorie elektromagnetischer Schwingungen, Berlin: Springer-Verlag, 1957.

58Fed   Fedorov, F.I.: Optika anisotropnych sred, (Optics of anisotropic media) in Russian, Minsk: Izdatelstvo Akademii Nauk BSSR, 1958.
58Mac   Macke, W.: Wellen, Leipzig: Akademische Verlagsgesellschaft Geest & Portig, 1958.
58Shu   Shubnikov, A.V.: Osnovy opticeskoj kristallografii (Principles of optical crystallography, in Russian), Moskva: Izdatelsvo Akademii nauk, 1958.

61Hoe   Hönl, H., Maue, A.W., Westpfahl, K.: Theorie der Beugung, in: Flügge, S. (ed.), Handbuch der Physik, Vol. 25/1, Berlin: Springer-Verlag, 1961.
61Ram   Ramachandran, G.N., Ramaseshan, S.: Crystal optics, in: Handbuch der Physik, Vol. **25**/1, Flügge, S. (ed.), Berlin: Springer-Verlag, 1961, p. 1.

62Lan   Landolt-Börnstein: Zahlenwerte und Funktionen, 6th edition, Vol. 2: Eigenschaften der Materie in ihren Aggregatzuständen, part 8: Optische Konstanten, Hellwege, K.H., Hellwege, A.M. (eds.), Berlin: Springer-Verlag 1962.
62Shu   Shurcliff, W.A.: Polarized light, Cambridge, Mass.: Harvard University Press, 1962.

64Bro   Brouwer, W.: Matrix methods in optical instrument design, New York: W.A. Benjamin, 1964.
64Kog   Kogelnik, H.: Coupling and conversion coefficients for optical modes, in: Quasi-optics, Brooklyn: Polytechnic Press, 1964.
64Lun   Luneburg, R.K.: Mathematical theory of optics, Berkeley: University of California Press, 1964.
64Wol   Wolf, E., Marchand, E.W.: J. Opt. Soc. Am. **54** (1964) 587.

66Kog1 Kogelnik, H., Li, T.: Appl. Opt. **5** (1966) 1550.
66Kog2 Kogelnik, H., Li, T.: Proc. IEEE **54** (1966) 1312.
66Rub   Rubinowicz, A.: Die Beugungswelle in der Kirchhoffschen Theorie der Beugung, Warszawa: Polnischer Verlag der Wissenschaften PWN, 1966.

67Loh   Lohmann, A.W., Paris, D.P.: Appl. Opt. **6** (1967) 1739.
67Rud   Rudolph, D., Schmahl, G.: Umschau Wiss. Tech. **67** (1967) 225.
67Sok   Sokolov, A.V.: Optical properties of metals, New York: American Elsevier, 1967.

68Goo   Goodman, J.W.: Introduction to Fourier optics, San Francisco: McGraw-Hill, 1968.
68Pap   Papoulis, A: Systems and transforms with applications in optics, New York: McGraw Hill, 1968.
68Ves   Veselago, V.G.: Sov. Phys. Usp. **10** (1968) 509.

69Arn   Arnaud, J.A., Kogelnik, H.: Appl. Opt. **8** (1969) 1687.
69Cam   Campbell, J.P., DeShazer, L.G.: J. Opt. Soc. Am. **59** (1969) 1427.
69Kog   Kogelnik, H.: Bell Syst. Tech. J. **48** (1969) 2909.
69Str   Stroke, G.W.: Diffraction gratings, in: Flügge, S. (ed.): Handbuch der Physik, Vol. 29, Berlin: Springer-Verlag, 1969, p. 425.

70Abr   Abramowitz, M., Steguhn, I.A.: Handbook of mathematical functions, New York: Dover Publications, 1970.
70Ber   Berek, M.: Grundlagen der praktischen Optik, Berlin: Verlag Walter de Gruyter, 1970.
70Col   Collins, S.A.: J. Opt. Soc. Am. **60** (1970) 1168.

71Col   Collier, R.J., Burckhardt, C.B., Lin, L.H.: Optical holography, New York: Academic Press, 1971.
71Pan   Pankove, J.I.: Optical processes in semiconductors, New York: Dover, 1971.
71Sch   Schell, R.G., Tyras, G.: J. Opt. Soc. Am. **61** (1971) 31.

72Mar   Marcuse, D.: Light transmission optics, New York: Van Nostrand Reinhold Company, 1972.
72Sta   Stavroudis, O.N.: The optics of rays, wavefronts and caustics, New York: Academic Press, 1972.
72Woo   Wooten, F.: Optical properties of solids, New York: Academic Press, 1972.

73Men   Menzel, E., Mirande, W., Weingärtner, I.: Fourier-Optik und Holographie, Wien: Springer-Verlag, 1973.

74Jun   Junghans, J., Keller, M., Weber, H.: Appl. Opt. **13** (1974) 2793.
74Sle   Slevogt, H.: Technische Optik, Berlin: Walter de Gruyter, 1974.
74Wel   Welford, W.T.: Aberrations of the symmetrical optical system, London: Academic Press, 1974.

75Ger   Gerrard, A., Burch, J.M.: Introduction to matrix methods in optics, London: John Wiley & Sons, 1975.
75Jac   Jackson, J.D.: Classical electrodynamics, New York: John Wiley & Sons, 1975.
75Lax   Lax, M., Louisell, W.H., McKnight, W.B.: Phys. Rev. A **11** (1975) 1365.

76Arn   Arnaud, J.: Beam and fiber optics, New York: Academic Press, 1976.
76Fed   Fedorov, F.I., Fillipov, V.V.: Otrazheniye i prelomlenie sveta prozrachnymi kristallami, (Reflection and refraction of light at transparent crystals) in Russian, Minsk: Izdatelstvo Nauka i Technika, 1976.
76Jen   Jenkins, F.A., White, H.E.: Fundamentals of optics, New York: McGraw Hill, 1976.

77Azz   Azzam, R.M.A., Bashara, N.M.: Ellipsometry and polarized light, Amsterdam: North Holland, 1977.
77Gon   Gontcharenkov, A.M.: Gaussovy putchki sveta (Gaussian light beams) in Russian, Minsk: Nauka i Technika, 1977.

78Dri	Driscoll, W.G., Vaughan, W. (eds.): Handbook of optics, New York: McGraw-Hill, 1978.
78Gas	Gaskill, J.D.: Linear systems, Fourier transforms and optics, New York: John Wiley & Sons, 1978.
78Gra	Grau, G.K.: Quantenelektronik, Braunschweig: Vieweg, 1978.
78Loh	Lohmann, A.W.: Optical information processing, Erlangen: University Print, 1978.
78Mar	Marchand, E.W.: Gradient index optics, New York: Academic Press, 1978.
79Agr	Agrawal, G.P., Pattanyak, D.N.: J. Opt. Soc. Am. **69** (1979) 575.
79Dav	Davis, L.W.: Phys. Rev. A **19** (1979) 1177.
79Wah	Wahlstrom, E.E.: Optical crystallography, New York: John Wiley & Sons, 1979.
80Gra	Grau, G.K., Leminger, O.G., Sauter, E.G.: Int. J. Electron. Commun. (AEÜ) **34** (1980) 259.
80Hof	Hofmann, Chr.: Die optische Abbildung, Leipzig: Akademische Verlagsgesellschaft Geest & Portig, 1980.
80Pet	Petit, R. (ed.): Electromagnetic theory of gratings, Berlin: Springer-Verlag, 1980.
80Sch	Schott-Katalog, Optisches Glas, Mainz, 1980.
81Gue	Güther, R.: Opt. Appl. **11** (1981) 97.
81Kam	Kaminski, A.A.: Laser crystals, their physics and properties, Berlin: Springer-Verlag, 1981.
81Rus	Russell, P.St.J.: Phys. Rep. **71** (1981) 209.
81Sol	Solymar, L., Cooke, D.J.: Volume holography and volume gratings, London: Academic Press, 1981.
82Bru	Brunner, W., Junge, K.: Wissensspeicher Lasertechnik, Leipzig: Fachbuchverlag, 1982.
82Car	Carter, W.H.: Appl. Opt. **21** (1982) 1989.
82Hac	Hack, H., Neuroth, N.: Appl. Opt. **21** (1982) 3239.
82Hut	Hutley, M.: Diffraction gratings, London: Academic Press, 1982.
82Wag	Wagner, E.E., Tomlinson, W.J.: Appl. Opt. **21** (1982) 2671.
83Mac	Macutow, B., Arsenault, H.H.: J. Opt. Soc. Am. **73** (1983) 1360.
83Ste	Steward, E.G.: Fourier Optics: An introduction, Chichester: Horwood, 1983.
84Agr	Agranovich, V.M., Ginzburg, V.L.: Crystal optics with spatial dispersion and excitons, Berlin: Springer-Verlag, 1984.
84Haf	Haferkorn, H., Richter, W.: Synthese optischer Systeme, Berlin: Deutscher Verlag der Wissenschaften, 1984.
84Hau	Haus, H.: Waves and fields in optoelectronics, Englewood Cliffs: Prentice Hall, 1984.
84Rus	Rusinov, M.M. (ed.): Vycislitelnaja optika, spravocnik (Numerical optics, handbook, in Russian), Leningrad: Maschinostroenije, 1984.
84Sch	Schreier, D. (ed.): Synthetische Holografie, Leipzig: Fachbuchverlag, 1984.
84Yar	Yariv, A., Yeh, P.: Optical waves in crystals, New York: John Wiley & Sons, 1984.
85Bic	Bickel, G., Häusler, G., Maul, M.: Opt. Eng. **24** (1985) 975.
85Iiz	Iizuka, K.: Engineering optics, Berlin: Springer-Verlag, 1985.
85Koh	Kohlrausch, F.: Praktische Physik, Hahn, D., Wagner, S. (eds.), Vol. 1, Stuttgart: B.G. Teubner, 1985.
85Sud	Sudarshan, E.C.G., Mukunda, N., Simon, R.: Opt. Acta **32** (1985) 855.
85Wan	Wang, S.: Opt. Quantum Electron. **17** (1985) 1.

86Haf   Haferkorn, H.: Bewertung optischer Systeme, Berlin: Deutscher Verlag der Wissenschaften, 1986.
86Lan   Landau, L.D., Lifschitz, E.M.: Quantenmechanik, Berlin: Akademie-Verlag, 1986.
86Mah   Mahajan, V.N.: J. Opt. Soc. Am. A **3** (1986) 470.
86Sie   Siegman, A.E.: Lasers, Mill Valley: University Science Books, 1986.
86Sol   Solimeno, S., Crosignani, B., DiPorto, P.: Guiding, diffraction and confinement of optical radiation, Orlando: Academic Press, 1986.
86Sta   Stamnes, J.J.: Waves in focal regions, Bristol: Adam Hilger, 1986.

87Chr   Chrisp, M.P.: Aberration-corrected holographic gratings and their mountings, in: Shannon, R.R., Wyant, J.C. (eds.): Applied optics and optical engineering, Vol. 10, San Diego: Academic Press, 1987.
87Dur   Durnin, J.: J. Opt. Soc. Am. A **4** (1987) 651.
87EMI   Electronic Materials Information Service (EMIS): Series on semiconductor materials, beginning 1987, for example No. 16: Properties of Gallium Arsenide, London: INSPEC, IEE 1996, the chapters: optical properties.
87Gor   Gori, F., Guattari, G., Padovani, C.: Opt. Commun. **64** (1987) 491.
87Hil   Hillerich, B.: Opt. Quantum Electron. **19** (1987) 209.
87Nau   Naumann, H., Schröder, G.: Bauelemente der Optik, München: Hanser-Verlag, 1987.
87Ree   Rees, W.G.: Eur. J. Phys. **8** (1987) 44.

88Bey   Beyer, H. (ed.): Handbuch der Mikroskopie, Berlin: Verlag der Technik, 1988.
88Boi   Boiko, B.B., Petrov, N.S.: Otrashenie sveta ot ysilivajuschich i nelineinych sred, (Reflection of light at amplifying and nonlinear media) in Russian, Minsk. Izdatelstvo Nauka i Technika, 1988.
88Kle   Klein, M.V., Furtak, T.E.: Optik, Berlin: Springer-Verlag, 1988.
88Mil   Milonny, P.W., Eberly, J.H.: Lasers, New York: John Wiley & Sons, 1988.
88Neu   Neumann, E.G.: Single-mode fibers, Berlin: Springer-Verlag, 1988.
88Par   Parker, S.P. (ed.):Optics source book, New York: McGraw-Hill, 1988.
88Sim   Simon, R., Mukunda, N., Sudarshan, E.C.G.: Opt. Commun. **65** (1988) 322.
88Yeh   Yeh, P.: Optical waves in layered media, New York: John Wiley & Sons, 1988.

89Ars   Arsenault, H.H., Szoplik, T., Macutow, B. (eds.): Optical processing and computing, Boston: Academic Press, 1989.
89Gha   Ghatak, A., Thyagarajan, K.: Optical electronics, Cambridge: Cambridge University Press, 1989.
89Hil   Hillerich, B.: IEEE J. Lightwave Technol. **7** (1989) 77.
89Kor   Koronkevich, V.P.: Computer synthesis of diffraction optical elements, in: Arsenault, H.H., Szoplik, T., Macutow, B. (eds.): Optical processing and computing, Boston: Academic Press, 1989, p. 277.
89Som   Sommerfeld, A.: Vorlesungen über Theoretische Physik: Optik, Frankfurt: H. Deutsch, 1989.

90Gec   Geckeler, S.: Lichtwellenleiter für die optische Nachrichtenübertragung, Berlin: Springer-Verlag, 1990.
90Kli   Kliger, D.S., Lewis, J.W., Randall, C.E.: Polarized light in optics and spectroscopy, Boston: Academic Press, 1990.
90Roe   Röseler, A.: Infrared spectroscopic ellipsometry, Berlin: Akademie-Verlag, 1990.

91Dmi   Dmitriev, V.G., Gurzadan, G.G., Nykogosyan, D.N.: Handbook of nonlinear optical crystals, Berlin: Springer-Verlag, 1991.
91Eic   Eichler, J., Eichler, H.-J.: Laser, Berlin: Springer-Verlag, 1991.

91Gra	Grau, G., Freude, W.: Optische Nachrichtentechnik, Berlin: Springer-Verlag, 1991.
91Ish	Ishimaru, A.: Electromagnetic wave propagation, radiation, and scattering, Englewood Cliffs: Prentice Hall, 1991.
91Klo	Klocek, P. (ed.): Handbook of infrared optical materials, New York: Marcel Dekker, 1991.
91Lan	Landau, L.D., Lifschitz, E.M. (eds.): Lehrbuch der theoretischen Physik, Vol. 4: Berestetzki, W.B., Lifschitz, E.M., Pitaevski, L.P.: Quantenelektrodynamik, Berlin: Akademie-Verlag, 1991.
91Lee	Lee, H.S., Steward, B.W., Will, D., Fenichel, H.: Appl. Phys. Lett. **59** (1991) 3096.
91Mah	Mahajan, V.N.: Aberration theory made simple, Bellingham: SPIE Optical Engineering Press, 1991.
91Nie	Nieto-Vesperinas, M.: Scattering and diffraction in physical optics, New York: John Wiley & Sons, 1991.
91Sal	Saleh, B.E.A., Teich, M.C.: Fundamentals of photonics, New York: John Wiley & Sons, 1991.
91Spl	Splett, A., Majd, M., Petermann, K.: IEEE Photonics Technology Letters **3** (1991) 466.
91Wu	Wu, H.-D., Barnes, F.S. (eds.): Microlenses, coupling light to optical fibers, New York: IEEE Press, 1991.
91Wyr	Wyrowski, F., Bryngdahl, O.: Rep. Prog. Phys. **54** (1991) 1481–1571.
92Col	Collett, E.: Polarized light: Fundamentals and applications, New York: Marcel Dekker, 1992.
92Cor	Corning: Optical Glass Catalog, Corning, SA, July 1992.
92Lan	Landau, L.D., Lifschitz, E.M.: Klassische Feldtheorie, Berlin: Akademie Verlag, 1992.
92Lug	Vander Lugt, A.: Optical signal processing, New York: John Wiley & Sons, 1992.
92Ung	Unger, H.G.: Optische Nachrichtentechnik, Vol. II, Heidelberg: Hüthig Buch Verlag, 1992.
92Wue	Wünsche, A.: J. Opt. Soc. Am. A **6** (1992) 765.
93Sto	Stößel, W.: Fourieroptik, Berlin: Springer-Verlag, 1993.
94Fel	Felsen, L.B., Marcuvitz, N.: Radiation and scattering of waves, Englewood Cliffs: Prentice-Hall, 1994.
94Kri	Krivoshlykov, S.G.: Quantum-theoretical formalism for inhomogeneous graded-index waveguides, Berlin: Akademie-Verlag, 1994.
94Leg	Leger, J.R., Chen, D., Wang, Z.: Opt. Lett. **19** (1994) 108.
94Leh	Lehner, G.: Elektromagnetische Feldtheorie, Berlin: Springer-Verlag, 1994.
94Roe	Römer, H.: Theoretische Optik, Weinheim: VCH Verlagsgesellschaft, 1994.
94Soi	Soifer, V.A., Golub, M.: Laser beam mode selection by computer generated holograms, Boca Raton: CRC Press, 1994.
95Bac	Bach, H., Neuroth, N.: The properties of optical glass, Berlin: Springer-Verlag, 1995.
95Bas	Bass, M. (ed.): Handbook of optics, Vol. I and Vol. II, New York: McGraw-Hill, 1995.
95Gae	Gaebe, C.: Proc. SPIE (Int. Soc. Opt. Eng.) **2610** (1995) 184.
95Gou	Gousbet, G., Lock, J.A., Grehan, G.: Appl. Opt. **34** (1995) 2133.
95Joa	Joannopoulos, J.D., Meade, R.D., Winn, J.N.: Photonic crystals, Princeton: Princeton University Press, 1995.
95Kli	Klingshirn, C.F.: Semiconductor optics, Berlin: Springer-Verlag, 1995.
95Mae	März, R.: Integrated optics, design and modeling, Boston: Artech House, 1995.
95Man	Mandel, L., Wolf, E.: Optical coherent and quantum optics, Cambridge: Cambridge University Press, 1995.
95Sny	Snyder, A.W., Love, J.D.: Optical waveguide theory, London: Chapman & Hall, 1995.

References for 3.1

95Wil	Wilson, R.G.: Fourier series and optical transform techniques in contemporary optics, New York: John Wiley & Sons, 1995.
96Die	Diels, J.-C., Rudolph, W.: Ultrashort laser pulse phenomena, San Diego: Academic Press, 1996.
96Don	MacDonald, R.P., Boothroyd, S.A., Okamoto, T., Chrostowski, J., Syrett, B.A.: Opt. Commun. **122** (1996) 169.
96For	Forbes, G.W.: J. Opt. Soc. Am. A **13** (1996) 1816.
96Gro	Gross, H.: Propagation höhermodiger Laserstrahlung und deren Wechselwirkung mit optischen Systemen, Stuttgart: B.G. Teubner, 1996.
96Hal	Hall, D.G.: Opt. Lett. **21** (1996) 9.
96Har	Hariharan, P.: Optical holography, Cambridge: Cambridge UP, 1996.
96Koe	Koechner, W.: Solid-state laser engineering, Berlin: Springer-Verlag, 1996.
96Lan	Landolt-Börnstein: Numerical data and functional relationships in science and technology, Madelung, O., Martienssen, W. (eds.): Comprehensive index, Berlin: Springer-Verlag, 1996, pp. 103, 104.
96Oha	Ohara-GmbH: Melcher, P.: Optischer Glas-Katalog, Hofheim a. Ts.: Verlag Gutenberg, 1996.
96Ped	Pedrotti, F.L., Pedrotti, L.S., Bausch, W., Schmidt, H.: Optik: Eine Einführung, München: Prentice Hall, 1996.
96Sch	Schott-Katalog, Optisches Glas, Nr. 10.000 0992, Mainz, 1996; updated version: www.schott.com/optics_devices/german/download.
96Tov	Tovar, A.A., Casperson, L.W.: J. Opt. Soc. Am. A **13** (1996) 2239.
96Yar	Yariv, A.: Optical electronics, Philadelphia: Holt, Rinehart and Winston, Inc., 1996.
97For	Forbes, G.W., Butler, D.J., Gordon, R.L., Asatryan, A.A.: J. Opt. Soc. Am. A **14** (1997) 3300.
97Gal	Gale, M.T.: Direct writing of continuous-relief micro-optics, in: Herzig, H.P. (ed.): Micro-optics, London: Taylor & Francis, 1997, p. 87.
97Her	Herzig, H.P. (ed.): Micro-optics, London: Taylor & Francis, 1997.
97Hua	Huard, S.: Polarization of light, Chichester: John Wiley & Sons, 1997.
97Kle	Kley, E.-B.: Microelectron. Eng. **34** (1997) 261.
97Leg	Leger, J.R.: Diffractive laser resonators, in: Turunen, J., Wyrowski, F. (eds.), Diffractive optics for industrial and commercial applications, Berlin: Akademie-Verlag, 1997, p. 189.
97Loe	Loewen, E.G., Popov, E.: Diffraction gratings and applications, New York: Marcel Dekker, 1997.
97Nik	Nikogosyan, D.N.: Properties of optical and laser-related materials: A Handbook, Chichester: John Wiley & Sons, 1997.
97Tur	Turunen, J., Wyrowski, F.: Diffractive optics for industrial and commercial applications, Berlin: Akademieverlag, 1997.
97Zen	Zeng, X., Liang, C., An, Y.: Appl. Opt. **36** (1997) 2042.
98Hec	Hecht, E.: Optics, Reading: Addison-Wesley, 1998.
98Hoy	Hoya Optical Glass Catalogue 1998.
98Mah	Mahajan, V.N.: Optical imaging and aberrations, part 1: Ray geometrical optics, Bellingham: SPIE Optical Engineering Press, 1998.
98Pal	Palik, E.D. (ed.): Handbook of optical constants of solids, Vols. 1, 2, 3, San Diego: Academic Press, 1998.
98Pu	Pu, J., Zhang, H.: Opt. Commun. **151** (1998) 331.
98Sta	Stamnes, J.J., Eide, H.A.: J. Opt. Soc. Am. A **15** (1998) 1285; 1292; 1308.
98Sve	Svelto, O.: Principles of lasers, New York: Plenum Press, 1998.

99Bor   Born, M., Wolf, E.: Principles of optics, Oxford: Pergamon Press, 1999.
99Gao   Gao, C.: Characterization and transformation of astigmatic laser beams, Berlin: Verlag Wissenschaft und Technik, 1999.
99Gue   Güther, R.: Gaußstrahlen in der angewandten Optik, part 1, 2, in: Prenzel, W.-D.: Jahrbuch für Optik und Feinmechanik, Berlin: Schiele & Schön, Vol. 45 (1998) 1; Vol. 46 (1999) 12.
99Lau   Lauterborn, W., Kurz, T., Wiesenfeldt, M.: Coherent optics, Berlin: Springer-Verlag, 1999.
99Pau   Paul, H. (ed.): Lexikon der Optik, Heidelberg: Spektrum-Verlag, 1999.
99Zei   Zeitner, U.D., Aagedal, H., Wyrowski, F.: Appl. Opt. **38** (1999) 980.

00Gou   Gousbet, G., Mees, L., Grehan, G., Ren, K.-H.: Appl. Opt. **39** (2000) 1008.
00Mey   Meyreuis, P.: Digital diffractive optics, Weinheim: Wiley-VCH, 2000.
00Pen   Pendry, J.B.: Phys. Rev. Lett. **85** (2000) 3966.
00Tur   Turunen, J., Kuittinen, M., Wyrowski, F.: Diffractive optics: Electromagnetic approach, in: Wolf, E. (ed.), Progress of Optics, Vol. 40, Amsterdam: Elsevier, 2000, p. 343–388.

01Ben   Bennett, C.V., Kolner, B.H.: IEEE J. Quantum Electron. **37** (2001) 20.
01Iff   Iffländer, R.: Solid-state lasers for materials processing, Berlin: Springer-Verlag, 2001.
01Jah   Jahns, J.: Photonik, München: Oldenbourg, 2001.
01Mah   Mahajan, V.N.: Optical imaging and aberrations, part 2: Wave diffraction optics, Bellingham: SPIE Optical Engineering Press, 2001.
01Sak   Sakoda, K.: Optical properties of photonic crystals, Berlin: Springer-Verlag, 2001.

02Gom   Gomez-Reino, C., Perez, M.V., Bao, C.: Gradient-Index Optics: Fundamentals and Applications, Berlin: Springer-Verlag, 2002.

03DIN   DIN 1335, Geometrische Optik – Bezeichnungen und Definitionen, Berlin: Beuth-Verlag, 2003.
03Wad   Wadsworth, W.J., Percival, R.M., Bouwmans, G., Knight, J.C., Russell, P.S.J.: Optics Express **11** (2003) 48.
03Wal   Walz, G. (ed.): Lexikon der Mathematik, Vol. 5, Heidelberg: Spektrum Akademischer Verlag, 2003.

04Ber   Bergmann, L., Schaefer, C.: Lehrbuch der Experimentalphysik, Band 3: Optik, Niedrig, H. (ed.), Berlin: Walter de Gruyter, 2004.
04Bus   Busch, K., Lölkes, S., Wehrspohn, R.B., Föll, H. (eds.): Photonic Crystals, Weinheim: Wiley-VCH, 2005.

05Ele   Eleftheriades, G.V., Balmain, K.G. (eds.): Negative-refraction metamaterials, Hoboken: John Wiley & Sons, 2005.
05Gro1 Gross, H. (ed.): Handbook of Optical Systems, Vol. 1: Gross, H.: Fundamentals of Technical Optics, Weinheim: Wiley-VCH, 2005.
05Gro2 Gross, H. (ed.): Handbook of Optical Systems, Vol. 2: Singer, W., Totzeck, M., Gross, H.: Physical Image Formation, Weinheim: Wiley-VCH, 2005.
05Hod   Hodgson, N., Weber, H.: Laser Resonators and Beam Propagation, New York: Springer, 2005.
05Ram   Ramakrishna, S.A.: Rep. Prog. Phys. **68** (2005) 449.

Part 4

Nonlinear optics

4.1 Frequency conversion in crystals

G.G. Gurzadyan

4.1.1 Introduction

4.1.1.1 Symbols and abbreviations

4.1.1.1.1 Symbols

η	conversion efficiency
η (energy)	energy conversion efficiency
η (power)	power conversion efficiency
η (quantum)	quantum conversion efficiency
τ_p, τ	pulse duration
α	angle between interacting beams
$\Delta\lambda$	wavelength bandwidth
$\Delta\nu$	frequency bandwidth
$\Delta\theta$	angular bandwidth
E	energy
f	laser pulse repetition rate
I_0	pump intensity
I_{thr}	threshold intensity
φ_{pm}	phase-matching angle in the XY plane from X axis
L	crystal length
λ	wavelength
n	refractive index
n_o	ordinary refractive index
n_e	extraordinary refractive index
ν	wave number, frequency
P	power
θ_{pm}	phase-matching angle from Z axis
ρ	birefringence (walk-off) angle
T, T_{pm}	crystal temperature
Type I	o + o → e or e + e → o
Type II	o + e → e or o + e → o
ooe	o + o → e or e → o + o
eeo	e + e → o or o → e + e
eoe	e + o → e or e → e + o
oeo	o + e → o or o → e + o

4.1.1.1.2 Abbreviations

av	average
cw	continuous wave
DFG	difference frequency generation
DROPO	doubly resonant OPO
ERR	external ring resonator
FIHG	fifth harmonic generation
FOHG	fourth harmonic generation
ICDFG	intracavity difference frequency generation
ICSHG	intracavity second harmonic generation
IR	infrared
mid IR	middle infrared
NC	noncollinear
NCSHG	noncollinear second harmonic generation
OPA	optical parametric amplifier
OPO	optical parametric oscillator
SFG	sum frequency generation
SH	second harmonic
SHG	second harmonic generation
SIHG	sixth harmonic generation
SP OPO	synchronously pumped OPO
SROPO	singly resonant OPO
SRS	stimulated Raman scattering
THG	third harmonic generation
TROPO	triply resonant OPO
TWOPO	traveling-wave OPO
UV	ultraviolet

4.1.1.1.3 Crystals

Chemical formula	Symbol	Crystal name
Ag_3AsS_3		Proustite
$AgGaS_2$		Silver Thiogallate
$AgGaSe_2$		Silver Gallium Selenide
Ag_3SbS_3		Pyrargyrite
$Ba_2NaNb_5O_{15}$		Barium Sodium Niobate (Banana)
$\beta-BaB_2O_4$	BBO	Beta-Barium Borate
$CdGeAs_2$		Cadmium Germanium Arsenide
CdSe		Cadmium Selenide
CsB_3O_5	CBO	Cesium Borate
CsH_2AsO_4	CDA	Cesium Dihydrogen Arsenate
$CsLiB_6O_{10}$	CLBO	Cesium Lithium Borate
$C_6H_6N_2O_3$	POM	3-Methyl-4-Nitro-Pyridine-1-Oxide
$C_8H_8O_3$	MHBA	4-Hydroxy-3-Methoxy-Benzaldehyde (Vanillin)
$C_{10}H_{11}N_3O_6$	MAP	Methyl N-(2,4-Dinitrophenyl)-L-Alaninate
$C_{10}H_{13}N_3O_3$	DAN	N-[2-(Dimethylamino)-5-Nitrophenyl]-Acetamide
$C_{11}H_{14}N_2O_3$	NPP	N-(4-Nitrophenyl)-(L)-Propinol
CsD_2AsO_4	DCDA	Cesium Dideuterium Arsenate
GaSe		Gallium Selenide

$HgGa_2S_4$		Mercury Thiogallate
$\alpha-HIO_3$		$\alpha-$Iodic Acid
$KB_5O_8\ 4D_2O$	DKB5	Potassium Pentaborate Tetradeuterate
$KB_5O_8\ 4H_2O$	KB5	Potassium Pentaborate Tetrahydrate
KD_2AsO_4	DKDA	Potassium Dideuterium Arsenate
KD_2PO_4	DKDP	Potassium Dideuterium Phosphate
KH_2PO_4	KDP	Potassium Dihydrogen Phosphate
$KNbO_3$		Potassium Niobate
$KTiOAsO_4$	KTA	Potassium Titanyl Arsenate
$KTiOPO_4$	KTP	Potassium Titanyl Phosphate
LiB_3O_5	LBO	Lithium Triborate
$LiCOOH\ H_2O$	LFM	Lithium Fomate
$LiIO_3$		Lithium Iodate
$LiNbO_3$		Lithium Niobate
$LiNbO_3{:}MgO$		Mg:O-doped Lithium Niobate
$(NH_2)_2CO$		Urea
$NH_4H_2AsO_4$	ADA	Ammonium Dihydrogen Arsenate
$NH_4H_2PO_4$	ADP	Ammonium Dihydrogen Phosphate
$NO_2C_6H_4NH_2$	mNA	meta-Nitroaniline
RbH_2AsO_4	RDA	Rubidium Dihydrogen Arsenate
RbH_2PO_4	RDP	Rubidium Dihydrogen Phosphate
$RbTiOAsO_4$	RTA	Rubidium Titanyl Arsenate
Te		Tellurium
Tl_3AsSe_3		Thallium Arsenic Selenide
$ZnGeP_2$		Zinc Germanium Phosphide

4.1.1.2 Historical layout

The pioneering work of *Franken et al.* [61Fra] on second harmonic generation of ruby laser radiation in quartz and invention of the phase-matching concept [62Gio, 62Mak] generated a new direction in the freshly born field of nonlinear optics: frequency conversion in crystals. Sum frequency generation by mixing the outputs of two ruby lasers in quartz was already realized in 1962 [62Mil, 62Bas]. *Zernike and Berman* [65Zer] were the first to demonstrate difference frequency mixing. Optical parametric oscillation was experimentally realized in 1965 by *Giordmaine and Miller* [65Gio]. First monographs on nonlinear optics by *Akhmanov and Khokhlov* [64Akh] and *Bloembergen* [65Blo] greatly stimulated development of the nonlinear frequency converters. At present the conversion of laser radiation in nonlinear crystals is a powerful method for generating widely tunable radiation in the ultraviolet, visible, near, mid, and far IR regions.

For theoretical and experimental details of nonlinear frequency conversions in crystals, see monographs by *Zernike and Midwinter* [73Zer], *Danelyus, Piskarskas et al.* [83Dan], *Dmitriev and Tarasov* [87Dmi], *Shen* [84She], Handbook of nonlinear optical crystals (by *Dmitriev, Gurzadyan, Nikogosyan*) [91Dmi, 99Dmi], Handbook of nonlinear optics (by *Sutherland*) [96Sut]. For frequency conversion of femtosecond laser pulses, see also [88Akh]. For linear and nonlinear optical properties of the crystals, see [77Nik, 79Kur, 84Jer, 87Nik, 87Che, 96Sut, 99Dmi, 00Cha, 00Sas]. For related nonlinear phenomena, see [96Sut]. For the historical perspective of the nonlinear frequency conversion over the first forty years, see [00Bye]. In the following section, Sect. 4.1.2, we present some basic equations which may be useful for simple calculations of frequency converters.

4.1.2 Fundamentals

4.1.2.1 Three-wave interactions

Dielectric polarization $\boldsymbol{P}$ (dipole moment of unit volume of the substance) is related to the field $\boldsymbol{E}$ by the material equation of the medium [64Akh, 65Blo] (Chap. 1.1):

$$\boldsymbol{P}(\boldsymbol{E}) = \varepsilon_0 \left(\chi^{(1)} \boldsymbol{E} + \chi^{(2)} \boldsymbol{E}^2 + \chi^{(3)} \boldsymbol{E}^3 + \ldots \right) \tag{4.1.1}$$

with

$\varepsilon_0 = 8.854 \times 10^{-12}$ CV^{-1}m^{-1} : dielectric permittivity of free space,
$\chi^{(1)} = n^2 - 1$: the linear, and $\chi^{(2)}$, $\chi^{(3)}$ etc.: the nonlinear dielectric susceptibilities.

In the present chapter, Chap. 4.1, we consider only three-wave interactions in crystals with square nonlinearity ($\chi^{(2)} \neq 0$). The following nonlinear frequency conversion processes are considered:

Second Harmonic Generation (SHG):

$$\omega + \omega = 2\omega , \tag{4.1.2}$$

Sum-Frequency Generation (SFG) or up-conversion:

$$\omega_1 + \omega_2 = \omega_3 , \tag{4.1.3}$$

Difference-Frequency Generation (DFG) or down-conversion:

$$\omega_3 - \omega_2 = \omega_1 , \tag{4.1.4}$$

Optical Parametric Oscillation (OPO):

$$\omega_3 = \omega_2 + \omega_1 . \tag{4.1.5}$$

For efficient frequency conversion *phase matching* should be fulfilled:

$$\boldsymbol{k}_1 + \boldsymbol{k}_2 = \boldsymbol{k}_3 \tag{4.1.6}$$

with

$\boldsymbol{k}_i$: the wave vectors for ω_1, ω_2, ω_3, respectively.

Two types of phase matching are introduced:

type I: o + o → e or e + e → o ,
type II: o + e → e or o + e → o ,

or with shortened notations:

ooe: o + o → e or e → o + o ,
eeo: e + e → o or o → e + e ,
eoe: e + o → e or e → e + o ,
oeo: o + e → o or o → e + o .

In the shortened notation (ooe, eoe, ...) applies: $\omega_1 < \omega_2 < \omega_3$, i.e. the first symbol refers to the longest-wavelength radiation, and the latter to the shortest-wavelength radiation. Here, *o-beam*, or ordinary beam, is the beam with polarization normal to the principal plane of the crystal, i.e.

the plane containing the wave vector $\mathbf{k}$ and crystallophysical axis Z (or optical axis, for uniaxial crystals). The *e-beam*, or extraordinary beam, is the beam with polarization in the principal plane.

The methods of angular and temperature phase-matching tuning are used in frequency converters. Angular tuning is rather simple and more rapid than temperature tuning. Temperature tuning is generally used in the case of 90° phase matching, i.e., when the birefringence angle is zero. This method is mainly used in crystals with a strong temperature dependence of phase matching: $LiNbO_3$, LBO, $KNbO_3$, and $Ba_2NaNb_5O_{15}$.

4.1.2.2 Uniaxial crystals

For uniaxial crystals the difference between the refractive indices of the ordinary and extraordinary beams, *birefringence* Δn, is zero along the optical axis (crystallophysical axis Z) and maximum in the normal direction. The refractive index of the ordinary beam does not depend on the direction of propagation, however, the refractive index of the extraordinary beam $n_e(\theta)$ is a function of the polar angle θ between the Z axis and the vector $\mathbf{k}$ (but not of the azimuthal angle φ) (Fig. 4.1.1):

$$n_e(\theta) = n_o \left[\frac{1 + \tan^2 \theta}{1 + \left(\frac{n_o}{n_e}\right)^2 \tan^2 \theta} \right]^{\frac{1}{2}}, \qquad (4.1.7)$$

where n_o and n_e are the refractive indices of the ordinary and extraordinary beams in the plane normal to the Z axis and termed as corresponding *principal values*. Note that if $n_o > n_e$, the crystal is *negative*, and if $n_o < n_e$, it is *positive*. For an o-beam the indicatrix of the refractive indices is a sphere with radius n_o, and an ellipsoid of rotation with semiaxes n_o and n_e for an e-beam (Fig. 4.1.2). In the crystal the beam, in general, is divided into two beams with orthogonal polarizations; the angle between these beams ρ is the *birefringence* (or *walk-off*) angle.

Equations for calculating phase-matching angles in uniaxial crystals are given in Table 4.1.1 [86Nik, 99Dmi].

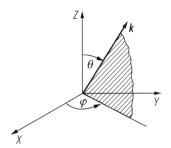

Fig. 4.1.1. Polar coordinate system for description of refraction properties of uniaxial crystals ($\mathbf{k}$ is the light propagation direction, Z is the optic axis, θ and φ are the coordinate angles).

4.1.2.3 Biaxial crystals

For biaxial crystals the optical indicatrix has a bilayer surface with four points of interlayer contact which correspond to the directions of two optical axis. In the simple case of light propagation in

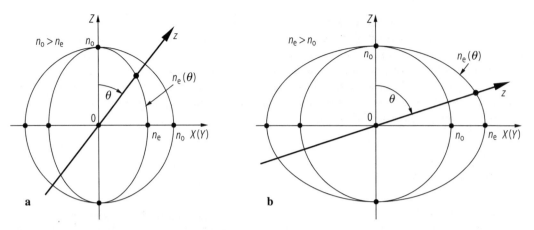

Fig. 4.1.2. Dependence of refractive index on light propagation direction and polarization (index surface) in uniaxial crystals: (**a**) negative: $n_o > n_e$ and (**b**) positive: $n_e > n_o$.

Table 4.1.1. Equations for calculating phase-matching angles in uniaxial crystals [86Nik, 99Dmi].

Negative uniaxial crystals	Positive uniaxial crystals
$\tan^2 \theta_{\text{pm}}^{\text{ooe}} = (1-U)/(W-1)$	$\tan^2 \theta_{\text{pm}}^{\text{eeo}} \approx (1-U)/(U-S)$
$\tan^2 \theta_{\text{pm}}^{\text{eoe}} \approx (1-U)/(W-R)$	$\tan^2 \theta_{\text{pm}}^{\text{oeo}} = (1-V)/(V-Y)$
$\tan^2 \theta_{\text{pm}}^{\text{oee}} \approx (1-U)/(W-Q)$	$\tan^2 \theta_{\text{pm}}^{\text{eoo}} = (1-T)/(T-Z)$

Notations:
$U = (A+B)^2/C^2$; $W = (A+B)^2/F^2$; $R = (A+B)^2/(D+B)^2$;
$Q = (A+B)^2/(A+E)^2$; $S = (A+B)^2/(D+E)^2$; $V = B^2/(C-A)^2$;
$Y = B^2/E^2$; $T = A^2/(C-B)^2$; $Z = A^2/D^2$;
$A = n_{o1}/\lambda_1$; $B = n_{o2}/\lambda_2$; $C = n_{o3}/\lambda_3$;
$D = n_{e1}/\lambda_1$; $E = n_{e2}/\lambda_2$; $F = n_{e3}/\lambda_3$.

These expressions can be generalized to noncollinear phase matching. In this case, for example, the phase-matching angle $\theta_{\text{pm}}^{\text{ooe}}$ is determined from the above presented equation using the new coefficients U and W:

$$U = (A^2 + B^2 + 2AB\cos\gamma)/C^2, \quad W = (A^2 + B^2 + 2AB\cos\gamma)/F^2,$$

where γ is the angle between the wave vectors $\mathbf{k}_1$ and $\mathbf{k}_2$.

the principal planes XY, YZ, and XZ the dependencies of refractive indices on the direction of light propagation represent a combination of an ellipse and a circle (Fig. 4.1.3). Thus in the principal planes a biaxial crystal can be considered as a uniaxial crystal, e.g. a biaxial crystal with $n_Z > n_Y > n_X$ in the XY plane is similar to a negative uniaxial crystal with $n_o = n_Z$ and

$$n_e(\varphi) = n_Y \left(\frac{1 + \tan^2\varphi}{1 + (n_Y/n_X)^2 \tan^2\varphi} \right)^{\frac{1}{2}}. \tag{4.1.8}$$

The angle V_Z between the optical axis and Z axis for the case $n_Z > n_Y > n_X$ can be found from:

$$\sin V_Z = \frac{n_Z}{n_Y} \left(\frac{n_Y^2 - n_X^2}{n_Z^2 - n_X^2} \right)^{\frac{1}{2}} \tag{4.1.9}$$

and for the case $n_X > n_Y > n_Z$:

$$\cos V_Z = \frac{n_X}{n_Y} \left(\frac{n_Y^2 - n_Z^2}{n_X^2 - n_Z^2} \right)^{\frac{1}{2}}. \tag{4.1.10}$$

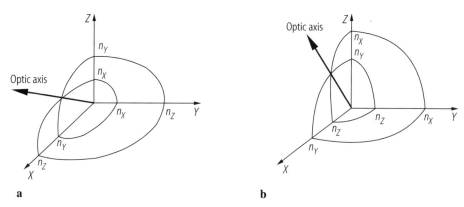

Fig. 4.1.3. Dependence of refractive index on light propagation direction and polarization (index surface) in biaxial crystals: (a) $n_X < n_Y < n_Z$, (b) $n_X > n_Y > n_Z$.

For a positive biaxial crystal the bisectrix of the acute angle between optical axes coincides with $n_{\max}$ and for a negative one the bisectrix coincides with $n_{\min}$.

Equations for calculating phase-matching angles upon propagation in principal planes of biaxial crystals are given in Table 4.1.2 [87Nik, 99Dmi].

4.1.2.4 Effective nonlinearity

Miller delta formulation [64Mil]:

$$\varepsilon_0 E_i(\omega_3) = \delta_{ijk} P_j(\omega_1) P_k(\omega_2) ,\qquad(4.1.11)$$

where the Miller coefficient,

$$\delta_{ijk} = \frac{1}{2\varepsilon_0} \frac{\chi^{(2)}_{ijk}(\omega_3)}{\chi^{(1)}_{ii}(\omega_1)\chi^{(1)}_{jj}(\omega_2)\chi^{(1)}_{kk}(\omega_3)} ,\qquad(4.1.12)$$

has small dispersion and is almost constant for a wide range of crystals.

For anisotropic media the coefficients $\chi^{(1)}$ and $\chi^{(2)}$ are, in general, the second- and third-rank tensors, respectively. In practice, the tensor

$$d_{ijk} = \frac{1}{2}\chi_{ijk}\qquad(4.1.13)$$

is used instead of χ_{ijk}. Usually, the "plane" representation of d_{ijk} in the form d_{il} is used, the relation between l and jk is:

jk		l
11	$\leftrightarrow$	1
22	$\leftrightarrow$	2
33	$\leftrightarrow$	3
23 or 32	$\leftrightarrow$	4
31 or 13	$\leftrightarrow$	5
12 or 21	$\leftrightarrow$	6

Table 4.1.2. Equations for calculating phase-matching angles in biaxial crystals upon light propagation in the principal planes [87Nik, 99Dmi].

(a) $n_X < n_Y < n_Z$

Principal plane	Type of interaction	Equations	Notations
XY	ooe	$\tan^2\varphi = \dfrac{1-U}{W-1}$	$U = \left(\dfrac{A+B}{C}\right)^2 ; W = \left(\dfrac{A+B}{F}\right)^2 ; A = \dfrac{n_{Z1}}{\lambda_1} ; B = \dfrac{n_{Z2}}{\lambda_2} ; C = \dfrac{n_{Y3}}{\lambda_3} ; F = \dfrac{n_{X3}}{\lambda_3}$
	eoe	$\tan^2\varphi \approx \dfrac{1-U}{W-R}$	$U = \left(\dfrac{A+B}{C}\right)^2 ; W = \left(\dfrac{A+B}{F}\right)^2 ; R = \left(\dfrac{A+B}{D+B}\right)^2 ; A = \dfrac{n_{Y1}}{\lambda_1} ; B = \dfrac{n_{Z2}}{\lambda_2} ; C = \dfrac{n_{Y3}}{\lambda_3} ; D = \dfrac{n_{X1}}{\lambda_1} ; F = \dfrac{n_{X3}}{\lambda_3}$
	oee	$\tan^2\varphi \approx \dfrac{1-U}{W-Q}$	$U = \left(\dfrac{A+B}{C}\right)^2 ; W = \left(\dfrac{A+B}{F}\right)^2 ; Q = \left(\dfrac{A+B}{A+E}\right)^2 ; A = \dfrac{n_{Z1}}{\lambda_1} ; B = \dfrac{n_{Y2}}{\lambda_2} ; C = \dfrac{n_{Y3}}{\lambda_3} ; E = \dfrac{n_{X2}}{\lambda_2} ; F = \dfrac{n_{X3}}{\lambda_3}$
YZ	eeo	$\tan^2\theta \approx \dfrac{1-U}{U-S}$	$U = \left(\dfrac{A+B}{C}\right)^2 ; S = \left(\dfrac{A+B}{D+E}\right)^2 ; A = \dfrac{n_{Y1}}{\lambda_1} ; B = \dfrac{n_{Y2}}{\lambda_2} ; C = \dfrac{n_{X3}}{\lambda_3} ; D = \dfrac{n_{Z1}}{\lambda_1} ; E = \dfrac{n_{Z2}}{\lambda_2}$
	oeo	$\tan^2\theta = \dfrac{1-V}{V-Y}$	$V = \left(\dfrac{B}{C-A}\right)^2 ; Y = \left(\dfrac{B}{E}\right)^2 ; A = \dfrac{n_{X1}}{\lambda_1} ; B = \dfrac{n_{Y2}}{\lambda_2} ; C = \dfrac{n_{X3}}{\lambda_3} ; E = \dfrac{n_{Z2}}{\lambda_2}$
	eoo	$\tan^2\theta = \dfrac{1-T}{T-Z}$	$T = \left(\dfrac{A}{C-B}\right)^2 ; Z = \left(\dfrac{A}{D}\right)^2 ; A = \dfrac{n_{Y1}}{\lambda_1} ; B = \dfrac{n_{X2}}{\lambda_2} ; C = \dfrac{n_{X3}}{\lambda_3} ; D = \dfrac{n_{Z1}}{\lambda_1}$
XZ, $\theta < V_Z$	ooe	$\tan^2\theta = \dfrac{1-U}{W-1}$	$U = \left(\dfrac{A+B}{C}\right)^2 ; W = \left(\dfrac{A+B}{F}\right)^2 ; A = \dfrac{n_{Y1}}{\lambda_1} ; B = \dfrac{n_{Y2}}{\lambda_2} ; C = \dfrac{n_{X3}}{\lambda_3} ; F = \dfrac{n_{Z3}}{\lambda_3}$
	eoe	$\tan^2\theta \approx \dfrac{1-U}{W-R}$	$U = \left(\dfrac{A+B}{C}\right)^2 ; W = \left(\dfrac{A+B}{F}\right)^2 ; R = \left(\dfrac{A+B}{D+B}\right)^2 ; A = \dfrac{n_{X1}}{\lambda_1} ; B = \dfrac{n_{Y2}}{\lambda_2} ; C = \dfrac{n_{X3}}{\lambda_3} ; D = \dfrac{n_{Z1}}{\lambda_1} ; F = \dfrac{n_{Z3}}{\lambda_3}$
	oee	$\tan^2\theta \approx \dfrac{1-U}{W-Q}$	$U = \left(\dfrac{A+B}{C}\right)^2 ; W = \left(\dfrac{A+B}{F}\right)^2 ; Q = \left(\dfrac{A+B}{A+E}\right)^2 ; A = \dfrac{n_{Y1}}{\lambda_1} ; B = \dfrac{n_{X2}}{\lambda_2} ; C = \dfrac{n_{X3}}{\lambda_3} ; E = \dfrac{n_{Z2}}{\lambda_2} ; F = \dfrac{n_{Z3}}{\lambda_3}$
XZ, $\theta > V_Z$	eeo	$\tan^2\theta \approx \dfrac{1-U}{U-S}$	$U = \left(\dfrac{A+B}{C}\right)^2 ; S = \left(\dfrac{A+B}{D+E}\right)^2 ; A = \dfrac{n_{X1}}{\lambda_1} ; B = \dfrac{n_{X2}}{\lambda_2} ; C = \dfrac{n_{Y3}}{\lambda_3} ; D = \dfrac{n_{Z1}}{\lambda_1} ; E = \dfrac{n_{Z2}}{\lambda_2}$
	oeo	$\tan^2\theta = \dfrac{1-V}{V-Y}$	$V = \left(\dfrac{B}{C-A}\right)^2 ; Y = \left(\dfrac{B}{E}\right)^2 ; A = \dfrac{n_{Y1}}{\lambda_1} ; B = \dfrac{n_{X2}}{\lambda_2} ; C = \dfrac{n_{Y3}}{\lambda_3} ; E = \dfrac{n_{Z2}}{\lambda_2}$
	eoo	$\tan^2\theta = \dfrac{1-T}{T-Z}$	$T = \left(\dfrac{A}{C-B}\right)^2 ; Z = \left(\dfrac{A}{D}\right)^2 ; A = \dfrac{n_{X1}}{\lambda_1} ; B = \dfrac{n_{Y2}}{\lambda_2} ; C = \dfrac{n_{Y3}}{\lambda_3} ; D = \dfrac{n_{Z1}}{\lambda_1}$

(continued)

Table 4.1.2 continued.
(b) $n_X > n_Y > n_Z$

Principal plane	Type of interaction	Equations	Notations
XY	eeo	$\tan^2\varphi \approx \dfrac{1-U}{U-S}$	$U = \left(\dfrac{A+B}{C}\right)^2$; $S = \left(\dfrac{A+B}{D+E}\right)^2$; $A = \dfrac{n_{Y1}}{\lambda_1}$; $B = \dfrac{n_{Y2}}{\lambda_2}$; $C = \dfrac{n_{Z3}}{\lambda_3}$; $D = \dfrac{n_{X1}}{\lambda_1}$; $E = \dfrac{n_{X2}}{\lambda_2}$
	oeo	$\tan^2\varphi = \dfrac{1-V}{V-Y}$	$V = \left(\dfrac{B}{C-A}\right)^2$; $Y = \left(\dfrac{B}{E}\right)^2$; $A = \dfrac{n_{Z1}}{\lambda_1}$; $B = \dfrac{n_{Y2}}{\lambda_2}$; $C = \dfrac{n_{Z3}}{\lambda_3}$; $E = \dfrac{n_{X2}}{\lambda_2}$
	eoo	$\tan^2\varphi = \dfrac{1-T}{T-Z}$	$T = \left(\dfrac{A}{C-B}\right)^2$; $Z = \left(\dfrac{A}{D}\right)^2$; $A = \dfrac{n_{Y1}}{\lambda_1}$; $B = \dfrac{n_{Z2}}{\lambda_2}$; $C = \dfrac{n_{Z3}}{\lambda_3}$; $D = \dfrac{n_{X1}}{\lambda_1}$
YZ	ooe	$\tan^2\theta = \dfrac{1-U}{W-1}$	$U = \left(\dfrac{A+B}{C}\right)^2$; $W = \left(\dfrac{A+B}{F}\right)^2$; $A = \dfrac{n_{X1}}{\lambda_1}$; $B = \dfrac{n_{X2}}{\lambda_2}$; $C = \dfrac{n_{Y3}}{\lambda_3}$; $F = \dfrac{n_{Z3}}{\lambda_3}$
	eoe	$\tan^2\theta \approx \dfrac{1-U}{W-R}$	$U = \left(\dfrac{A+B}{C}\right)^2$; $W = \left(\dfrac{A+B}{F}\right)^2$; $R = \left(\dfrac{A+B}{D+B}\right)^2$; $A = \dfrac{n_{Z1}}{\lambda_1}$; $B = \dfrac{n_{X2}}{\lambda_2}$; $C = \dfrac{n_{Y3}}{\lambda_3}$; $D = \dfrac{n_{Z1}}{\lambda_1}$; $E = \dfrac{n_{Z2}}{\lambda_2}$
	oee	$\tan^2\theta \approx \dfrac{1-U}{W-Q}$	$U = \left(\dfrac{A+B}{C}\right)^2$; $W = \left(\dfrac{A+B}{F}\right)^2$; $Q = \left(\dfrac{A+B}{A+E}\right)^2$; $A = \dfrac{n_{X1}}{\lambda_1}$; $B = \dfrac{n_{Y2}}{\lambda_2}$; $C = \dfrac{n_{Y3}}{\lambda_3}$; $E = \dfrac{n_{Z2}}{\lambda_2}$
XZ $\theta < V_Z$	eeo	$\tan^2\theta = \dfrac{1-U}{U-S}$	$U = \left(\dfrac{A+B}{C}\right)^2$; $S = \left(\dfrac{A+B}{D+E}\right)^2$; $A = \dfrac{n_{Y1}}{\lambda_1}$; $B = \dfrac{n_{X2}}{\lambda_2}$; $C = \dfrac{n_{Y3}}{\lambda_3}$; $D = \dfrac{n_{Z1}}{\lambda_1}$; $E = \dfrac{n_{Z2}}{\lambda_2}$
	oeo	$\tan^2\theta = \dfrac{1-V}{V-Y}$	$V = \left(\dfrac{B}{C-A}\right)^2$; $Y = \left(\dfrac{B}{E}\right)^2$; $A = \dfrac{n_{Y1}}{\lambda_1}$; $B = \dfrac{n_{X2}}{\lambda_2}$; $C = \dfrac{n_{Y3}}{\lambda_3}$; $E = \dfrac{n_{Z2}}{\lambda_2}$
	eoo	$\tan^2\theta = \dfrac{1-T}{T-Z}$	$T = \left(\dfrac{A}{C-B}\right)^2$; $Z = \left(\dfrac{A}{D}\right)^2$; $A = \dfrac{n_{X1}}{\lambda_1}$; $B = \dfrac{n_{Y2}}{\lambda_2}$; $C = \dfrac{n_{Y3}}{\lambda_3}$; $D = \dfrac{n_{Z1}}{\lambda_1}$
XZ $\theta > V_Z$	ooe	$\tan^2\theta = \dfrac{1-U}{W-1}$	$U = \left(\dfrac{A+B}{C}\right)^2$; $W = \left(\dfrac{A+B}{F}\right)^2$; $A = \dfrac{n_{Y1}}{\lambda_1}$; $B = \dfrac{n_{Y2}}{\lambda_2}$; $C = \dfrac{n_{X3}}{\lambda_3}$; $F = \dfrac{n_{Z3}}{\lambda_3}$
	eoe	$\tan^2\theta \approx \dfrac{1-U}{W-R}$	$U = \left(\dfrac{A+B}{C}\right)^2$; $W = \left(\dfrac{A+B}{F}\right)^2$; $R = \left(\dfrac{A+B}{D+B}\right)^2$; $A = \dfrac{n_{X1}}{\lambda_1}$; $B = \dfrac{n_{Y2}}{\lambda_2}$; $C = \dfrac{n_{X3}}{\lambda_3}$; $D = \dfrac{n_{Z1}}{\lambda_1}$; $F = \dfrac{n_{Z3}}{\lambda_3}$
	oee	$\tan^2\theta \approx \dfrac{1-U}{W-Q}$	$U = \left(\dfrac{A+B}{C}\right)^2$; $W = \left(\dfrac{A+B}{F}\right)^2$; $Q = \left(\dfrac{A+B}{A+E}\right)^2$; $A = \dfrac{n_{Y1}}{\lambda_1}$; $B = \dfrac{n_{X2}}{\lambda_2}$; $C = \dfrac{n_{X3}}{\lambda_3}$; $E = \dfrac{n_{Z2}}{\lambda_2}$; $F = \dfrac{n_{Z3}}{\lambda_3}$

Kleinman symmetry conditions [62Kle]: $d_{21} = d_{16}$, $d_{24} = d_{32}$, $d_{31} = d_{15}$, $d_{13} = d_{35}$, $d_{14} = d_{36}$, $d_{25} = d_{12} = d_{26}$, $d_{32} = d_{24}$ are valid in the case of non-dispersion of electron nonlinear polarizability. The equations for calculating the conversion efficiency include the effective nonlinear coefficients d_{eff}, which comprise all summation operations along the polarization directions of the interacting waves and thus reduce the calculation to one dimension. Effective nonlinearities d_{eff} for different crystal point groups under valid *Kleinman symmetry* conditions are presented in Table 4.1.3.

The conversion factors for SI and CGS-esu systems are given in Table 4.1.4.

Table 4.1.3. Expressions for d_{eff} in nonlinear crystals when *Kleinman symmetry* relations are valid.
(a) Uniaxial crystals

Point group	Type of interaction ooe, oeo, eoo	eeo, eoe, oee
4, 4mm	$d_{15} \sin \theta$	0
6, 6mm	$d_{15} \sin \theta$	0
$\bar{6}m2$	$d_{22} \cos \theta \sin(3\varphi)$	$d_{22} \cos^2 \theta \cos \varphi$
3m	$d_{15} \sin \theta - d_{22} \cos \theta \sin(3\varphi)$	$d_{22} \cos^2 \theta \cos(3\varphi)$
$\bar{6}$	$(d_{11} \cos(3\varphi) - d_{22} \sin(3\varphi)) \cos \theta$	$(d_{11} \sin(3\varphi) + d_{22} \cos(3\varphi)) \cos^2 \theta$
3	$(d_{11} \cos(3\varphi) - d_{22} \sin(3\varphi)) \cos \theta + d_{15} \sin \theta$	$(d_{11} \sin(3\varphi) + d_{22} \cos(3\varphi)) \cos^2 \theta$
32	$d_{11} \cos \theta \cos(3\varphi)$	$d_{11} \cos^2 \theta \sin(3\varphi)$
$\bar{4}$	$(d_{14} \sin(2\varphi) + d_{15} \cos(2\varphi)) \sin \theta$	$(d_{14} \cos(2\varphi) - d_{15} \sin(2\varphi)) \sin(2\theta)$
$\bar{4}2m$	$d_{36} \sin \theta \sin(2\varphi)$	$d_{36} \sin(2\theta) \cos(2\varphi)$

(b) Biaxial crystals (assignments of crystallophysical and crystallographic axes: for $mm2$ and 222 point groups: $X, Y, Z \to a, b, c$; for 2 and m point groups: $Y \to b$)

Point group	Principal plane	Type of interaction ooe, oeo, eoo	eeo, eoe, oee
2	XY	$d_{23} \cos \varphi$	$d_{36} \sin(2\varphi)$
	YZ	$d_{21} \cos \theta$	$d_{36} \sin(2\theta)$
	XZ	0	$d_{21} \cos^2 \theta + d_{23} \sin^2 \theta - d_{36} \sin(2\theta)$
m	XY	$d_{13} \sin \varphi$	$d_{31} \sin^2 \varphi + d_{32} \cos^2 \varphi$
	YZ	$d_{31} \sin \theta$	$d_{13} \sin^2 \theta + d_{12} \cos^2 \theta$
	XZ	$d_{12} \cos \theta - d_{32} \sin \theta$	0
mm2	XY	0	$d_{31} \sin^2 \varphi + d_{32} \cos^2 \varphi$
	YZ	$d_{31} \sin \theta$	0
	XZ	$d_{32} \sin \theta$	0
222	XY	0	$d_{36} \sin(2\varphi)$
	YZ	0	$d_{36} \sin(2\theta)$
	XZ	0	$d_{36} \sin(2\theta)$

Table 4.1.4. Units and conversion factors.

Nonlinear coefficient	MKS or SI units		CGS or electrostatic units
$\chi_{ij}^{(1)}$	1 (SI, dimensionless)	=	$\dfrac{1}{4\pi}$ (esu, dimensionless)
d_{ij} or $\chi_{ijk}^{(2)}$	1 V^{-1}m	=	$\dfrac{3 \times 10^4}{4\pi}$ (erg^{-1} cm^3)$^{\frac{1}{2}}$
δ_{ij}	1 C^{-1}m^2	=	$\dfrac{4\pi}{3 \times 10^5}$ (erg^{-1} cm^3)$^{\frac{1}{2}}$

Note that in SI units $\boldsymbol{P}^{(n)} = \varepsilon_0 \boldsymbol{\chi}^{(n)} \boldsymbol{E}^n$ (with $\boldsymbol{P}^{(n)}$ expressed in C m^{-2}), whereas in CGS or esu units $\boldsymbol{P}^{(n)} = \boldsymbol{\chi}^{(n)} \boldsymbol{E}^n$ (with $\boldsymbol{P}^{(n)}$ expressed in esu).

4.1.2.5 Frequency conversion efficiency

4.1.2.5.1 General approach

The conversion efficiency of a three-wave interaction process for the case of square nonlinearity

$$\boldsymbol{P}_{\text{nl}} = \varepsilon_0 \chi^{(2)} \boldsymbol{E}^2 \qquad (4.1.14)$$

can be determined from the wave equation derived from Maxwell's equations [64Akh, 65Blo, 73Zer, 99Dmi], see also (1.1.4)–(1.1.7),

$$\nabla \times \nabla \times \boldsymbol{E} + \frac{(1 + \chi^{(1)})}{c^2} \frac{\partial^2 \boldsymbol{E}}{\partial t^2} = -\frac{1}{\varepsilon_0 c^2} \frac{\partial^2 \boldsymbol{P}_{\text{nl}}}{\partial t^2} \qquad (4.1.15)$$

with the initial and boundary conditions for the electric field $\boldsymbol{E}$.

An exact calculation of the nonlinear conversion efficiency for SHG, SFG, and DFG generally requires a numerical calculation. In some simple cases analytical expressions are available. In order to choose the proper method, the contribution of different effects in the nonlinear mixing process should be determined. For this purpose the following approach is introduced [99Dmi]:

– Consider the *effective lengths* of the interaction process:
  1. Aperture length L_{a}:

     $$L_{\text{a}} = d_0 \rho^{-1}, \qquad (4.1.16)$$

     where d_0 is the beam diameter.
  2. Quasistatic interaction length L_{qs}:

     $$L_{\text{qs}} = \tau \nu^{-1}, \qquad (4.1.17)$$

     where τ is the radiation pulse width and ν is the mismatch of reverse group velocities. For SHG

     $$\nu = u_\omega^{-1} - u_{2\omega}^{-1}, \qquad (4.1.18)$$

     where u_ω and $u_{2\omega}$ are the group velocities of the corresponding waves ω and 2ω.
  3. Diffraction length L_{dif}:

     $$L_{\text{dif}} = k d_0^2. \qquad (4.1.19)$$

  4. Dispersion-spreading length L_{ds}:

     $$L_{\text{ds}} = \tau^2 g^{-1}, \qquad (4.1.20)$$

where g is the dispersion-spreading coefficient

$$g = \frac{1}{2}\left(\frac{\partial^2 k}{\partial \omega^2}\right) . \qquad (4.1.21)$$

5. Nonlinear interaction length L_{nl}:

$$L_{nl} = (\sigma\, a_0)^{-1} . \qquad (4.1.22)$$

Here σ is the nonlinear coupling coefficient:

$$\sigma_{1,2} = 4\pi k_{1,2}\, n_{1,2}^{-2}\, d_{\text{eff}} , \qquad (4.1.23)$$

$$\sigma_3 = 2\pi k_3\, n_3^{-2}\, d_{\text{eff}} , \qquad (4.1.24)$$

and

$$a_0 = \left(a_1^2(0) + a_2^2(0) + a_3^2(0)\right)^{\frac{1}{2}} , \qquad (4.1.25)$$

where $a_n(0)$ are the wave amplitudes of interacting waves λ_1, λ_2, and λ_3 at the input surface of the crystal.
- The length of the crystal L should be compared with L_{eff} from above equations. If $L < L_{\text{eff}}$ the respective effect can be neglected.

4.1.2.5.2 Plane-wave fixed field approximation

When the conditions $L < L_{nl}$ and $L < L_{\text{eff}}$ are fulfilled, the so-called fixed-field approximation is realized. For SHG, $\omega + \omega = 2\omega$ and $\Delta k = 2k_\omega - k_{2\omega}$, the conversion efficiency η is determined by the equation:

$$\eta = P_{2\omega}/P_\omega = \frac{2\pi^2 d_{\text{eff}}^2 L^2 P_\omega}{\varepsilon_0\, c\, n_\omega^2\, n_{2\omega}\, \lambda_2^2\, A}\, \text{sinc}^2\left(\frac{|\Delta k|\, L}{2}\right) . \qquad (4.1.26)$$

For SFG, $\omega_1 + \omega_2 = \omega_3$ and $\Delta k = k_1 + k_2 - k_3$, the conversion efficiency η is:

$$\eta = P_3/P_1 = \frac{8\pi^2 d_{\text{eff}}^2 L^2 P_2}{\varepsilon_0\, c\, n_1\, n_2\, n_3\, \lambda_3^2\, A}\, \text{sinc}^2\left(\frac{|\Delta k|\, L}{2}\right) . \qquad (4.1.27)$$

For DFG, $\omega_1 = \omega_3 - \omega_2$ and $\Delta k = k_1 + k_2 - k_3$, the conversion efficiency η is:

$$\eta = P_1/P_3 = \frac{8\pi^2 d_{\text{eff}}^2 L^2 P_2}{\varepsilon_0\, c\, n_1\, n_2\, n_3\, \lambda_1^2\, A}\, \text{sinc}^2\left(\frac{|\Delta k|\, L}{2}\right) . \qquad (4.1.28)$$

Note that all the above equations are for the SI system, i.e. $[d_{\text{eff}}] = \text{m/V}$; $[P] = \text{W}$; $[L] = \text{m}$; $[\lambda] = \text{m}$; $[A] = \text{m}^2$; $\varepsilon_0 = 8.854 \times 10^{-12}\,\text{A s/(V m)}$; $c = 3 \times 10^8\,\text{m/s}$.

When the powers of the mixing waves are almost equal, the conversion efficiency is for THG, $\omega + 2\omega = 3\omega$:

$$\eta = \frac{P_{3\omega}}{(P_{2\omega} P_\omega)^{\frac{1}{2}}} ; \qquad (4.1.29)$$

for FOHG in the case of $\omega + 3\omega = 4\omega$:

$$\eta = \frac{P_{4\omega}}{(P_{3\omega} P_\omega)^{\frac{1}{2}}},\qquad(4.1.30)$$

or for $2\omega + 2\omega = 4\omega$:

$$\eta = \frac{P_{4\omega}}{P_{2\omega}};\qquad(4.1.31)$$

for SFG, $\omega_1 + \omega_2 = \omega_3$:

$$\eta = \frac{P_3}{(P_1 P_2)^{\frac{1}{2}}};\qquad(4.1.32)$$

for DFG, $\omega_1 = \omega_3 - \omega_2$:

$$\eta = \frac{P_1}{(P_2 P_3)^{\frac{1}{2}}}.\qquad(4.1.33)$$

In some cases (mentioned additionally) the conversion efficiency is calculated from the power (energy) of fundamental radiation, e.g. for fifth harmonic generation, $\omega + 4\omega = 5\omega$:

$$\eta = \frac{P_{5\omega}}{P_\omega}.\qquad(4.1.34)$$

Corresponding equations are valid for energy conversion efficiencies by substituting the pulse energy instead of power in the above equations.

The efficiency η in the case of OPO is calculated by the equation

$$\eta = \frac{E_\text{OPO}}{E_0},\qquad(4.1.35)$$

where E_OPO is the total OPO radiation energy (signal + idler) and E_0 is the energy of the pump radiation. Conversion efficiency can also be determined in terms of *pump depletion*:

$$\eta = 1 - \frac{E_\text{unc}}{E_\text{pump}},\qquad(4.1.36)$$

where E_unc is the energy of unconverted pumping beam after the OPO crystals. Pump depletions are usually significantly greater than the ordinary η values.

The quantum conversion efficiency (for the ratio of converted and mixing quanta) in the case of exact phase-matching ($\Delta k = 0$) for sum-frequency generation, $\omega_1 + \omega_2 = \omega_3$, is determined by the following equation (SI system):

$$\eta = \frac{P_3 \lambda_3}{P_1 \lambda_1} = \sin^2\left(2\pi d_\text{eff} L \sqrt{\frac{2 P_2}{\varepsilon_0 c\, n_1 n_2 n_3 \lambda_1 \lambda_3 A}}\right),\qquad(4.1.37)$$

and for difference-frequency generation, $\omega_1 = \omega_3 - \omega_2$:

$$\eta = \frac{P_1 \lambda_1}{P_3 \lambda_3} = \sin^2\left(2\pi d_\text{eff} L \sqrt{\frac{2 P_2}{\varepsilon_0 c\, n_1 n_2 n_3 \lambda_1 \lambda_3 A}}\right).\qquad(4.1.38)$$

In the presence of linear absorption all the above equations for conversion efficiencies should be multiplied by the factor

$$\exp(-\alpha L) \approx 1 - \alpha L,\qquad(4.1.39)$$

where α is the linear absorption coefficient of the crystal.

4.1.2.5.3 SHG in "nonlinear regime" (fundamental wave depletion)

Analytical equation for SHG power conversion efficiency for the case of fundamental power depletion in the plane-wave approximation and for exact phase matching ($\Delta k = 0$) is given below [99Dmi]:

$$\eta = \frac{P_{2\omega}}{P_\omega} = \tanh^2\left(\frac{L}{L_{\text{nl}}}\right) . \tag{4.1.40}$$

In order to calculate

$$L_{\text{nl}} = (\sigma\, a_0)^{-1} \tag{4.1.41}$$

one should determine a_0 [V cm^{-1}]:

$$a_0 = \left[\frac{752\, P_\omega}{\pi\,\zeta^2\, n}\right]^{\frac{1}{2}} \tag{4.1.42}$$

from input radiation power P_ω [W] and the characteristic radius of the beam ζ [cm], and the parameter σ [V^{-1}]

$$\sigma = \frac{8\,\pi^2\, d_{\text{eff}}}{n\,\lambda_\omega} ; \tag{4.1.43}$$

where λ_1 is in m, d_{eff} in mV^{-1}.

4.1.3 Selection of data

Literature up to the end of 1998 is compiled in this chapter. Attempts were made to select the most reliable and recent data.

Tables in Sect. 4.1.4–4.1.8 present data on second, third, fourth, fifth, and sixth harmonic generation of Nd:YAG laser (including intracavity and in external resonant cavities), harmonic generation of iodine, ruby, Ti:sapphire, semiconductor, dye, argon, He–Ne, NH$_3$, CO, and CO$_2$ lasers, sum-frequency mixing (including up-conversion of IR radiation into the visible), difference-frequency generation, optical parametric oscillation (cw, nanosecond, picosecond, and femtosecond in the UV, visible, near and mid IR regions) and picosecond continuum generation.

Second harmonic generation of Nd:YAG laser was realized with conversion efficiency of $\eta = 80$ % in KDP and KTP, THG with $\eta = 80$ % in KDP, FOHG with $\eta = 80 - 90$ % (calculated from SH) in ADP and KDP, FIHG in KDP, ADP (upon cooling) and BBO and urea (at room temperature). Second harmonic generation of Ti:sapphire laser with $\eta = 75$ % was achieved in LBO, minimum pulse durations for SH were as short as 10–16 fs (BBO, LBO). Third and fourth harmonics of Ti:sapphire laser were generated in BBO, thus covering the range of wavelengths 193–285 nm. Second harmonic of CO$_2$ laser with $\eta = 50$ % was obtained in ZnGeP$_2$.

Sum-frequency generation (mixing) is used, in particular, for extending the range of generating radiation into the ultraviolet. By use of SFG the shortest wavelengths in VUV were achieved with KB5 crystal (166 nm), LBO (172.7 nm), CBO, CLBO (185 nm), BBO, KDP and ADP (189, 190, and 208 nm, respectively). At present, $\lambda = 166$ nm is the minimum wavelength achieved by frequency conversion in crystals. Sum-frequency generation is also used for up-conversion of near IR (1–5 μm) and CO$_2$ laser radiation into the visible. Maximum conversion efficiencies up to 40–60 % were obtained for the latter case in AgGaS$_2$, CdSe, and HgGa$_2$S$_4$ crystals.

Difference-frequency generation makes it possible to produce IR radiation in the near IR (up to 7.7 μm, in LiIO$_3$), mid IR (up to 18–23 μm, in AgGaSe$_2$, GaSe, CdSe, Ag$_3$AsS$_3$) and far IR (0.05–30 mm, in LiNbO$_3$ and GaP).

Optical parametric oscillation is a powerful method for generating continuously tunable radiation in the UV (up to 314–330 nm, in LBO and urea), visible, and IR regions (up to 16–18 μm, in CdSe and GaSe). Singly resonant OPO, or SROPO, uses resonant feedback at only the signal or idler frequency. Doubly resonant OPO, or DROPO, uses resonant feedback of both signal and idler frequencies. Exotic triply resonant OPO, or TROPO, with resonant feedback also at pump frequency, and quadruply resonant OPO, or QROPO, with SHG inside the OPO cavity and resonant feedback also at the second harmonic, are used very seldom.

Different OPO schemes and their energetic, temporal, spectral, and spatial characteristics are considered in detail in [73Zer, 78Dmi, 83Dan, 87Dmi] and in the three special issues of the Journal of the Optical Society of America B (Vol. 10, No. 9 and 11, 1993 and Vol. 12, No. 11, 1995) devoted to optical parametric oscillators. In Tables 4.1.30–4.1.33 we list only the main OPO parameters realized in practice: pump wavelengths, phase-matching angles, pump thresholds (peak intensity and/or average power), tuning ranges, OPO pulse durations, and conversion efficiencies for OPO experiments in the UV, visible, and near IR spectral ranges. The column headed *notes* gives data on the OPO type, pump intensities, crystal lengths, phase-matching temperatures, and output characteristics of OPO radiation (energy, power, bandwidth).

High conversion efficiencies were obtained with resonant schemes of cw OPO ($\eta = 40 - 80$ % with LiNbO$_3$:MgO crystal), nanosecond ($\eta = 60$ % with BBO), traveling-wave and synchronously pumped picosecond OPO ($\eta = 45-75$ % with KDP, KTP, KTA, BBO), and synchronously pumped femtosecond OPO ($\eta = 50$ % with BBO). Minimum pulse durations were 13 fs in SP OPO with BBO crystal, pumped by the second harmonic of a Ti:sapphire laser. Very low power thresholds (0.4 mW) were achieved with LiNbO$_3$:MgO containing quadruply resonant OPO. In general, in the case of OPO the total conversion efficiencies to both, idler and signal wavelengths, are presented. In most cases the conversion efficiency corresponds to the maximum for the range of wavelengths.

The picosecond continuum, first detected in media with cubic nonlinearity (D$_2$O, H$_2$O, etc.), was also observed in crystals with square nonlinearity (KDP, LiIO$_3$, LiNbO$_3$, etc.).

We don't pretend to comprehend all directions of frequency conversion in crystals. Some special aspects, e.g. second harmonic generation in layers and films, waveguides and fibers, periodically poled crystals, liquid crystals, as well as different design configurations of frequency converters have been beyond our consideration. For "justification" we refer to Artur L. Schawlow's famous saying: "To do successful research, you don't need to know everything. You just need to know of one thing that isn't known".

4.1.4 Harmonic generation (second, third, fourth, fifth, and sixth)

Table 4.1.5. Second harmonic generation of Nd:YAG laser radiation (1.064 → 0.532 μm).

Crystal	Type of interaction	θ_{pm} [deg]	I_0 [W cm^{-2}]	τ_p [ns]	L [mm]	Conversion efficiency [%]	Ref.	Notes
KDP	ooe	41	10^9	0.15	25	32 (energy)	[75Att]	
	ooe	41	–	0.05	25	60	[76Att]	
	ooe	41	8×10^9	0.03	14	82 (energy)	[78Mat]	
	ooe	41	7×10^9	0.03	20	81 (energy)	[78Mat]	
	ooe	41.35	–	0.1 ms	40	0.38 (energy)	[93Dim]	$\lambda = 946$ nm
DKDP	eoe	53.5	10^8	18	30	50 (power)	[76Mac]	
	eoe	53.5	3×10^9	0.25	40	70 (power)	[76Mac]	
	eoe	53.5	8×10^7	20	30	50 (energy)	[78Kog]	$P_{2\omega} = 10$ W
	ooe	36.6	3×10^8	8	20	40 (energy)	[91Bor1]	
	eoe	53.7	3×10^8	8	20	50 (energy)	[91Bor1]	
CDA	ooe	90	2×10^8	10	17.5	57 (power)	[74Kat2]	$T = 48$ °C
	ooe	90	4×10^9	0.007	13	25 (energy)	[72Rab]	
DCDA	ooe	90	8×10^7	20	21	40 (energy)	[78Kog]	$T = 90\ldots100$ °C
	ooe	90	3×10^8	20	16	40 (energy)	[78Kog]	$P_{2\omega} = 10$ W
	ooe	90	2×10^8	10	13.5	45 (power)	[74Kat2]	$T = 112$ °C
	ooe	90	9×10^7	–	29	50 (power)	[74Amm]	
	ooe	90	–	15	20	57	[76Hon]	$P_{2\omega} = 6$ W
RDA	ooe	50	–	10	–	34 (power)	[75Kat2]	$T = 25$ °C
RDP	ooe	50.8	2×10^8	10	15.3	36 (power)	[74Kat5]	
LiIO$_3$	ooe	30	7×10^7	–	18	44 (power)	[73Dmi]	
	ooe	30	3×10^9	0.04	5	50	[84Van]	
LiNbO$_3$	ooe	90	3.7×10^7	10	20	40	[81Bye]	$T = 120$ °C
	ooe [a]	90			9–30	50	[88Amm]	$P_{2\omega} = 1$ W
LFM	ooe	55.1	2×10^8	–	15	36	[75And]	
KTP	eoe	26 [b]	–	10	–	22	[78Har]	
	eoe	26 [b]	–	0.04	5	18	[83Joh]	
	eoe	25.2 [b]	–	0.07	7.2	52	[85Ale]	
	eoe	25 [b]	2.5×10^8	15	4	60	[85Bel]	
	eoe	30 [b]	2×10^7	35	9	40 (energy)	[86Dri]	multimode
	eoe	30 [b]	9×10^7	35	4	45 (energy)	[86Dri]	multimode
	eoe	30 [b]	10^8	30	5.1	60 (energy)	[86Dri]	two-pass
	eoe	30 [b]	10^8	30	8	50 (energy)	[86Dri]	Gaussian
	eoe	26 [b]	–	0.2	5	55	[87Moo1]	$E_{2\omega} = 0.19$ J
	eoe	23 [b]	2.5×10^8	10	3	30	[86Lav]	
	eoe	23 [b]	3.2×10^8	8.5	4.5	55 (power)	[93Bol]	

(continued)

Table 4.1.5 continued.

Crystal	Type of inter-action	θ_{pm} [deg]	I_0 [W cm^{-2}]	τ_p [ns]	L [mm]	Conversion efficiency [%]	Ref.	Notes
KTP	eoe	–	–	8	7	80 (energy)	[92Bro]	$E_{2\omega} = 0.72$ J, $T = 55$ °C
	eoe	–	9×10^7	17	10	97 (energy)	[97Coo]	Multistage system with 3 SHG crystals, $E_{2\omega} = 0.2$ J
KNbO$_3$	ooe	19	4.7×10^7	11	4.8	40 (energy)	[92See]	Nd:YLF (1.047 μm)
BBO	ooe	–	1.9×10^8	14	6	47	[87Adh]	$P_{2\omega} = 4.5$ W
	ooe	–	1.67×10^8	14	6	38	[87Adh]	$P_{2\omega} = 8.5$ W
	ooe	–	2.53×10^8	14	6	37	[87Adh]	$P_{2\omega} = 36$ W
	ooe	21	2×10^9	1	6.8	68 (energy)	[86Che]	
	ooe	21	2.5×10^8	8	6.8	58 (energy)	[86Che]	
	ooe	22.8	1.4×10^8	–	7	32 (power)	[90Bha2]	
	ooe	22.8	1.6×10^8	8	7.5	55–60 (energy)	[91Bor1]	
LBO	ooe	0 [b]	10^9	0.035	15	65 (energy)	[91Hua]	$T = 148.5 \pm 0.5$ °C
	ooe	0 [b]	5×10^8	10	12.5	60 (energy)	[90Lin2]	$T = 149$ °C
	ooe	12 [b]	$(5-8) \times 10^8$	9	14	70 (energy)	[91Xie]	
	ooe	12 [b]	1.4×10^8	8	17	55–60 (energy)	[91Bor1]	
CLBO	ooe	29.4	10^{11}	0.0015	7	53 (energy)	[96Sha]	
	type II	41.9	3×10^8	7	12	55 (energy)	[98Yap]	$E_{2\omega} = 1.55$ J

[a] LiNbO$_3$ grown from congruent melt.
[b] φ_{pm}.

Table 4.1.6. Second harmonic generation of Nd:YAG laser radiation in organic crystals.

Crystal	Type of inter-action	d_{eff} / d_{36}(KDP)	θ_{pm} [deg]	φ_{pm} [deg]	η [%]	Ref.	Notes
POM	eeo	13.6	18.1 (1.32 μm)	90	50	[88Jos]	$L = 7$ mm, $\tau_p = 160$ ps
MAP	eoe	38.3	2.2	0	30	[77Oud]	$L = 1$ mm
MAP	oeo	37.7	11	90	40	[77Oud]	$L = 1.7$ mm
mNA	ooe	37.7	90	55	15	[74Dav]	$L = 2.5$ mm, $\Delta\theta = 2.9$ mrad
mNA	ooe	6.8	90	8.5	85	[80Kat3]	NCSHG in the XY plane, $L = 3$ mm
DAN	–	–	40	0	20	[87Nor]	$L = 2$ mm
MHBA	–	30	–	–	59	[93Zha2]	$L = 3$ mm

Table 4.1.7. Intracavity SHG of Nd:YAG laser radiation (1.064 → 0.532 µm).

Crystal	θ_{pm} [deg]	L [mm]	Mode of Nd:YAG laser operation	$P_{2\omega}$ [W]	η [%]	Ref.
LiIO$_3$	29	–	Q-switched	0.3	100	[69Des]
	29	20	cw	4	40 (0.12 [a])	[81Dmi]
	29	–	Continuous pump, mode-locked, $\tau = 800$ ps	5	40 (0.13 [a])	[82Gol]
	29	15	$\tau = 180$ µs, $f = 50$ Hz	100 (peak)	0.06 [a]	[80Koe]
	34	4	Diode-laser pumped cw Nd:YAG laser, $\lambda = 946$ nm	0.52	–	[97Kel]
LiNbO$_3$	90	–	Continuous pump, Q-switched	0.31	100	[72Dmi]
	90	1	$\tau = 60$ ns, $f = 400$ Hz	100 (peak)	–	[68Smi]
Banana	90	3	cw	1.1	100	[68Geu]
	90	–	Continuous pump, Q-switched	0.016	100	[70Che]
	90	5	–	0.3…0.5	–	[74Gul]
KTP	26	3.5	Q-switched	5.6	–	[84Liu]
	–	4.6	Acoustooptic modulation, $f = 4…25$ kHz	28	54 (0.6 [a])	[87Per]
	–	–	Diode-laser pumped cw Nd:YAG laser	0.03…0.1	6 [a]	[92Ant]
	–	15	Diode-laser pumped mode-locked Nd:YAG laser, $\tau = 120$ ps, $f = 160$ MHz	3	56 (1.3 [a])	[92Mar]
	–	15	Diode-laser pumped cw Nd:YAG laser	2.8	47 (0.04 [a])	[92Mar]
	90	4.4	Q-switched Nd:YalO$_3$ laser, $\lambda = 1.08$ µm	15	–	[86Gar]
	–	5	Diode-laser pumped Nd:YVO$_4$ laser	0.07	9.1 [a]	[94Tai]
KNbO$_3$	90	5	cw	0.366	90	[77Fuk]
	60	3.7	Diode-laser pumped cw Nd:YAG laser, $\lambda = 946$ nm	0.0031	0.74 [a]	[89Ris]
	0	6.2	Diode-laser pumped cw Nd:YAG laser	0.002	1 [a]	[89Bia]
	90	1.3	Ti-sapphire laser pumped Nd:YalO$_3$ laser, $\lambda = 946$ nm	0.015	–	[95Zar]
BBO	25	4	Diode-laser pumped cw Nd:YAG laser, $\lambda = 946$ nm	0.55	–	[97Kel]
LBO	$\varphi = 11.4$	9	Diode-laser pumped Q-switched Nd:YAG laser, $\tau = 60$ ns	4	10 [a]	[94Han]

[a] Conversion efficiency calculated with respect to the energy of pumping flash lamps or diode lasers.

Table 4.1.8. Second harmonic generation of Nd:YAG laser radiation (1.064 → 0.532 μm) in external resonant cavities.

Crystal	θ_{pm} [deg]	T_{pm} [°C]	L [mm]	Mode of laser operation	$P_{2\omega}$ [W]	η [%]	Ref.
LiNbO$_3$:MgO	90	–	12.5	Diode-laser pumped, cw	0.03	56	[88Koz]
	90	110	12	Diode-laser pumped, cw (monolithic ring frequency doubler)	0.2	65	[91Ger]
	90	107	–	Diode-laser pumped, cw (monolithic ring frequency doubler)	0.005	50	[93Fie]
	90	110	7.5	Diode-laser pumped, cw (monolithic ring frequency doubler)	0.1	82	[94Pas]
LiNbO$_3$	90	233.7	10	Injection-locked Nd:YAG laser	1.6	69	[91Jun]
KTP	90	63	10	cw YAlO$_3$:Nd laser ($\lambda = 1.08$ μm)	0.6	85	[92Ou]
LBO	90 (θ), 0 (φ)	149.5	6	Injection-locked cw Nd:YAG laser	6.5	36	[91Yan]
	90	167	12	Diode-laser pumped mode-locked Nd:YLF laser ($\lambda = 1.047$ μm, $\tau = 12$ ps, $f = 225$ MHz)	0.75	54	[92Mal]

Table 4.1.9. Third harmonic generation of Nd:YAG laser radiation (1.064 → 0.355 μm).

Crystal	Type of interaction	θ_{pm} [deg]	τ_p [ns]	L [mm]	η [%]	Ref.	Notes
KDP	eoe	58	0.15	12	32 (energy)	[75Att]	$I_0 = 1$ GW cm^{-2}
KDP	eoe [a]	58	25	–	6 (energy)	[79And1]	$P = 40$ MW
KDP	eoe	58	0.05	–	10 (energy)	[72Kun]	
DKDP	eoe	59.5	8	20	17 (energy)	[91Bor1]	$I_0 = 0.25$ GW cm^{-2}
RDA	ooe	66.2	8	14.8	12 (power)	[75Kat2]	$\Delta\theta L = 1.0$ mrad cm
RDP	ooe	61.2	–	15.3	44 (power)	[74Kat5]	$I_0 = 0.2$ GW cm^{-2}
RDP	ooe	61.2	8	15.3	21 (power)	[74Kat1]	
LiIO$_3$	ooe	47	0.8	8	0.7 (power)	[85Bog]	$P_{av} = 4.5$ mW
LiIO$_3$	ooe	47.5	–	4	4 (power)	[71Oka]	
BBO	eoe	64	8	5.5	23 (energy)	[86Che]	$I_0 = 0.25$ GW cm^{-2}
BBO	ooe	31.3	8	7.5	20 (energy)	[91Bor1]	$I_0 = 0.19$ GW cm^{-2}
BBO	ooe	31.3	9	6	35 (quantum)	[93Wu]	Intracavity THG, $P = 0.2$ W
CBO	type II	40.3 [b]	0.035	5	80 [c]	[97Wu]	$I_0 = 5$ GW cm^{-2}
LBO	type I	38.1 [b]	8	12.2	22 (energy)	[91Bor1]	$I_0 = 0.19$ GW cm^{-2}
LBO	type II	41	8	12.6	60 (energy)	[89Wu]	

[a] Neodymium silicate glass laser.
[b] φ_{pm}.
[c] Conversion efficiency from 0.532 μm.

Table 4.1.10. Fourth harmonic generation of Nd:YAG laser radiation (1.064 → 0.266 μm).

Crystal	Type of interaction	θ_{pm} [deg]	I_0 [W cm^{-2}]	τ_p [ns]	L [mm]	Conversion efficiency (from 532 nm) [%]	Ref.	Notes
KDP	ooe	78	–	7	–	30...35	[77Aba]	
DKDP	ooe	90	8×10^9	0.03	4	75	[77Rei]	
DKDP	ooe	90	5×10^7	25	20	40	[76Liu]	$T = 60\,°C$, $P = 2.5$ MW
DKDP	ooe	90	–	600	50	3.4	[85Per]	$T = 49.8\,°C$, $P_{av} = 0.5$ W
ADP	ooe	90	8×10^9	0.03	4	85	[77Rei]	
ADP	ooe	90	–	8	30	15 [a]	[75Kat3]	$T = 51.2\,°C$, $P_{av} = 5$ W
BBO	ooe	48	–	5	–	16	[88Lag]	$E = 80$ mJ
BBO	ooe	48	1.6×10^8	1	5	52	[86Che]	
BBO	ooe	57.8	–	80 μs	6.6	0.17	[93Dim]	Nd:YAG laser cooled to 253 K, $\lambda = 946$ nm
CBO [b]	ooe	52.3	4×10^9	0.035	10	60	[97Wu]	
CLBO	ooe	62	–	10	9	30	[95Mor1]	$E = 110$ mJ
CLBO	ooe	61.6	–	7	10	50	[96Yap]	$E = 500$ mJ
CLBO	ooe	62.5	10^{11}	1.5 ps	10	24	[96Sha]	
CLBO	–	–	–	0.014	6	38	[97Sri]	
CLBO	ooe	–	1.7×10^8	46	15	20	[01Koj]	$T = 140\,°C$, $P_{av} = 20.5$ W
Li$_2$B$_4$O$_7$	ooe	66	–	10	35	20	[97Kom]	$E = 160$ mJ

[a] Efficiency of conversion from 1.064 μm.
[b] 1.064 + 0.355 → 0.266 μm; conversion efficiency from 0.355 μm.

Table 4.1.11. Fifth harmonic (1.064 → 0.2128 μm) and sixth harmonic generation (1.064 → 0.1774 μm) of Nd:YAG laser radiation.

Crystal	θ_{pm} [deg]	Type of interaction	Crystal temperature [°C]	Output parameters	τ_p [ns]	Ref.
KDP	90	ooe [a]	−70	$E = 0.1$ mJ	−	[69Akm]
KDP	90	ooe	−35	$P_{av} = 2.6$ mW, $f = 120$ kHz	30	[78Mas]
KDP	90	ooe	−40	$P_{av} = 2$ mW, $f = 6$ kHz	30	[79Jon]
KDP	84	ooe [b]	20	$E = 0.45$ mJ	0.015	[88Gar, 89Aru]
ADP	90	ooe	−40	$P_{av} = 5\ldots 7$ mW, $f = 10$ Hz	10	[76Mas1]
ADP	90	ooe [c]	−67.5	$E = 20$ J	0.5	[88Beg]
KB5	$53 \pm 1(\varphi)$	eeo	20	$E = 0.7$ mJ	6	[76Kat1]
KB5	$53 \pm 1(\varphi)$	eeo	20	$E = 0.1$ mJ	0.02	[82Tan]
KB5	$52.1(\varphi)$	eeo	20	$E = 0.3$ mJ	0.03	[80Aru]
Urea	72	eeo	20	$E = 30$ mJ	10	[80Kat1]
BBO	55	ooe	20	$E = 20$ mJ	5	[86Che, 88Lag]
CLBO	−	ooe	20	$E = 35$ mJ	10	[95Mor1]
CLBO	67.3	ooe	20	$E = 230$ mJ	7	[96Yap]
$Li_2B_4O_7$	80	ooe	20	$E = 70$ mJ	10	[97Kom]
KB5 [d]	$90(\theta), 68.5(\varphi)$	eeo	20	$P_{av} = 6$ mW	6	[96Ume]
KB5 [d]	$80(\theta), 90(\varphi)$	ooe	20		6	[96Ume]

[a] Neodymium silicate glass laser.
[b] Nd:YAlO$_3$ laser.
[c] Nd:YLF laser.
[d] Sixth harmonic generation, $\omega + 5\omega = 6\omega$.

Table 4.1.12. Generation of harmonics of Nd:YAG laser radiation with $\lambda = 1.318$ μm.

Number of harmonic	λ [nm]	Crystal	θ_{ooe} [deg]	L [mm]	τ_p [ns]	Output parameters	Energy conversion efficiency [%]	Ref.
2	659.4	LiNbO$_3$	44.67	16	40	85 kW	10	[81Akm]
3	439.6	KDP	42.05	30	40	3.4 kW	0.4	[81Akm]
4	329.7	KDP	53.47	30	40	6 kW	0.6	[81Akm]
5 [a]	263.8	KDP	55.33	30	30	0.2 kW	0.02	[81Akm]
6 [b]	219.3	KB5	78 (eeo)	15	45	3 kW	0.5	[87Aru]
2	659.4	DCDA	70.38	13.5	25	1.4 MW	40	[76Kat2]
2 [c]	659.4	LiIO$_3$	22	10	30	1 W (av.)	100	[81Kaz]
2	659.4	LiNbO$_3$	90 ($T = 300$ °C)	19	50	60 mJ	48	[83Kaz]
2	659.4	LiNbO$_3$	90	20	50	10 mJ	21	[83Kaz]
2 [c,d]	659	LBO	$\varphi = 3.7$	−	2	0.3 W (av.)	−	[94Lin]
2 [d]	659	LBO	along Z axis	16	76	0.85 mJ	40	[95Mor2]
3	439.6	KDP	42.05 ($T = 300$ °C)	40	50	1.4 mJ	3	[83Kaz]
3	439.6	LiIO$_3$	−	8	50	1.4 mJ	1.2	[83Kaz]

[a] $\omega + 4\omega = 5\omega$.
[b] $3\omega + 3\omega = 6\omega$.
[c] Intracavity SHG.
[d] Nd:YLF laser.

Table 4.1.13. Generation of harmonics of high-power Nd:glass laser radiation in KDP crystals.

Fundamental radiation			Second harmonic					Third and fourth harmonics					Ref.
λ [μm]	I_0 [10^9 W cm^{-2}]	τ_P [ns]	λ [μm]	Type of interaction	η [%]	Crystal length [mm]	E [J]	λ [μm]	Type of interaction	η [%]	Crystal length [mm]	E [J]	
1.054	2.5	0.14	0.53	eoe [a]	67	12	9	0.35	eoe	80	12	11	[80Sek]
1.054	3.5	0.7	0.53	eoe	67	12	25	0.35	eoe	80	12	30	[80Sek]
1.064	2.5	0.1	0.532	eoe	67	8	17	0.266	ooe	30	7	4	[80Lot]
1.064	9.5	0.7	0.532	ooe	83	10	346						[82Lin]
1.064	2.0	0.7	0.532	eoe	67	12	–	0.355	eoe	55	10	41	[82Lin]
1.064	1.2	0.7	0.532	ooe	–	10	–	0.266	ooe	51	10	50	[82Lin]
1.06	0.2	25	0.53	ooe	80	40	60						[82Ibr]
1.06	2.7	0.5	0.53	ooe	90	30	20						[83Gul]
1.06	2.7	0.5	0.53	eoe [a]	67	18	–	0.35	eoe	81	18	10...20	[83Gul]
1.053	1.5	0.6	0.53	eoe	70	16	80	0.26	ooe	46	7	53	[85Bru]
1.054	5	0.5	0.53	eoe	87	17.5	–	0.264	ooe	92 [b]	10	–	[88Beg]

[a] The angle between the polarization vector of the fundamental radiation and o-ray is 35°.
[b] Conversion efficiency from 0.527 μm.

Table 4.1.14. Generation of harmonics of iodine laser radiation: $\lambda = 1.315$ μm ($\tau_P = 1$ ns) [80Fil, 81Wit, 83Fil, 83Bre].

	SHG		THG		FOHG		FIHG		SIHG	
	$\omega + \omega = 2\omega$		$\omega + 2\omega = 3\omega$		$2\omega + 2\omega = 4\omega$		$2\omega + 3\omega = 5\omega$		$3\omega + 3\omega = 6\omega$	
Wavelength [nm]	657.6		438.4		328.8		263.0		219.2	
Crystal	DKDP	KDP	DKDP	KDP	KDP		KDP		KB5	
Crystal length [mm]	19	20	20	10	40		–		10	
Type of interaction	eoe	eoe	eoe [a]	ooe	ooe		ooe		eeo	
θ_{pm} [deg]	51.3	61.4	48	42.2	53.6		74		80.5 (φ_{pm})	
Conversion efficiency [%] at										
$I_0 = (1...1.5) \times 10^9$ W cm^{-2}	30	16	30	6	15		–		–	
$I_0 = 3 \times 10^9$ W cm^{-2}	70	–	50	–	30		9		3	

Table 4.1.15. Second harmonic generation of ruby laser radiation (694.3 → 347.1 nm).

Crystal	Type of interaction	θ_{pm} [deg]	I_0 [W cm^{-2}]	L [cm]	Power conversion efficiency [%]	Ref.	Notes
RDA	ooe	80.3 (90)	1.5×10^8	1.45	58	[74Kat3]	$T = 20\,°C$ (90 °C), $L\Delta\theta = 4.37$ mrad cm
RDP	ooe	67	1.8×10^8	1.0	37	[74Kat4]	$T = 20\,°C$, $L\Delta\theta = 2.4$ mrad cm
LiIO$_3$	ooe	52	1.3×10^8	1.1	40	[70Nat]	$L\Delta\theta = 0.2$ mrad cm

Table 4.1.16. Harmonic generation of Ti:sapphire (Ti:Al$_2$O$_3$) laser radiation.

(a) Second harmonic generation.

Crystal	$\lambda_{2\omega}$ [nm]	τ	θ_{pm} [deg]	L [mm]	Output power $P_{2\omega}$ [mW]	η [%]	Ref.	Notes
KDP	390	150 fs	43	3...40	300	50	[95Kry]	
LiIO$_3$	360...425	1.5 ps	43	10	700	50	[91Neb]	
BBO	360...425	1.5 ps	30	8	450	27	[91Neb]	
BBO	430	54 fs	27.5	0.055	230	75 (5.2 [a])	[92Ell]	ICSHG
BBO	383...407	–	ooe	5	170	7.4 [a]	[93Poi]	ICSHG
BBO	425	16 fs	28	0.1...1	40	–	[95Ash]	
BBO	438	10 fs	26.7; ooe	0.04	3.6	1	[98Ste]	
BBO	400	150 fs	ooe	0.5	150	38	[98Zha]	
LBO	400	150 fs	–	3	130	32	[98Zha]	
LBO	350...450	12...25 ns	90 (θ), 22...40 (φ)	5	25 mJ	30	[91Skr]	
LBO	360...425	1.5 ps	90 (θ), 32 (φ)	8	350	20	[91Neb]	
LBO	410	cw	90 (θ), 31.8 (φ)	10.7	410	21.6	[93Bou]	ERR
LBO	416	14 fs	90 (θ), 29 (φ)	0.1	30	–	[94Bac]	ICSHG
LBO	400	1.5 ps	type I	10	1280	75	[94Wat]	ERR
LBO	398	cw	90 (θ), 31.7 (φ)	8	650	70	[95Zho, 96Zho]	ICSHG
KNbO$_3$	430...470	35 ns	along a axis	7.9	7.8 kW	45 (2 [a])	[90Wu]	ICSHG
KNbO$_3$	430	cw	–	6	650	48	[91Pol]	ERR

[a] Total conversion efficiency from the pump source.

(b) Third harmonic generation: $\omega + 2\omega = 3\omega$.

Crystal	$\lambda_{3\omega}$ [nm]	τ	θ_{pm} [deg]	L [mm]	Output power $P_{3\omega}$ [mW]	η [%]	Ref.	Notes
BBO	240...285	1.8 ps	50, ooe	6.5...12	150	30	[91Neb, 92Neb]	$f = 82$ MHz
LBO	266...283	1 ps	90 (θ), 70 (φ)	7	35	10	[91Neb]	$f = 82$ MHz
BBO	252...267	180 fs	58, eoe	0.3	18	6	[93Rin]	$f = 1$ kHz

(c) Fourth harmonic generation.

Crystal	$\lambda_{4\omega}$ [nm]	τ	θ_{pm} [deg]	L [mm]	Output power $P_{4\omega}$ [mW]	η [%]	Ref.	Notes
BBO [a]	205...213	1 ps	ooe	8	10	4	[91Neb]	$f = 82$ MHz
BBO [b]	193...210	1...2 ps	75, ooe	6.9	10	4	[92Neb]	$f = 82$ MHz
BBO [b]	193...210	165 fs	65, ooe	0.1	6	3	[98Rot]	$f = 82$ MHz
BBO [b]	193...210	340 fs	65, ooe	0.3	15	–	[98Rot]	$f = 82$ MHz
BBO [b]	189...200	180 fs	71, ooe	0.1	4	1	[93Rin]	NC, $f = 1$ kHz
BBO [b]	186	10 ns	81 (θ), 30 (φ), ooe	5	0.008	–	[99Kou]	$T = 91$ K

[a] $2\omega + 2\omega = 4\omega$.
[b] $\omega + 3\omega = 4\omega$.

Table 4.1.17. Second harmonic generation of semiconductor laser radiation in $KNbO_3$.

λ_ω [nm]	Phase-matching conditions	L [mm]	$P_{2\omega}$ [mW]	η [%]	Ref.	Notes
842	$T = -23$ °C	5	24	14	[89Gol]	external resonator
865	along a axis	5	0.215	1.7	[89Dix]	External Ring Resonator (ERR)
842	along a axis	5	6.7	0.57	[90Hem]	ERR, cw
856	along a axis, $T = 15$ °C	7	41	39	[90Koz]	external resonator
972	along b axis	5	1.2	4.8	[92Zim]	distributed Bragg reflection semiconductor laser
858	–	12.4	62	1.1	[93Gol]	
858	–	12.4	80	–	[95Gol]	THG in LBO, 90 (θ), 31.8 (φ); 15 mm, $\lambda = 286$ nm, 0.05 mW
972	along b axis	6.5	156	–	[95Zim]	ERR, FOHG in BBO (14 mm) in ERR, $\lambda = 243$ nm, 2.1 mW
860	along a axis	10	50	60	[97Lod]	
858	along a axis	10	90	–	[98Mat]	FOHG in BBO ($\theta = 71$°): $\lambda = 214.5$ nm, 0.1 mW

Table 4.1.18. Second harmonic generation of dye laser radiation.

Crystal	$\lambda_{2\omega}$ [nm]	Parameters of output radiation (energy, power, pulse duration); conversion efficiency	Ref.	Notes
KDP	267.5–310	0.1 kW (peak), $\eta = 1$ %	[76Str]	
KDP	280–310	50 mJ	[77Hir]	
KDP	280	90 mW, $\eta = 10$ %	[95Nie]	$L = 55$ mm, external cavity
ADP	280–310	50 mJ, $\eta = 8.4$ %	[77Hir]	
ADP [a]	290–315	up to 1 mW, $\eta = 0.03$ %	[72Gab]	
ADP [a]	250–260	120 µW	[80Web]	$\theta_{ooe} = 90$°, $T = 200...280$ K
ADP [a]	293	0.13 mW, $\eta = 0.08$ %, $\tau = 3$ ps	[80Yam]	$L = 3$ mm
ADP [a]	295	$\eta = 10^{-4}$, $\tau = 3...4$ ps	[80Wel]	$L = 1...3$ mm
RDP	313.8–318.5	3.6 MW, $\tau = 8$ ns, $\eta = 52$ % in power	[75Kat1]	$\theta = 90$°, $T = 20...98$ °C, $I_0 = 36$ MW cm^{-2}, $L = 25$ mm
RDP	310–335	3.2 MW, $\tau = 10$ ns, $f = 10$ Hz, $\eta = 36$ %	[77Kat2]	$\theta = 90$°

(continued)

Table 4.1.18 continued.

Crystal	$\lambda_{2\omega}$ [nm]	Parameters of output radiation (energy, power, pulse duration); conversion efficiency	Ref.	Notes
ADA	292–302	30 mW	[77Fer]	$\theta = 90°$
ADA [a]	285–315	400 mW (single-mode regime), 50 mW (multimode regime)	[76Fro]	$\theta = 90°$, temperature tuning, $L = 30$ mm
DKDA	310–355	$0.8\ldots3.2$ MW, $\tau = 10$ ns, $f = 10$ Hz, $\eta = 9\ldots36\%$	[77Kat2]	$\theta = 90°$, $L = 15$ mm
LiIO$_3$[a]	295	$\eta = 10^{-4}$, $\tau = 2.1$ ps	[80Wel]	$L = 0.3$ mm
LiIO$_3$[a]	293–312	0.37 mW, cw regime	[86Bue]	$L = 10$ mm
LiIO$_3$	293–330	15 mW, cw regime	[83Maj]	$L = 1$ mm
LiIO$_3$	293	3 kW, $\eta = 30\%$	[76Str]	$L = 6$ mm
LiIO$_3$	293–310	4 mW, cw regime, $\eta = 0.4\%$	[75Bet]	$L = 6$ mm, $\Delta\lambda = 0.03$ nm
LiIO$_3$	293–310	21 mW, cw regime, $\eta = 2\%$	[75Bet]	$L = 6$ mm, $\Delta\nu = 30$ MHz
BBO	204.8–215	100 kW, $\tau = 8$ ns, $\eta = 4\ldots17\%$	[86Kat]	$\theta = 70°\ldots90°$
BBO	205–310	50 kW, $\tau = 9\ldots22$ ns, $\eta = 1\ldots36\%$	[86Miy]	$L = 6$ and 8 mm
BBO [a]	315	20 mW, $\tau = 43$ fs	[88Ede]	$\theta = 38°$, $\varphi = 90°$, $L = 55$ μm
BBO	230–303	$0.02\ldots0.18$ mJ, $\tau = 17$ ns	[90Mue]	$\theta_{ooe} = 40°\ldots60°$, $L = 7$ mm
BBO [a]	243	30 mW, cw regime	[91Kal]	$\theta_{ooe} = 55°$, $L = 8$ mm, $\Delta\nu = 200$ Hz
KB5	217.3–234.5	0.3 kW, $\tau = 7$ ns, $\eta = 1\%$	[75Dew]	XY plane, eeo
KB5	217.1–240	$5\ldots6$ μJ, $\tau = 3\ldots4$ ns, $\eta = 10\%$	[76Dew]	XY plane, $\theta_{ooe} = 90°\ldots0°$
KB5	217.1–315.0	$5\ldots6$ μJ, 5 ns, 10%	[76Dew]	XY plane, $\varphi_{eeo} = 90°\ldots31°$, $L = 10$ mm
KB5	217–250	$0.1\ldots5$ μJ, $\eta = 0.2\ldots5\%$	[76Zac]	XY plane, $\varphi_{eeo} = 90°\ldots65°$
DKB5	216.15	2 μJ, $\tau = 3$ ns, $\eta = 5\%$	[78Pai]	$\theta = 90°$, $\varphi = 90°$
LFM	230–300	$\eta = 2\%$	[73Dun]	XZ plane, $\theta_{ooe} = 35°\ldots45°$, $L = 10$ mm
LFM [a]	290–315	$\eta = 10^{-4}$	[72Gab]	XZ plane, $\theta_{ooe} = 45°$ (590 nm)
LFM [a]	238–249	70 μJ (244 nm), cw regime	[80Bas]	XZ plane, $\theta_{ooe} = 39°$ (486 nm)
LFM	237.5–260	20 W, nanosecond regime, $\eta = 0.7\%$	[76Str]	
LFM [a]	243	1.4 mW, cw regime	[84Foo]	$\theta_{ooe} = 36.8°$, $L = 15$ mm
LFM	285–310	4 μJ, cw regime	[75Bet]	
KNbO$_3$	425–468	400 kW, $\eta = 43\%$	[79Kat]	angular tuning in XY and YZ planes, temperature tuning ($20\ldots220$ °C) along the a axis
KNbO$_3$	419–475	12 μW, cw regime, $\eta = 0.065\%$	[83Bau]	along the a axis, T from -36 °C to $+180$ °C, $L = 9$ mm
KNbO$_3$[a]	425–435	21 mW, cw regime, $\eta = 1.1\%$	[85Bau]	along the a axis, $T = 0\ldots50$ °C, $L = 9$ mm
Urea	238–300	–	[79Hal]	$\theta_{eeo} = 90°\ldots45°$, $L = 2$ mm
Urea	298–370	–	[79Hal]	$\theta_{eoo} = 90°\ldots50°$, $L = 2$ mm

[a] Intracavity SHG.

Table 4.1.19. Second harmonic generation of gas laser radiation.

Type of laser	Crystal	λ [μm]	θ_{pm} [deg]	T [°C]	Ref.
Argon laser	KDP [a]	0.5145	90	−13.7	[67Lab]
	ADP	0.4965	90	−93.2	[73Jai]
	ADP	0.5017	90	−68.4	[73Jai]
	ADP	0.5145	90	−10.2	[73Jai]
	ADP [a]	0.5145	90	−10	[82Ber]
	KB5	0.4579	67.2 (φ_{pm})	20	[76Che]
	KB5	0.4765	60.2 (φ_{pm})	20	[76Che]
	KB5	0.4880	56.6 (φ_{pm})	20	[76Che]
	KB5	0.5145	50.2 (φ_{pm})	20	[76Che]
	BBO	0.5145	49.5	20	[86Xin]
	BBO	0.4965	52.5	20	[86Xin]
	BBO	0.4880	54.5	20	[86Xin]
	BBO	0.4765	57.0	20	[86Xin]
	BBO [a]	0.4880	55	20	[89Zim]
	BBO [a]	0.5145	−	20	[92Tai]
He-Ne laser	LiIO$_3$ [a]	1.152 ... 1.198	25	20	[83Kac]
	LiNbO$_3$	1.152	90	169	[74Ant]
	LiNbO$_3$	1.152	90	281	[75Kus]
	AgGaS$_2$	3.39	33	20	[75Bad]
NH$_3$ laser	Te	12.8	−	−	[80Sha]
	CdGeAs$_2$	11.7	35.7	−	[87And3]
CO laser	ZnGeP$_2$	5.2 ... 6.3	47.5	−	[87And2]

[a] Intracavity SHG.

Table 4.1.20. Harmonic generation of CO_2 laser radiation.

Crystal	λ [μm]	Nonlinear process	Type of interaction, θ_{pm} [deg]	I_0 [W cm^{-2}]	L [mm]	η (power) [%]	Ref.
Ag$_3$AsS$_3$	10.6	SHG	ooe, 22.5	1.1×10^7	4.4	2.2	[75Nik2]
AgGaSe$_2$	10.6	SHG	ooe, 57.5	1.7×10^6	15.3	2.7	[74Bye]
AgGaSe$_2$	10.25	SHG	ooe, 52.7	$< 10^7$	21	35	[85Eck]
AgGaSe$_2$	10.6	SHG	ooe, 53	−	20	0.1 [a]	[97Sto]
ZnGeP$_2$	9.19 ... 9.7; 10.15 ... 10.8	SHG	eeo, 76	−	−	5	[84And]
ZnGeP$_2$	8.6	SHG	eeo, 55.8	−	−	10.1	[87And4]
ZnGeP$_2$	10.6	SHG	eeo, 76	10^9	3	49	[87And1]
ZnGeP$_2$	10.26 ... 10.61	SHG	eeo	4.4×10^7	7.2	11.3	[93Bar2]
ZnGeP$_2$	9.6	SHG	eeo, 70	5.5×10^7	10	8.1	[94Mas]
ZnGeP$_2$	10.78	SHG	eeo, 90	−	10	−	[97Kat]
CdGeAs$_2$	10.6	SHG	oeo, 48.4	1.4×10^7	9	15	[74Kil]
CdGeAs$_2$	10.6	SHG	eeo, 32.5	−	13	21	[76Men]
CdGeAs$_2$	10.6	SHG	eeo, 32.5	−	13	0.44 [a]	[76Men]

(continued)

Table 4.1.20 continued.

Crystal	λ [μm]	Nonlinear process	Type of interaction, θ_{pm} [deg]	I_0 [W cm^{-2}]	L [mm]	η (power) [%]	Ref.
Tl$_3$AsSe$_3$	9.6	SHG	–	–	3.7	10.9	[87Pas]
Tl$_3$AsSe$_3$	9.6	SHG	ooe, 19	10^7	5…6	28	[89Auy]
Tl$_3$AsSe$_3$	10.6	SHG	ooe	6.3×10^8	4.57	57	[91Suh]
Tl$_3$AsSe$_3$	9.25	SHG	ooe, 19	2×10^7	46	20	[96Suh]
GaSe	9.3…10.6	SHG	ooe, 12.8…14.4	2×10^7	6.5	9	[89Abd]
GaSe	9.2…11.0	SHG	ooe, 13	–	2.5	–	[95Bha]
CdGeAs$_2$	–	THG	oeo, 45	–	4.5	1.5	[79Men]
Tl$_3$AsSe$_3$	9.6	THG	ooe, 21	10^7	5…6	–	[89Auy]
ZnGeP$_2$	10.6	FOHG	eeo, 47.5	–	10	14 [b]	[87And1]
ZnGeP$_2$	–	FOHG	eeo, 47.5	–	5	2	[85And]
ZnGeP$_2$	10.6	FOHG	eeo, 47.8	–	10	–	[97Sto] [a]
ZnGeP$_2$	9.55	FOHG	eeo, 49	–	10	10	[98Cho]
Tl$_3$AsSe$_3$	9.6	FOHG	ooe, 27	10^7	5…6	27 [b]	[89Auy]
Tl$_3$AsSe$_3$	9.6	FIHG	ooe, 28	10^7	5…6	45 [c]	[89Auy]

[a] Continuous-wave regime.
[b] Conversion efficiency from 2ω.
[c] Conversion efficiency from 4ω.

4.1.5 Sum frequency generation

Table 4.1.21. Sum frequency generation of UV radiation in KDP.

λ_{SF} [nm]	Sources of interacting radiation	τ_p [ns]	Conversion efficiency, power, energy	Ref.
190–212	SRS of 1.064 μm + sum frequency radiation (220–250 nm) [83Tak]	0.02	20–40 μJ	[85Tak]
215–223	2ω of dye laser + Nd:YAG laser	10	10 kW	[76Mas1]
215–245	SRS of 266 nm (4ω of Nd:YAG laser) + OPO (0.9–1.4 μm)	0.02	100 μJ	[83Tak]
217–275	2ω of dye laser + Nd:YAG laser (1.064 μm)	25–30	50–55 %, 10 mW (average)	[83Kop]
217–226	OPO (1.1–1.5 μm) + 4ω of Nd:YAG laser (266 nm)	0.02	100 kW	[82Tan]
218–244	(269–315 nm) [79Ang] + Nd:YAG laser	0.03	0.1 mJ	[79Ang]
239	Nd:YAG laser (1.064 μm) + XeCl laser (308 nm)	0.7	50%	[81Lyu]
240–242	2ω of ruby laser (347 nm) + dye laser	30	1 MW	[78Sti3]
257–320	Dye laser + argon laser	cw regime	0.2 mW	[77Bli]
269–315	SRS of 532 nm (2ω of Nd:YAG laser) + 532 nm	0.03	1–3 mJ	[79Ang]
269–287	OPO (1.29–3.6 μm) + 3ω of Nd:YAG laser (355 nm)	0.02	100 kW	[82Tan]
271	Two copper vapor lasers (511 and 578 nm)	35	1.5%, 100 mW (average)	[89Cou]
288–393 [a]	OPO (0.63–1.5 μm) + 2ω of Nd:YAG laser (0.532 nm)	0.02	100 kW	[82Tan]
360–415	Dye laser + Nd:YAG laser	25–30	60–70%	[79Dud]
362–432	Dye laser + Nd:YAG laser	0.03	20%	[76Moo]

[a] DKDP crystal was used.

Table 4.1.22. Sum frequency generation of UV radiation in ADP.

λ_{SF} [nm]	Sources of interacting radiation	τ_p [ns]	Conversion efficiency, power, energy	Ref.
208–214	2ω of dye laser + Nd:YAG laser, $\theta = 90°$, $T = -120°\ldots0°C$	10	1.7 μJ	[76Mas1]
222–235	2ω of dye laser + Nd:YAG laser	10	10%	[76Mas1]
240–248	Dye laser + 2ω of ruby laser, $\theta = 90°$, $T = -20\ldots+80°C$	30	4%, 1 MW	[78Sti3]
243–247 [a]	Dye laser + argon laser (363.8 nm)	cw regime	4 mW	[91Kal, 83Cou]
243 [a]	Dye laser + argon laser (351 nm), $\theta = 90°$, $T = 8°C$	cw regime	0.3 mW	[83Hem1]
247.5	Dye laser + krypton laser (413.1 nm), $\theta = 90°$, $T = -103°C$	cw regime	–	[79Mar]
246–259	Dye laser + 2ω of Nd:YAG laser, $\theta = 90°$, $T = -120\ldots0°C$	10	1%, 3 μJ	[76Mas1]
252–268 [a]	Dye laser + argon laser (477, 488, 497 nm), $\theta_{ooe} = 90°$	cw regime	8 mW	[82Liu]
270–307	Dye laser + 2ω of Nd:YAG laser, $\theta_{ooe} = 81°$	ps regime	–	[76Moo]

[a] ADP crystal was placed in an external resonator.

Table 4.1.23. Sum frequency generation of UV radiation in BBO.

λ_{SF} [nm]	Sources of interacting radiation	τ_p [ns]	Conversion efficiency, power, energy	Ref.
188.9–197	Dye laser (780–950 nm) + 2ω of another dye laser (248.5 nm)	10	up to 0.1 mJ	[88Mue]
190.8–196.1	Ti:sapphire laser (738–825 nm) + 2ω of Ar laser (257 nm)	–	tens of nW	[91Wat]
193	Dye laser + KrF laser (248.5 nm)	9	0.2%, 2 μJ	[88Mue]
193	Dye laser (707 nm) + 4ω of Nd:YAG laser	90–250 fs	10 μJ (250 fs)	[92Hof]
193.3	Dye laser (724 nm, 5 ps) + 4ω of Nd:YLF laser (263 nm, 25 ps)	0.01	1.7%, 4 μJ (2.5 mJ) [a]	[92Tom]
193.4	FOHG of dye laser radiation (774 nm, 300 fs), $\omega + 3\omega = 4\omega$	800 fs	0.5 μJ (1.5 mJ) [a]	[92Rin]
194	Ti:sapphire laser + 2ω of Ar laser (257 nm), three crystal configuration with external cavity	–	0.016 mJ	[92Wat]
194	Diode laser (792 nm) + 2ω of Ar laser (257 nm)	cw	2 mW	[97Ber]
195.3	THG of dye laser (T [crystal] = 95 K)	17	5%, 8 μJ	[88Lok]
196–205	Dye laser + 2ω of another dye laser	5	0.1 mJ	[92Hei]
197.7–202	THG of dye laser	0.008	1%, 1–4 mW	[88Gus]
198–204	THG of dye laser	5	20%, 1.7 mJ	[87Gla]
271	Two copper vapor lasers (511 and 578 nm)	35	0.9%, 64 mW	[89Cou]
362.6–436.4	Dye laser + Nd:YAG laser, noncollinear SFG (NCSFG), $\alpha = 4.8\ldots21.3°$	–	1%, 0.065 mJ	[90Bha1]
369	Diode laser (1310 nm) + Ar laser (515 nm)	–	1.3 μW	[91Sug]
370.6	Dye laser (568.6 nm) + Nd:YAG laser, NCSFG, $\alpha = 6.3°$	–	8–18%	[92Bha]

[a] After amplification in an ArF excimer gain module.

Table 4.1.24. Sum frequency generation of UV radiation in LBO.

λ_{SF} [nm]	Sources of interacting radiation	τ_p [ns]	Conversion efficiency, power, energy	Ref.
170–185 [a]	OPO (1.6–2.5 µm) + 4ω of Ti-sapphire laser (189–210 nm), $\theta = 66$–$90°$, ooe	100 fs	4	[98Pet3]
172.7–187	OPO (1.65–2.15 µm) + 4ω of Ti-sapphire laser (190–203.75 nm), $\theta = 90°$, $\varphi = 73°$, ooe	130 fs	50 nJ	[94Sei3]
185–187.5 [b]	OPO + 5ω of Nd:YAG laser (212.8 nm), $\theta = 62$–$74°$	–	–	[95Kat]
194 [b]	OPO + 5ω of Nd:YAG laser (212.8 nm), $\theta = 51.2°$, $\varphi = 90°$	5	2.2 %	[00Kag]
185 [c]	OPO + 5ω of Nd:YAG laser (212.8 nm), $\theta = 64°$	–	–	[97Ume]
194 [c]	OPO + 5ω of Nd:YAG laser (212.8 nm), $\theta = 53°$, $\varphi = 0°$	5	1 %	[00Kag]
195–210 [c] 226–265	Nd:YAG laser + 2ω of dye laser, 2ω or 3ω of dye laser	10	14 %	[00Bha]
188–195	OPO (1.6–2.3 µm) + 5ω of Nd:YAG laser (212.8 nm), $\theta = 90°$, $\varphi = 90$–$52°$, ooe	6	0.2–2 %, 2–40 µJ	[91Bor2]
187.7–195.2	OPO (1.591–2.394 µm) + 5ω of Nd:YAG laser, $\theta = 90°$, $\varphi = 88$–$50°$, ooe	8	3 kW (peak)	[92Wu]
191.4	SRS in H_2 (1.908 µm) + 5ω of Nd:YAG laser, $\theta = 90°$, $\varphi = 88$–$50°$, ooe	8	10 %, 67 kW (peak), 2 mW (average)	[92Wu]
218–242	OPO (1.2–2.6 µm) + 4ω of Nd:YAG laser (266 nm), $\theta = 90°$, $\varphi = 90$–$33°$, ooe	6	0.2–2 %, 20–400 µJ	[91Bor2]
232.5–238	Nd:YAG laser + 2ω of dye laser	10	–	[90Kat]
240–255	Nd:YAG laser + 2ω of dye laser, NCSFG	10	8 %, 0.12 mJ	[93Bha]

[a] $Li_2B_4O_7$ crystal was used.
[b] CBO crystal was used.
[c] CLBO crystal was used.

Table 4.1.25. Sum frequency generation of UV radiation in KB5.

λ_{SF} [nm]	Sources of interacting radiation	τ_p [ns]	Conversion efficiency, power, energy	Ref.
208–217	Two dye lasers, $\theta = 90°$, $\varphi = 90°$, eeo	10	0.025 %, 1 W	[76Dun]
196.6	Dye laser + 2ω of Nd:YAG	8	0.1 %, 0.5 mJ	[77Kat1]
207.3–217.4	Ruby laser (694.3 nm) + 2ω of dye laser	3	0.3 %, 0.8 mJ	[77Kat2]
201–212	Nd:YAG + 2ω of dye laser	20	10 %, 2–10 µJ	[77Sti]
185–200	Dye laser (740–910 nm) + 2ω of dye laser (237 nm), $\theta = 90°$, eeo	30	10 %, up to 10 µJ	[78Sti2]
211–216	Dye laser + Ar laser (351.1 nm)	cw regime	10^{-6}, 50–100 nW	[78Sti1]
196.7–226	OPO + 3ω and 4ω of Nd:YAG laser, $\theta = 90°$, $\varphi = 65°$, eeo	0.02	20 kW	[82Tan]
194.1–194.3	Dye laser + 2ω of Ar laser (257 nm)	cw regime	2 µW	[83Hem2]
200–222	OPO + 3ω and 4ω of Nd:YAG laser	0.045	2×10^{-5}, 1 µJ	[83Pet]
166–172	OPO (1.15–1.6 µm) + 4ω of Ti-sapphire laser, $\theta = 90°$, $\varphi = 90°$, eeo	200 fs	0.05–0.4 MW	[98Pet2]

Table 4.1.26. Up-conversion of near IR radiation into the visible.

Crystal	λ_{IR} [μm]	Pump source	η [%]	Ref.
LiIO$_3$	3.39	0.694 μm, mode-locked ruby laser	100	[73Gur]
	3.2...5	1.064 μm, Nd:YAG laser	0.001	[74Gur]
	2.38	0.488 μm, argon laser	4×10^{-8}	[75Mal2]
	1.98, 2.22, 2.67	0.694 μm, mode-locked ruby laser	0.14...0.28	[75Mal1]
	3.39	0.5145 μm, argon laser	2.4×10^{-2}	[80See]
	1...2	0.694 μm, ruby laser	18	[71Cam]
LiNbO$_3$	1.69...1.71	0.694 μm, Q-switched ruby laser	1	[67Mid]
	1.6...3.0	0.694 μm, Q-switched ruby laser	100	[75Aru]
	1.6	0.694 μm, ruby laser	10^{-5}	[68Mid]
	3.3913	0.633 μm, cw He-Ne laser	10^{-5}	[67Mil]
	3.3922	0.633 μm, cw He-Ne laser	5×10^{-5}	[73Bai]
KTP	1.064	0.809 μm, diode laser	68	[93Kea]
	1.54	0.78 μm, diode laser	7×10^{-4}	[93Wan1]
	1.064	0.824 μm, dye laser (intracavity SFG)	0.26	[90Ben]
	1.064	0.809 μm, diode laser	55	[92Ris]
	1.064	0.805 μm, diode laser	24	[92Kea]
	1.319; 1.338	0.532 μm, 2ω of Q-switched Nd:YAG laser	10	[89Sto]

[a] The angle between the polarization vector of the fundamental radiation and o-ray is 35°.

Table 4.1.27. Up-conversion of CO_2 laser radiation by sum-frequency generation.

Crystal	Pump source	λ_{pump} [μm]	Type of interaction	θ_{pm} [deg]	I_0 [W cm^{-2}]	L [mm]	η [%]	Ref.
Ag_3AsS_3	ns Nd:YAG laser, 740 W	1.064	eoe	20	–	6	0.84	[72Tse]
	Ruby laser, 1 ms	0.694	–	–	10^4	10	0.14	[72Luc]
	ns Nd:YAG laser	1.064	eoe	20	400	6	0.5	[73Alc]
	Nd:YAG laser	1.064	eoe	20	–	14	1.5	[74Vor]
	Ruby laser, 25 ps	0.694	ooe	25.2	10^8	5	10.7	[75Nik1]
	ns Nd:YAG laser	1.064	eoe	20	–	–	30 [a]	[79Jaa]
	ns Nd:YAG laser	1.064	eoe	20	$(0.5...1.2) \times 10^6$	–	8 [b]	[81And]
$AgGaS_2$	Nd:YAG laser	1.064	oee	40	6×10^5	3	40 [a]	[75Vor]
	Dye laser, 3 ns	0.598	ooe	90	–	5	40	[77Jan]
	Ruby laser, 30 ns	0.694	eoe	55	–	3.3	9	[77And1]
	ns Nd:YAG laser	1.064	oee	40	–	–	30	[78Vor]
	ns Nd:YAG laser	1.064	oee	40	$(0.5...1.2) \times 10^6$	–	14 [b]	[81And]
$HgGa_2S_4$	ns Nd:YAG laser	1.064	ooe	41.6	$(0.5...1.2) \times 10^6$	3.6	60 (20) [b]	[80And, 81And]
$ZnGeP_2$	Nd:YAG laser	1.064	oeo	82...89	–	10	1.4	[71Boy]
	ns Nd:YAG laser	1.064	oeo	82.9	$(0.5...1.2) \times 10^6$	–	6 [b]	[81And]
	Nd:YAG laser, 30 ns	1.064	oeo	82.5	3×10^6	3	5	[79And2]
CdSe	Nd:YAG laser	1.833	oeo	77	2.4×10^7	10	35 [a]	[71Her]
	HF laser, 250 ns	2.72	oeo	70.5	6×10^6	30	40	[76Fer]

[a] Power-conversion efficiency.
[b] Power-conversion efficiency for two cascades:
$10.6 + 1.064 \rightarrow 0.967$ μm,
$0.967 + 1.064 \rightarrow 0.507$ μm.

4.1.6 Difference frequency generation

Table 4.1.28. Generation of IR radiation by DFG.

(a) Crystal: LiIO$_3$

λ [µm]	Sources of interacting radiations, crystal parameters	Conversion efficiency, energy, power, τ_p	Ref.
4.1–5.2	Dye laser + ruby laser, ICDFG, $L = 12$ mm	100 W (peak)	[72Mel]
1.25–1.60; 3.40–5.65	Dye laser + Q–switched Nd:YAG laser (1.064 and 0.532 µm), ICDFG, $\theta_{ooe} = 21\text{–}28.5$ °	0.5–70 W (peak), $\Delta \nu = 0.1$ cm^{-1}, 60 ns	[75Gol]
2.6–7.7	Dye laser + 2ω of Nd:YAG laser, $\theta_{ooe} = 22$ °	2 nJ–50 µJ, 10 ns	[95Cha2]
2.3–4.6	Dye laser + argon laser (514 and 488 nm)	0.5–4 µW, cw	[76Wel]
4.3–5.3	Dye laser + 2ω of Nd:YAG laser, $\theta_{ooe} = 24.3$ °	–	[77Dob]
0.7–2.2	Dye laser + nitrogen laser, $\theta_{ooe} = 51\text{–}31$ °	3 ns	[78Koe]
3.8–6.0	Dye laser + copper vapor laser (511 nm), $\theta_c = 21\text{–}24$ °	10–100 µW, 20 ns	[82Ata]
3.5–5.4	Dye laser + 2ω of Nd:YAG laser, $\theta_{ooe} = 20$ °	0.8 mJ, 10 ns	[83Man]
1.2–1.6	Two dye lasers, $\theta_{ooe} = 29$ °	1.5–5 ps	[84Cot]
4.4–5.7	Dye laser + Nd:YAG laser, $\theta_{ooe} = 20\text{–}22$ °	550 kW, 8 ns	[85Kat]
~ 5	Two dye lasers, $\theta_{ooe} = 20$ °, $L = 3$ mm	10 %, 10 nJ, 400 fs	[91Els]
2.5–5.3	Signal and idler pulses of OPO, $\theta_{ooe} = 21$ °	0.2 mW, $f = 82$ MHz, 200 fs	[94Loh]
6.8–7.7	Dye laser + 2ω of Nd:YAG laser, $\theta = 28\text{–}29$ °	100 mW (peak)	[95Cha1]

(b) Crystal: LiNbO$_3$

λ [µm]	Sources of interacting radiations, crystal parameters	Conversion efficiency, energy, power, τ_p	Ref.
3–4	Dye laser + ruby laser	1 %, 6 kW	[71Dew]
2.2–4.2	Dye laser + argon laser	1 µW, cw	[74Pin]
2–4.5	Dye laser (1.2 ps) + argon laser (100 ps), $\theta = 90$ °, $T = 200\ldots 400$ °C	25 µW (average), 1.2 ps, $f = 138$ MHz	[84Rud, 85Ree]
2–4	Dye laser + Nd:YAG laser, $\theta_{ooe} = 46\ldots 57$ °	60 %, 1.6 MW	[80Kat2]
2.04	Two dye lasers, $\theta_{ooe} = 90$ °	50 %, $\Delta \lambda = 0.03$ nm	[77Sey]
1.7–4.0	CPM dye laser + subpicosecond continuum, $\theta_c = 55$ °, $L = 1$ mm	10 kW (peak), 0.2 ps, $\Delta \nu = 100$ cm^{-1}	[87Moo2]
4.043	Two Nd:YAG lasers (1.064 and 1.444 µm), $L = 25$ mm	5.5 %, 30 mJ, 14 ns	[94Won]
1.6–4.8	Nd:glass laser + OPO	6 %, 30 µJ, 1–3 ps	[95DiT]

(c) Crystal: BBO

λ [µm]	Sources of interacting radiations, crystal parameters	Conversion efficiency, energy, power, τ_p	Ref.
2.5	Dye laser (620 nm) + picosecond continuum (825 nm), $\theta_{ooe} = 20.3$ °, $L = 5$ mm	5 %, 4 µJ, 0.5 ps	[91Pla]
0.9–1.5	Dye laser + Nd:YAG laser, $\theta_{ooe} = 20.5\text{–}24.5$ °, $L = 10$ mm	23 %, 4.5 mJ, 8 ns	[93Ash]
2.04–3.42	Two dye lasers, NCDFG, $\theta_{ooe} = 12\text{–}17$ °, $L = 6$ mm	300–400 W (peak)	[91Bha]
1.23–1.76	Dye laser + Ti:sapphire laser	10 µW (average), 150 fs, $f = 80$ MHz	[93Sei]

(d) Crystal: KTP

λ [μm]	Sources of interacting radiations, crystal parameters	Conversion efficiency, energy, power, τ_p	Ref.
1.4–1.6	Dye laser + Nd:YAG laser, $\theta_{eoe} = 76$–$78°$, $\varphi = 0°$	8.4 kW, $f = 76$ MHz, 94 fs	[75Bri]
1.35–1.75	Dye laser + 2ω of Ti:sapphire laser, ICDFG	10 W (peak), 1.6 ps	[94Pet]
2.8–3.6	Ti:sapphire laser + OPO, $\theta_{eoe} = 90°$, $\varphi = 47°$	40–150 μW, 90–350 fs, $f = 82$ MHz	[95Gal1]
1.2–2.2	Nd:YAG laser + dye laser, $\theta_{eoe} = 90°$, $\varphi = 31°$	36 % (quantum), 1 mJ	[95Cha3]
1.05–2.8	Two Ti:sapphire lasers, dye laser + Ti:sapphire laser	20 μW, cw	[96Mom]
1.14–1.23	Dye laser (550–570 nm) + Nd:YAG laser, $\theta_{eeo} = 82$–$90°$, $\varphi = 0°$	22 % (quantum), 3.3 mJ	[96Bha]

(e) Crystal: KTA

λ [μm]	Sources of interacting radiations, crystal parameters	Conversion efficiency, energy, power, τ_p	Ref.
2.66–5.25	Ti:sapphire laser + Nd:YAG laser, $\theta_{eoe} = 40°$, $\varphi = 0°$	60 % (quantum), 1–15 mJ, 2 ns	[95Kun]

(f) Crystal: Ag$_3$AsS$_3$

λ [μm]	Sources of interacting radiations, crystal parameters	Conversion efficiency, energy, power, τ_p	Ref.
11–23	Two dye lasers	3 W (peak), 30 ns	[76Hoc]
3.7–10.2	OPO (1.06–1.67 μm) + 2ω of phosphate glass laser (527 nm)	25–50 μJ, 10 ps	[80Bar1]

(g) Crystal: AgGaS$_2$

λ [μm]	Sources of interacting radiations, crystal parameters	Conversion efficiency, energy, power, τ_p	Ref.
5.5–18.3	Two dye lasers, $\theta = 90°$	4 W, 4 ns	[76Sey]
5–11	Dye laser + Nd:YAG laser, $\theta_{eoe} = 38$–$52°$	180 kW, 12 ns	[84Kat]
3.9–9.4	Dye laser + Nd:YAG laser	1 %, 8 ps	[85Els]
4–11	OPO (2–4 μm) + radiation at $\lambda = 1.4$–2.13 μm	1 kW, 8 ns	[86Bet]
8.7–11.6	Two dye lasers, $\theta_{ooe} = 65$–$85°$	0.1 mW, 500 ns	[74Han]
4.6–12	Two dye lasers, $\theta_{ooe} = 45$–$83°$	300 mW, 10 ns	[73Han]
7–9	Dye laser + Ti:sapphire laser, $\theta_{ooe} = 90°$	1 μW, cw, $\Delta\nu = 0.5$ MHz	[92Can]
4.76–6.45	Dye laser + Ti:sapphire laser, $\theta_{ooe} = 90°$, $L = 45$ mm	20 μW, cw, $\Delta\nu = 1$ MHz	[92Hie]
∼ 4.26	GaAlAs laser (858 nm) + Ti:sapphire laser (715 nm), $\theta_{ooe} = 90°$	47 μW (cw), 89 μW (50 μs)	[93Sim2]
4.73; 5.12	Diode laser + Ti:sapphire laser, $\theta_{ooe} = 90°$	1 μW, cw	[93Sim1]
5.2–6.4	Nd:YAG laser + near IR (DFG in LiIO$_3$)	35 %, 23 ps	[88Spe]
3.4–7.0	Dye laser + Nd:YAG laser, $\theta_c = 53.2°$	17 μW (average), 2.16 ps, $f = 76$ MHz	[91Yod]
4–10	Dye laser (1.1–1.4 μm) + Nd:glass laser (1.053 μm)	2 %, 10 nJ ... 1 μJ, 1 ps	[93Dah]

(continued)

Table 4.1.28 (g) continued.

λ [μm]	Sources of interacting radiations, crystal parameters	Conversion efficiency, energy, power, τ_p	Ref.
4.5–11.5	Dye laser (870–1000 nm) + Ti:sapphire laser (815 nm), $\theta_c = 45°$, $L = 1$ mm	10 nJ, $f = 1$ kHz, 400 fs	[93Ham]
9	Ti:sapphire laser with dual wavelength output (50–70 fs), $\theta_c = 44°$, $L = 1$ mm	0.03 pJ, $f = 85$ MHz	[93Bar1]
3.1–4.4	Ti:sapphire laser + Nd:YAG laser, ICDFG, $\theta_c = 74°$	0.3 mW, cw	[95Can]
2.5–5.5	Signal and idler pulses of OPO, $\theta = 40°$	0.5 mW, $f = 82$ MHz, 200 fs	[94Loh]
6.2–9.7	Two Ti:sapphire lasers (696–804 nm and 766–910 nm)	3 μJ, 0.08 %, 13 ns	[96Aka]
6.8–12.5	Two diode lasers (766–786 nm and 830–868 nm)	1 μW, cw	[98Pet1]
2.4–12	Signal and idler waves of BBO based OPA	2.5 mW, 50 fs	[98Gol]
5–12	Signal and idler waves of LiNbO$_3$ based OPO (1.8–2.7 μm)	0.1 mJ, 6 ns	[99Hai]
~ 5	Two diode lasers, $\theta = 90°$, $L = 30$ mm	0.2 μW, cw	[96Sch]

(h) Crystal: AgGaSe$_2$

λ [μm]	Sources of interacting radiations, crystal parameters	Conversion efficiency, energy, power, τ_p	Ref.
7–15	OPO (1.5–1.7 μm) + Nd:YAG laser (1.32 μm), $\theta_{ooe} = 90$–$57°$	1.2 %	[74Bye]
12.2–13	CO laser (5.67–5.85 μm) + CO$_2$ laser, $\theta = 61°$	0.2 μW, cw	[73Kil]
8–18	Idler and signal waves of OPO	0.1 mJ, 3–6 ns	[03Boo]
5–18	Idler and signal waves of OPO, $\theta_{ooe} = 51°$	0.2 mJ, 8 ns	[98Abe]

(i) Crystal: CdGeAs$_2$

λ [μm]	Sources of interacting radiations, crystal parameters	Conversion efficiency, energy, power, τ_p	Ref.
11.4–16.8	CO laser + CO$_2$ laser	4 μW, cw	[74Kil]

(j) Crystal: GaSe

λ [μm]	Sources of interacting radiations, crystal parameters	Conversion efficiency, energy, power, τ_p	Ref.
9.5–18	Dye laser + ruby laser	300 W, 20 ns	[76Abd]
4–12	Idler and signal waves of OPO	60 W	[78Bia]
7–16	Nd:YAG laser + laser on F$_2^-$ colour centers, $\theta_{ooe} = 13$–$15°$, $\theta_{eoe} = 12$–$16°$	0.1–1 kW, 10 ns	[80Gus]
6–18	Dye laser (1.1–1.4 μm) + Nd:glass laser (1.053 μm)	10 nJ ... 1 μJ, 1 ps	[93Dah]
5.2–18	Idler and signal waves of OPO, $L = 1$ mm	2 mW, 3.3 %, $f = 76$ MHz, 120 fs	[98Ehr]

(k) Crystal: CdSe

λ [μm]	Sources of interacting radiations, crystal parameters	Conversion efficiency, energy, power, τ_p	Ref.
16	OPO signal wave (1.995 μm) + OPO idler wave (2.28 μm), $\theta = 62.22°$	0.5 kW, 20 Hz, 10 ns	[77And2]
9–22	OPO (2–4 μm) + radiation at $\lambda = 1.4$–2.13 μm	10–100 W, 8 ns	[86Bet]
10–20	OPO signal and idler waves, $\theta = 70°$, eoo	50 % (quantum), 5–40 μJ, 10 ps	[95Dhi]

(1) Crystal: Te

λ [μm]	Sources of interacting radiations, crystal parameters	Conversion efficiency, energy, power, τ_p	Ref.
10.9–11.1	CO_2 laser (10.2 μm) + cw spin-flip laser (5.3 μm), $\theta_{eeo} = 14°$	10 μW	[75Bri]

Table 4.1.29. Difference frequency generation in the far IR region.

Pump sources	Crystal	ν [cm^{-1}]	λ [mm]	Power, energy	Ref.
Nd:glass (1.06 μm)	LiNbO$_3$	100	0.1	–	[65Zer]
Ruby laser (0.694 μm)	LiNbO$_3$	29	0.33	–	[69Yaj]
Two ruby lasers (0.694 μm), 1 MW, 30 ns	Quartz, LiNbO$_3$	1.2–8.0	1.25–8.33	20 mW	[69Far]
Nd:glass (1.06 μm), 50 mJ, 10 ps	ZnTe, LiNbO$_3$	8–30	0.33–1.25	20 mW/cm^{-1}	[71Yaj]
Nd:glass (1.06 μm), 10 ps	LiIO$_3$	–	–	–	[72Tak]
Dye laser (0.73–0.93 μm), 11–15 ns, 4–13 MW	ZnTe, ZnSe, LiNbO$_3$	5–30	0.33–2.00	1 W (ZnTe)	[73Mat]
Nd:glass (1.064 μm), 10 ps	LiNbO$_3$	0.4–2.5	4–25	60 W	[76Ave]
Two ruby lasers (0.694 μm), 20 ns	LiNbO$_3$	1–3.3	3–10	0.5 W	[79Ave]
Ruby laser (0.694 μm)	LiNbO$_3$	1.67–3.3	3–6	–	[80Mak]
Two dye lasers: $\tau_1 = 1$–2 ps, $\lambda_1 = 589$ nm, $E_1 = 0.2$ mJ; $\tau_2 = 20$ ns, $\lambda_2 = 590$–596 nm, $E_2 = 20$ mJ	LiNbO$_3$	20–200	0.05–0.5	3 nJ	[85Ber]
Nd:YAG laser (45 ps) + OPO (35 ps)	LiNbO$_3$	10–200	0.05–1	10 kW	[95Qiu]
CO_2 laser at two frequencies	GaAs	2–100	0.1–5.0	–	[85Rya]
Two CO_2 lasers	ZnGeP$_2$	70–110	0.09–0.14	1.7 μW	[72Boy]
Two CO_2 lasers	ZnGeP$_2$	99–100	0.1–0.11	3.6 μJ	[96Apo]
Nd:YAG (1.064 μm), 30 ns	GaP	0.33–1	10–30	1 mW	[87Len]

4.1.7 Optical parametric oscillation

Table 4.1.30. Continuous wave (cw) and nanosecond OPO in the UV, visible, and near IR regions.

Crystal	θ_{pm}, type of interaction	λ_{pump} [μm]	I_{thr} [MW cm^{-2}]	λ_{OPO} [μm]	τ_p [ns]	η [%]	Ref.	Notes
KDP	eoe	0.532	1000–2000	–	–	40–42 [a]	[86Bar]	TWOPO, $L_1 = 4$ cm, $L_2 = 6$ cm, $E = 2$ J
	eoe	0.35	1000	0.45–0.6	0.5	41 [a]	[87Beg]	TWOPO, $L_1 = 2$ cm, $L_2 = 6$ cm, $E = 0.35$ J, $I_0 = 6$–8 GW cm^{-2}
ADP	–	0.527	1500	0.93–1.21	–	37 [a]	[84Akh]	TWOPO, $E = 2.3$ J, $I_0 = 10$ GW cm^{-2}
	ooe	0.266	–	0.42–0.73	2	25	[71Yar]	TWOPO, $T = 50$–105 °C
	ooe	0.266	250	–	14	30	[75Zhd]	$L = 6$ cm, $I_0 = 1$ GW cm^{-2}
LiIO$_3$	$\theta_{ooe} = 24$°	1.06	50	2.5–3.2	40	15	[84Ash]	SROPO, $L = 6$ cm, $E = 0.1$ J
	$\theta_{ooe} = 23.1$–22.4°	0.694	5	1.15–1.9	20	50 [a]	[71Cam, 72Cam]	DROPO, $L = 0.85$ cm, $P = 10$ kW
	$\theta_{ooe} = 25$–30°	0.53	10	0.68–2.4	15	8	[70Izr]	SROPO, $L = 1.6$ cm
	$\theta_{ooe} = 23$–30°	0.532	10	0.63–3.35	30	20	[77Dzh]	SROPO
LiNbO$_3$	$\theta_{ooe} = 90$°	1.06	–	2.13	100	8	[69Amm]	DROPO, $L = 3$ mm
	$\theta_{ooe} = 90$°	1.06	–	1.4–4.45	20	15	[74Her]	SROPO, $I_0 = 10$ MW cm^{-2}
	43.3°	0.93	8 mJ	1.48–1.8; 1.95–2.55	16	9.7	[97Raf]	SROPO, $L = 50$ mm, broad spectral bandwidth ($\Delta\lambda = 320$ nm)
	$\theta_{ooe} = 90$°	0.473–0.659	–	0.55–3.65	130–70C	46 (67 [a])	[70Wal]	SROPO, $T = 110$–430 °C, $P_{av} = 105$ mW
LiNbO$_3$:MgO	$\theta_{ooe} = 90$°	1.06	0.4 mW	1–1.14	cw	–	[93Sch]	Quadruply resonant OPO
	$\theta_{ooe} = 90$°	0.532	35 mW	1.01–1.13	cw	40 (60 [a])	[89Koz]	DROPO, $T = 107$–110 °C
	$\theta_{ooe} = 90$°	0.532	12 mW	1.007–1.129	cw	34 (78 [a])	[89Nab]	DROPO, $T = 107$–111 °C, $P = 8.15$ mW
	$\theta_{ooe} = 90$°	0.532	13 mW	0.966–1.185	cw	38 (73 [a])	[93Ger]	DROPO, $T = 113$–126 °C, $L = 15$ mm, $P = 100$ mW
	$\theta = 90$°	0.532	28 mW	1.0–1.12	cw	81	[95Bre]	DROPO, $P = 105$ mW, $L = 7.5$ mm
	$\theta_{ooe} = 90$°	0.532	80 mW	0.788–1.640	cw	–	[98Tsu]	DROPO, $T = 80$–180 °C, $L = 15$ mm

(continued)

Table 4.1.30 continued.

Crystal	θ_{pm}, type of interaction	λ_{pump} [μm]	I_{thr} [MW cm^{-2}]	λ_{OPO} [μm]	τ_P [ns]	η [%]	Ref.	Notes
BBO	$\theta_{ooe} = 21.7$–$21.9°$	0.532	278	0.94–1.22	12	10	[89Fan]	SROPO, $L = 9$ mm, $E = 1$ mJ
	ooe	0.355	130	0.45–1.68	8	9.4	[88Che]	SROPO, $L = 11.5$ mm, $E = 15$ mJ
	$\theta_{ooe} = 24$–$33°$	0.355	20	0.412–2.55	2.5	24	[88Fan]	SROPO, $L = 12$ mm, $P_{av} = 140$ mW
	ooe	0.355	27	0.42–2.3	8	32	[89Bos]	SROPO, $L_1 = 11.5$ mm, $L_2 = 9.5$ mm, $\Delta\lambda = 0.03$ nm
	$\theta_{ooe} = 33.7$–$44.4°$	0.355	38	0.48–0.63; 0.81–1.36	8	12	[90Bos]	SROPO, $L_1 = 17$ mm, $L_2 = 10$ mm, $\Delta\lambda = 0.05$–0.3 nm
	$\theta_{ooe} = 23$–$33°$	0.355	20–40	0.402–3.036	7	40–61	[91Fix, 93Fix]	SROPO, $L = 15$ mm, $E = 0.1$–0.2
	$\theta = 28°$	0.355	–	0.453–2.3	6	7–9	[95Joh]	SROPO, $\Delta\nu = 0.2$ cm^{-1}, $E = 100$ mJ, SHG in KDP and BBO (220–450 nm)
	$\theta = 23$–$33°$	0.355	20	0.465–1.5	10	40	[94Glo]	SROPO, $L = 12$ mm, collinear and noncollinear geometries
	$\theta = 33°$	0.355	3.2 mJ	0.44–1.76	10	37	[97Oie]	
	$\theta = 35.9°$	0.355	–	0.5–0.7	10	–	[97Wan]	Broad spectral bandwidth OPO ($\Delta\lambda > 100$ nm) with noncollinear geometry, $L = 18$ mm
	$\theta_{ooe} = 35.5$–$37°$	0.308	150	0.422–0.477	8	10	[88Kom]	SROPO, $L = 7$ mm, $E = 0.26$ mJ
	ooe	0.308	18	0.354–2.37	17	64^a	[91Rob]	SROPO, $L = 20$ mm, $E = 20$ mJ
	ooe	0.308	–	0.4–0.56	17	15	[93Rob]	SROPO, $L = 20$ mm, $\Delta\nu = 0.07$ cm^{-1} (with intracavity etalon)
	$\theta_{ooe} = 30$–$48°$	0.266	–	0.302–2.248	7	6.3	[91Fix]	SROPO
	$\theta_{ooe} = 38.3°$	0.266	58	0.3–2.34	4.5	15	[00Kon]	$L = 14$ mm
LBO	$\theta = 90°, \varphi = 0°$	0.78–0.81	360 mW	1.49–1.70	cw	40^a	[94Col1]	DROPO, $L = 2$ cm, $T = 130$–185 °C, $P = 30$ mW
	$\theta = 90°, \varphi = 0°$	0.5235	700	0.924–1.208	12	45	[93Hal2]	DROPO, $L = 12$ mm, $T = 156$–166 °C
	$\theta = 90°, \varphi = 0°$	0.5145	50 mW	0.966–1.105	cw	10	[93Col1]	TROPO, $L = 20$ mm, $T = (183 \pm 3)$ °C, $P = 90$ mW
	$\theta = 0°, \varphi = 0°$	0.5145	1 W	0.93–0.946	cw	15	[94Rob2]	SROPO, $P = 0.5$ W, $L = 25$ mm
	$\theta = 0°, \varphi = 90°$	0.364	115 mW	0.494–0.502; 1.32–1.38	cw	9.4	[93Col2, 94Col2]	SROPO and DROPO, $L = 20$ mm, $T = 18$–86 °C, $P = 103$ mW
	$\theta = 90°, \varphi = 24$–$42°$	0.355	14	0.435–1.922	10	22	[91Wan]	DROPO, $I_0 = 40$ MW cm^{-2}, $E = 2.7$ mJ

(continued)

Table 4.1.30 continued.

Crystal	θ_{pm}, type of interaction	λ_{pump} [μm]	I_{thr} [MW cm^{-2}]	λ_{OPO} [μm]	τ_p [ns]	η [%]	Ref.	Notes
LBO	$\theta = 0°, \varphi = 0°$	0.355	15	0.48–0.457; 1.355–1.59	12	27	[92Cui]	SROPO, T = 20–200 °C
	$\theta = 90°$, $\varphi = 27$–$42°$	0.355	60	0.455–0.655; 0.76–1.62	10	35 [a]	[93Cui]	SROPO, L = 16 mm
	$\theta = 90°$, $\varphi = 20.1$–$42.1°$	0.355	50	0.414–2.47	5	45	[94Sch]	SROPO, L = 15 mm
	$\theta = 90°$, $\varphi = 26$–$52°$	0.308	26	0.355–0.497; 0.809–2.34	17	28–40 [a]	[91Rob, 92Rob]	SROPO, L = 15 mm
	type II in XZ and YZ planes, $\theta = 0$–$9°$	0.308	30	0.381–0.387; 1.5–1.6	5	35	[91Ebr2]	L = 16 mm, I_0 = 0.1 GW cm^{-2}
	$\theta = 0°, \varphi = 0°$	0.266	10	0.314; 1.74	10	10	[92Tan]	SROPO, L = 16 mm, T = 20 °C
	$\theta = 90°$, $\varphi = 37$–$47°$	0.266	–	0.307–0.325	4	–	[94Sch]	SROPO, L = 15 mm
KTP	$\theta = 50$–$58°$, $\varphi = 0°$	1.064	–	1.8–2.4	10	10	[90Lin1]	DROPO, E = 0.1–0.5 mJ
	$\theta = 90°, \varphi = 53°$	1.064	80	3.2	10	5	[91Kat]	SROPO, L = 15 mm, P = 0.2 W
	$\theta = 90°, \varphi = 0°$	1.06	–	1.61	15	47 (66 [a])	[93Mar1]	Diode-pumped Nd:YAG laser
	–	1.047	0.5 mJ	1.54; 3.28	18	20	[94Ter]	
	$\theta = 63.4°, \varphi = 0°$	1.047	0.6 mJ	1.58–1.84	10	40	[97Tan]	NC SROPO, L = 25 mm
	$\theta = 90°, \varphi = 0°$	0.7–0.95	70	1.04–1.38; 2.15–3.09	10	20	[92Kat]	SROPO, L = 15 mm
	$\theta = 90°$	0.7–0.9	–	1.03–1.28; 2.18–3.03	20	55	[94Zen]	E = 49 mJ, L = 15 mm
	$\theta = 90°, \varphi = 0°$	0.769	6 mW	1.1; 2.54	cw	–	[95Sch]	TROPO, L = 12 mm
	$\theta = 54°, \varphi = 0°$	0.73–0.80	–	1.38–1.67	cw	0.001	[93Wan2]	L = 10 mm, P = 2 μW
	$\theta = 90°, \varphi = 0°$	0.532	1.4 W, SROPO; 30 mW, DROPO	1.039; 1.09	cw	35	[93Yan1, 93Yan2]	SROPO and DROPO, L = 10 mm, P = 1.07 W

(continued)

Table 4.1.30 continued.

Crystal	θ_{pm}, type of interaction	λ_{pump} [μm]	I_{thr} [MW cm^{-2}]	λ_{OPO} [μm]	τ_p [ns]	η [%]	Ref.	Notes
KTP	$\varphi = 0°$	0.532	80	0.7–0.9; 1.3–2.2	3.5	12	[93Bos, 92Bos]	$L = 15$ mm, $E = 3$ mJ, $\Delta\nu = 0.02$ cm^{-1}
	$\theta = 69°, \varphi = 0°$	0.532	–	0.75–1.04	4–6	27	[95Sri]	OPO-OPA, $L_1 = L_2 = 10$ mm, $E = 45$ mJ
	$\theta = 60°, \varphi = 0°$	0.532	–	0.75–0.87; 1.83–1.37	4	–	[95Hui]	$\Delta\nu = 200$ MHz (with Fabry-Perot etalon), $E = 0.6$ mJ, $L = 16$ mm
	$\theta = 90°, \varphi = 0°$	0.532	4.3 W	1.09; 1.039	cw	28 (64 ^{a})	[94Yan]	SROPO, $P = 1.9$ W, $L = 15$ mm
	$\theta = 90°$, $\varphi = 25.3°$	0.531	40 mW	1.0617	cw	30	[93Lee]	DROPO, $L = 8$ mm
	$\theta = 69°, \varphi = 0°$	0.532	7	0.76–1.04	6	30	[93Mar2]	$L = 15$ mm, ICSHG in BBO with $\eta = 40\%$ (380–520 nm)
KTA	$\theta = 53°, \varphi = 0°$	0.773–0.792	–	1.45; 1.7	300	0.3	[92Jan]	DROPO, $L = 7$ mm
	type II	–	–	1.11–1.20; 2.44–2.86	cw	90	[98Edw]	Intracavity (Ti:Sa) SROPO, $P = 1.46$ W, $L = 11.5$ mm
RTA	$\theta = 90°, \varphi = 0°$	0.77–0.83	70 mW	1.21–1.26; 2.1–2.4	cw	–	[97Sch]	SROPO, $L = 12$ mm, $P = 84$ mW, $\Delta\nu < 10$ MHz
Banana	$\theta_{ooe} = 90°$	0.532	–	0.75–1.82	10	5	[80Bar3]	SROPO, $T = 80$–220 °C
KNbO$_3$	along the b axis	0.532	3.5	0.88–1.35	10	32	[82Kat]	DROPO, $T = 180$–200 °C, $P = 12$ MW
Urea	$\theta_{oeo} = 81$–90°	0.355	55 (45 mW)	0.5–0.51; 1.17–1.22	7	20	[84Don]	SROPO, $L = 12.7$ mm, $I_0 = 90$ MW cm^{-2}
	$\theta_{oeo} = 50$–90°	0.355	–	0.5–1.23	7	23	[85Ros2, 85Ros1]	SROPO, $L = 23$ mm
	$\theta_{oeo} = 64$–90°	0.308	16–20	0.537–0.72	4–6	37	[89Ebr]	$L = 15$ mm
NPP	eeo	0.266	–	0.33–0.42	7	–	[85Ros1, 92Jos, 93Dou]	$L = 1.9$ mm
	$\theta = 9.5$–13°, $\varphi = 0°$	0.5927	30	0.9–1.7	1	5		
	$\theta = 30°$	0.583–0.59	0.5	1–1.5	7	–	[95Kho]	

^{a} Pump depletion.

Table 4.1.31. Picosecond OPO in the UV, visible, and near IR regions.

Crystal	θ_{pm}, type of interaction	λ_{pump} [μm]	I_{thr} [MW cm^{-2}]	λ_{OPO} [μm]	τ_p [ps]	η [%]	Ref.	Notes
KDP	eoe	0.532	–	0.8–1.67	40	25	[78Kry]	TWOPO, $E = 1$ mJ, $L_1 = L_2 = 4$ cm
	eoe	0.532	–	0.9–1.3	30	51	[78Dan2, 79Kab]	TWOPO, $\Delta \nu \Delta \tau = 0.7$, $L_1 = 4$ cm, $L_2 = 6$ cm, $I_0 = 15$–20 GW cm^{-2}
	eoe	0.527	–	0.82–1.3	0.3–0.5	2	[83Bar, 82Dan]	SP OPO, $E = 20$ μJ
	eoe	0.355	–	0.45–0.64, 0.79–1.69	45	15	[78Dan3]	TWOPO, $L_1 = L_2 = 4$ cm
ADP	$\theta_{ooe} = 90°$	0.266	–	0.44–0.68	10	10	[76Mas2]	TWOPO, $T = 50$–110 °C, $L_1 = L_2 = 5$ cm
CDA	$\theta_{ooe} = 90°$	0.532	–	0.854–1.41	10	30–60	[74Mas]	$L = 3$ cm, $T = 50$–70 °C, $I_0 = 0.3$ GW cm^{-2}
	$\theta_{ooe} = 90°$	0.53	1000	0.8–1.3	10	12.5	[87Ion]	SP OPO, $L = 4$ cm, $I_0 = 3$ GW cm^{-2}
LiIO$_3$	ooe	0.532	–	0.61–4.25	6	4	[77Dan, 78Dan1]	TWOPO, $L_1 = 1$ cm, $L_2 = 2.5$ cm, $I_0 = 2$ GW cm^{-2}
	$\theta_{ooe} = 25$–$30°$	0.53	3000	0.68–2.4	–	5	[77Kry]	TWOPO, $L_1 = L_2 = 4$ cm, $I_0 = 6$ GW cm^{-2}
LiNbO$_3$	ooe	1.06	–	1.4–4.0	3.5	10	[78Sei]	TWOPO, $\Delta \nu = 6.5$ cm^{-1}, $I_0 = 1$ GW cm^{-2}
	45–51°	1.064	–	1.37–4.83	40	17	[77Iva]	TWOPO
	47°	1.054	100	1.35–2.11	0.5	15	[90Lae1]	SP OPO, $L = 18$ mm, $I_0 = 0.14$ GW cm^{-2}
	84°	0.53	–	0.66–2.7	40	17	[77Iva]	TWOPO, $T = 46$–360 °C
	90°	0.532	–	0.68–0.76	20	9	[79Liu]	SP OPO
	90°	0.532	8	0.85–1.4	15	17.5	[86Pis]	SROPO, $P = 30$ kW, $f = 10$ kHz
	90°	0.532	< 30	0.65–3.0	10	7.2	[87Ion]	SP OPO, $L = 25$ mm
LiNbO$_3$:MgO	$\theta = 48.5°$	0.75–0.84	4000	2.6–4.5	2–3	18	[96Lin]	$L_1 = L_2 = 20$ mm
	$\theta_{ooe} = 60$–$84°$	0.532	–	0.7–2.2	30	5.4	[91He]	TWOPO, $\Delta \lambda = 0.3$ nm (0.7 μm) and 1.4 nm (2 μm)

(continued)

Table 4.1.31 continued.

Crystal	θ_{pm}, type of interaction	λ_{pump} [μm]	I_{thr} [MW cm^{-2}]	λ_{OPO} [μm]	τ_p [ps]	η [%]	Ref.	Notes
BBO	$\theta_{ooe} = 20.7-22.8°$	0.532	–	0.67–2.58	18	13	[92Zhu]	TWOPO, $L_1 = L_2 = 9$ mm, $I_0 = 2.5-3.8$ GW cm^{-2}, $E = 0.1-0.5$ mJ
	ooe	0.53	–	0.63–3.2	1.3	25	[93Dan]	TWOPO-OPA, $L_1 = L_2 = 8$ mm
	ooe	0.36	500	0.406–3.17	20	30	[90Bur]	SP OPO, $L = 12$ mm, $I_0 = 2$ GW cm^{-2}, $E = 3$ mJ, $\Delta\lambda = 0.24$ nm
	$\theta_{ooe} = 26-33°$	0.355	–	0.4–2.0	15	30	[90Hua]	OPO-OPA, $L_1 = 12$ mm, $L_2 = 6$ mm, $L_3 = 15$ mm, $I_0 = 3$ GW cm^{-2}, $\Delta\lambda = 0.3$ nm
	ooe	0.355	–	0.4–2.86	24	6.5	[90Suk]	TWOPO, $L_1 = L_2 = L_3 = 8$ mm, $I_0 = 5$ GW cm^{-2}, $\Delta\nu = 10$ cm^{-1}
	ooe	0.355	–	0.43–2.1	15	30	[93Zha1]	Injection seeding, $L = 15$ mm
	$\theta_{ooe} = 33°$	0.3547	–	0.42–2.8	30	61	[94Hua]	OPO-OPA, $P = 51$ MW
LBO	ooe, $\theta = 90°$, $\varphi = 0°$	0.8	700 (400 mW)	1.15–2.26	1–2.2	27 (48 [a])	[95Ebr1]	SP OPO, $P = 325$ mW, $L = 30$ mm, $T = 120-230$ °C
	ooe, $\theta = 90°$, $\varphi = 0°$	0.8	320	1.374–1.53; 1.676–1.828	0.52	7.5	[95Ebr3]	SP OPO, $P = 90$ mW, $L = 16$ mm
	type I	0.77–0.8	350	1.16–2.26	1	34	[96Fre]	SP SROPO, $P = 580$ mW, $L = 16$ mm
	$\theta = 81°$, $\varphi = 5°$	0.57–0.63	–	1.2–1.5	0.58	10	[91Bay]	Injection seeding by 1.08 μm
	$\theta = 85°$, $\varphi = 9°$	0.57–0.63	–	1.2–1.5	0.4	25	[92Akh]	Injection seeding by 1.08 μm (40 ps), $L = 9$ mm, $I_0 = 1$ TW cm^{-2}
	$\theta = 90°$, $\varphi = 0°$	0.532	–	0.75–1.8	35	20	[91Hua]	Injection seeding OPO, $T = 106.5-148.5$ °C
	$\theta = 90°$, ooe	0.53	–	0.65–2.5	15	24	[91Lin]	OPA, angle ($\varphi = 8.7-15.9°$) and temperature tuning ($T = 103-210$ °C), $E = 0.45$ mJ
	$\theta = 90°$, $\varphi = 0°$	0.532	1500	0.77–1.7	100	30	[93Zho]	SP SROPO, $L = 15$ mm, $T = 105-137$ °C, $\Delta\lambda = 0.14$ nm
	ooe, $\theta = 90°$, $\varphi = 11.5°$	0.532	–	0.68–2.44	25	8	[95Liu]	$L_1 = L_2 = 15$ mm
	$\theta = 0°$, $\varphi = 0°$	0.532	–	0.75–1.8	15	20	[95Wal]	$P = 200$ mW, $L = 15$ mm
	$\theta = 90°$, $\varphi = 0°$	0.5235	2500 (10 mW)	0.652–2.65	12	13	[92Ebr2, 93Ebr, 93Hal1]	SROPO, $L = 12$ mm, $T = 125-190$ °C

(continued)

Table 4.1.31 continued.

Crystal	θ_{pm}, type of interaction	λ_{pump} [μm]	I_{thr} [MW cm^{-2}]	λ_{OPO} [μm]	τ_p [ps]	η [%]	Ref.	Notes
LBO	$\theta = 90°, \varphi = 0°$	0.5235	1100 (4.5 mW)	0.909–1.235	33	50	[93Hal1, 92Ebr1]	DROPO, $T = 167$–180 °C
	$\theta = 90°, \varphi = 0°$	0.5235	15 (30 mW)	0.65–2.65	1.7	50	[93Hal1]	DROPO, $L = 12$ mm, $P = 0.21$ W
	ooe, $\theta = 90°, \varphi = 0°$	0.5235	47 mW	0.839–1.392	1.8	70[a]	[94Rob1]	SP SROPO, $P = 88$ mW, $L = 3$ mm
	ooe, $\theta = 90°, \varphi = 0°$	0.5235	170 mW	0.65–2.7	1.63	75[a]	[95But]	SP OPO, $P = 210$ mW, $L = 15$ mm
	$\theta = 90°, \varphi = 0°$	0.523	100	0.72–1.91	1	34	[93McC1]	SP SROPO, $L = 13$ mm, $T = 125$–175 °C, $P_{av} = 89$ mW
	$\theta = 90°, \varphi = 0°$	0.523	80 (70 mW)	0.8–1.5	1.2–1.5	27 (75[a])	[93But]	SROPO, $L = 12$ mm, $P_{av} = 78$ mW
	$\theta = 90°, \varphi = 27$–$43°$	0.355	–	0.46–1.6	15	30	[91Zha]	Injection seeding from OPO, $L = 16$ mm, $I_0 = 2.8$ GW cm^{-2}, $E = 0.3$ mJ
	$\theta = 90°, \varphi = 18$–$42°$	0.355	–	0.403–2.58	12	28	[92Kra]	TWOPO, $L_1 = L_2 = 15$ mm, $I_0 = 5$ GW cm^{-2}, $E = 0.1$–1 mJ
	$\theta = 0°, \varphi = 0°$	0.355	2300	0.4159–0.4826	30	38	[92Hua, 93Hua]	TWOPO, $L = 10$ mm, $T = 21$–450 °C, $I_0 = 18$ GW cm^{-2}, $\Delta\lambda = 0.15$ nm
	$\theta = 90°, \varphi = 30$–$42°$	0.355	1000	0.452–1.65	9	26	[93Agn]	DROPO, $L = 10.5$ mm, $E = 0.15$ mJ
KTP	$\theta = 82$–$90°, \varphi = 0°$	1.064	0.8 W	1.57–1.59; 3.21–3.30	2–3	15	[93Chu]	SROPO, $L = 10$ mm, $f = 75$ MHz, $\Delta\lambda = 1.5$ nm
	–	1.064	–	2.128	100	25	[93Lot]	SP OPO with 6 KTP (total length 58 mm), $P = 14$ W
	$\theta = 81$–$90°, \varphi = 0°$	1.053	5.8 W	1.55–1.56; 3.22–3.28	12	21	[93Gra]	SP OPO, $L = 6$ mm, $P = 2$ W
	$\theta = 40.6$–$45.2°, \varphi = 0°$, oeo	0.8	–	1.02–1.16; 2.6–3.7	2.6	–	[95Gra]	OPA, $E = 0.04$ mJ
	$\theta = 53°, \varphi = 0°$, oeo	0.72–0.85	0.8 W	1.44–1.64	1.1	20	[96Qia]	SP SROPO, $P = 200$ mW, $L = 7$ mm

(continued)

Table 4.1.31 continued.

Crystal	θ_{pm}, type of interaction	λ_{pump} [μm]	I_{thr} [MW cm^{-2}]	λ_{OPO} [μm]	τ_p [ps]	η [%]	Ref.	Notes
KTP	$\theta = 90°, \varphi = 0°$	0.72–0.853	150	1.052–1.214; 2.286–2.871	1.2	42	[93Neb]	SP OPO, $L = 6$ mm, $P = 0.7$ W
	$\theta = 54°, \varphi = 0°$, eoo	0.532	250	0.614–4.16	0.39	12	[94Umb]	$L = 14$ mm
	eoe, $\theta = 90°$, $\varphi = 35°$	0.527	0.9 W	0.851–0.938; 1.2–1.381	2.4–3.2	13	[95Che]	SROPO, $P = 80$–280 mW, $L = 5$ mm
	$\theta = 40$–70°, $\varphi = 90°$	0.526	–	0.6–2.0	30	10	[88Van]	$L = 20$ mm
	$\theta = 40$–80°, $\varphi = 0°$	0.526	–	0.6–4.3	30	10	[88Van]	$L = 20$ mm
	$\theta = 90°$, $\varphi = 10$–35°	0.523	57 (61 mW)	1.002–1.096	2.2	16 (79^a)	[92McC]	SP OPO, $L = 5$ mm, $P = 42$ mW
	–	0.5235	1000 (2 mW)	0.946–1.02; 1.075–1.172	8	10 (56^a)	[91Ebr1, 93Hal]	SP SROPO, $L = 5$ mm, $P = 2$ mW
	$\theta = 90°$	0.523	60 (61 mW)	0.938–1.184	1–2	16	[93McC2]	SROPO, $L = 5$ mm, $f = 125$ MHz, $P = 40$ mW
	$\theta = 90°$, $\varphi = 0$–33°	0.526	0.5 W	1.01–1.1	14	44	[93Gra]	SP OPO, $L = 6$ mm, $P = 0.58$ W
KTA	$\theta = 90°, \varphi = 0°$	1.064	–	1.54; 3.47	7	75	[98Ruf]	SP OPO, $L = 15$ mm
Banana	$\theta_{ooe} = 90°$	0.532	5	0.8–1.6	10	25	[83Oni]	SP OPO
	$\theta_{ooe} = 90°$	0.53	50	0.65–3	10	5.3	[87Ion]	SP OPO, $I_0 = 250$ MW cm^{-2}
	$\theta_{ooe} = 90°$	0.532	7–9	0.672–2.56	15–45	8.1	[89Pis, 90Pis]	SP SROPO, $L = 10$ mm, $f = 139$ MHz, $T = 75$–350 °C
α-HIO$_3$	eoe	0.532	–	0.7–2.2	30–45	10–12	[77Dan]	TWOPO, $L_1 = L_2 = 2$ cm, $I_0 = 4$–5 GW cm^{-2}
	eoe	0.527	60	–	5–6	10	[80Bar2]	SP OPO, $\Delta\nu\Delta\tau = 0.7$

^{a} Pump depletion.

Table 4.1.32. Femtosecond OPO in the UV, visible, and near IR regions.

Crystal	θ_{pm}, type of interaction	λ_{pump} [μm]	I_{thr} [MW cm^{-2}]	λ_{OPO} [μm]	τ_P [fs]	η [%]	Ref.	Notes
BBO	$\theta_{eoe} = 28°$	0.78	–	1.1–2.6	60	35	[94Nis]	TWOPO, $L_1 = L_2 = 4.8$ mm, $E = 0.15$ mJ
	$\theta_{ooe} = 20°$	0.8	–	1.2–1.3	70	5	[94Sei1]	TWOPO, $L = 4$ mm
	ooe	0.62	–	0.45–2.8	200	15	[91Joo]	TWOPO, $L_1 = 5$ mm, $L_2 = 7$ mm, $E = 20$ μJ
	ooe, eoe	0.6	–	0.75–3.1	180–250	23	[93Dan]	TWOPO-OPA, $L_1 = L_2 = 8$ mm, $I_0 = 70$ GW cm^{-2}
	$\theta_{ooe} = 19.5$–$21°$	0.53	2200	0.68–2.4	75	30	[88Bro]	SP SROPO, $L = 7.2$ mm, $I_0 = 2.2$ GW cm^{-2}, $E = 2$ mJ
	ooe	0.527	–	0.7–1.8	65–260	3	[90Lae2, 91Lae, 93Lae]	SP SROPO, $L = 5.8$ mm
	ooe	0.527	–	1.04–1.07	70	–	[92Dub]	OPA with gain ratio 2×10^4
	$\theta_{ooe} = 32°$	0.4	–	0.566–0.676	30	10	[94Dri]	SP OPO, $P = 100$ mW
	ooe	0.4	–	0.59–0.666	13	50	[95Gal2]	SP OPO, $P = 130$ mW
	ooe	0.395	–	0.55–0.69	14	–	[98Shi]	NC OPA, seeding with white light continuum
	$\theta_{ooe} = 32°$	0.39	–	0.5–0.7	11	–	[97Cer]	OPA, seeding with white light continuum, $L = 1$ mm
LBO	–	0.8	–	1.1–2.4	40	38	[95Kaf]	SP OPO, $L = 6$ mm, $P = 550$ mW
	–	0.77–0.8	320	1.374–1.530; 1.676–1.828	720	7.5	[95Ebr2]	SP OPO, $P = 90$ mW
	$\theta = 90°$, $\varphi = 0°$	0.605	–	0.85–0.97; 1.6–2.1	200	10–15	[93Dan, 93Ban]	TWOPO, $L_1 = L_2 = L_3 = 15$ mm, $T = 30$–85 °C, $I_0 = 25$ GW cm^{-2}
KTP	$\theta = 43°$, $\varphi = 0°$, eoo	0.83	325 mW	1.05–1.16; 2.9–4.0	175	15 [a]	[95McC]	SP OPO, $L = 2$ mm
	$\theta = 90°$, $\varphi = 0°$, eoo	0.816	–	2.5–2.9	160	55 [a]	[95Hol]	SP OPO-OPA, $L_1 = L_2 = 0.9$ mm, $E = 0.55$ μJ
	$\varphi = 0°$	0.765–0.815	–	1.22–1.37; 1.82–2.15	57–135	55 [a]	[92Pel]	$L = 1.15$ mm, $f = 90$ MHz, $P = 340$ mW (135 fs) and 115 mW (57 fs)

(continued)

Table 4.1.32 continued.

Crystal	θ_{pm}, type of interaction	λ_{pump} [μm]	I_{thr} [MW cm^{-2}]	λ_{OPO} [μm]	τ_P [fs]	η [%]	Ref.	Notes
KTP	$\theta = 67°, \varphi = 0°$	0.765	40000; (180 mW)	1.2–1.34; 1.78–2.1	62	–	[92Fu]	SP OPO, $L = 1.5$ mm, $f = 76$ MHz, $P = 175$ mW
		0.745	100 mW	0.53–0.585	200	29	[97Kar]	SP OPO with ICSHG (self-doubling OPO)
	$\theta = 45°, \varphi = 0°$	0.68	–	1.16–2.2; 0.58–0.657	57	60 [a]	[93Pow1]	$L = 1.5$ mm, $P = 0.68$ W, ICSHG in BBO ($L = 47$ μm)
	$\varphi = 0°$	0.645	110 mW	1.2–1.34	220	13	[92Mak]	SP OPO, $P = 30$ mW
	$\theta = 53°, \varphi = 0°$	0.61	–	0.755–1.04; 1.5–3.2	105–120	–	[90Wac, 91Wac]	SP OPO in CPM dye laser cavity, $L = 1.4$ mm
	$\theta = 62°, \varphi = 0°$, eoo	0.524	2000	1.2–1.6	260	10	[95Rau]	SP OPO, $L = 3$ mm
	$\theta = 62°, \varphi = 0°$	0.5235	–	1.2–1.7	300	–	[98Lae]	SP OPO, $L = 6$ mm, $E = 10$ nJ
KTA	$\varphi = 0°$, oeo	0.78	–	1.29–1.44; 1.83–1.91	85–150	10–15	[93Pow2]	$L = 1.47$ mm, P = 75 mW
RTA	$\theta = 53°, \varphi = 0°$, eoo	0.76–0.82	–	1.03–1.3; 2.15–3.65	58	25	[94Pow]	SP SROPO, $L = 1.8$ mm, $P = 250$ mW
	$\theta = 90°, \varphi = 0°$, eoo	0.78–0.86	50 mW	1.33	70	32	[95Rei]	$L = 2$ mm, $P = 185$ mW
	–	Ti:Sa	–	1.25; 2.25	78	33	[97Rei]	SP OPO, $f = 344$ MHz, $P = 0.6$ W
KNbO$_3$	$\theta = 38°, \varphi = 90°$	0.78	–	2.3–5.2	60–90	23	[95Spe, 96Spe]	$L = 1$ mm, $P = 170$–300 mW
NPP	–	0.62	–	0.8–1.6	150–290	–	[86Led, 87Led]	$L = 1.5$ mm

[a] Pump depletion.

Table 4.1.33. Optical parametric oscillation in the mid IR region.

Crystal	λ_{pump} [μm]	λ_{OPO} [μm]	τ_p	Conversion efficiency [%]	Ref.
Ag_3AsS_3	1.065	1.82–2.56	26 ns	1	[72Han]
	1.064	1.2–8	8 ps	0.01–1	[83Els]
$AgGaS_2$	1.064	1.2–10	8 ps	0.1–10	[84Els]
	1.06	1.4–4.0	18 ns	16	[84Fan]
	1.064	4.5–8.7	15–20 ps	5.4	[91Bak]
	1.064	1.16–12.9	19 ps	25	[93Kra]
	1.064	1.319; 5.505	45–80 ps	63 [a]	[94Che]
	1.047	2.6–7	0.5–2.6 ps	–	[98Lae]
	0.845	1.267; 2.535	cw	2	[98Dou]
	0.74–0.85	3.3–10	160 fs	20	[94Sei2]
$AgGaSe_2$	2.05	2.65–9.02	30 ns	> 18	[86Eck]
	2.06	~ 4.1	~ 30 ns	23	[93Bud]
	1.57	6–14	6 ns	20	[97Cha]
	1.34	1.6–1.7; 6.7–6.9	30 ns	> 18	[86Eck]
$ZnGeP_2$	2.94	5.51–5.38; 6.29–6.46	80 ps	5.3	[85Vod]
	2.94	5–5.3; 5.9–6.3	150 ps	17	[87Vod]
	2.79	5.3; 5.9	~ 100 ps	10	[93Vod2]
	2.8; 2.94	4–10	~ 100 ps	1–18	[91Vod, 93Vod1, 95Vod2]
GaSe	2.8; 2.94	3.5–18	~ 100 ps	1	[91Vod, 93Vod1, 95Vod1]
CdSe	1.833	9.8–10.4; 2.26–2.23	300 ns	40	[72Her]
	2.36	7.9–13.7	40 ns	15	[72Dav, 73Dav]
	2.87	4.3–4.5; 8.1–8.3	140 ns	15	[74Wei]
	2.87	14.1–16.4	–	–	[76Wen]

[a] Pump depletion.

4.1.8 Picosecond continuum generation

Table 4.1.34. Picosecond continuum generation in crystals.

Crystal	λ_{pump} [μm]	I_{pump} [10^9 W cm^{-2}]	λ_{cont} [μm]	η [%]	Cut angle of crystals	Ref.
KDP	1.054	50	0.3–1.1	10	$\theta = 49°$	[83Mur]
KDP	0.527	30–40	0.84–1.4	15	$\theta = 42°$	[82Bar]
$LiIO_3$	0.355	–	0.46–1.55	–	$\theta = 90°$	[85Pok]
$LiIO_3$	0.532	0.3	0.67–2.58	–	$\theta = 90°$	[85Pok]
$LiIO_3$	1.064	–	1.72–3.0	–	$\theta = 90°$	[85Pok]
$LiNbO_3$	1.064	–	1.92–2.38	3	$\theta = 44.7°$	[75Cam]
GaAs	9.3	100	3–14	–	–	[85Cor]

References for 4.1

61Fra   Franken, P.A., Hill, A.E., Peters, C.W., Weinreich, G.: Phys. Rev. Lett. **7** (1961) 118.

62Bas   Bass, M., Franken, P.A., Hill, A.E., Peters, C.W., Weinreich, G.: Phys. Rev. Lett. **8** (1962) 18.
62Gio   Giordmaine, J.A.: Phys. Rev. Lett. **8** (1962) 19.
62Kle   Kleinman, D.A.: Phys. Rev. **126** (1962) 1977.
62Mak   Maker, P.D., Terhune, R.W., Nisenoff, M., Savage, C.M.: Phys. Rev. Lett. **8** (1962) 21.
62Mil   Miller, R.C., Savage, A.: Phys. Rev. **128** (1962) 2175.

64Akh   Akhmanov, S.A., Khokhlov, R.V.: Problems of nonlinear optics, Moscow: VINITI, 1964; English Transl.: Akhmanov, S.A., Khokhlov, R.V.: Problems of nonlinear optics, New York: Gordon and Breach, 1971.
64Mil   Miller, R.C.: Appl. Phys. Lett. **5** (1964) 17–19.

65Blo   Bloembergen, N.: Nonlinear optics, New York: Benjamin, 1965.
65Gio   Giordmaine, J.A., Miller, R.C.: Phys. Rev. Lett. **14** (1965) 973.
65Zer   Zernike jr., F., Berman, P.R.: Phys. Rev. Lett. **15** (1965) 999.

67Lab   Labuda, E.F., Johnson, A.M.: IEEE J. Quantum Electron. **3** (1967) 164.
67Mid   Midwinter, J.E., Warner, J.: J. Appl. Phys. **38** (1967) 519.
67Mil   Miller, R.C., Nordland, W.A.: IEEE J. Quantum Electron. **3** (1967) 642.

68Geu   Geusic, J.E., Levinstein, H.J., Singh, S., Smith, R.G., Van Uitert, L.G.: Appl. Phys. Lett. **12** (1968) 306.
68Mid   Midwinter, J.E.: Appl. Phys. Lett. **12** (1968) 68.
68Smi   Smith, R.G., Geusic, J.E., Levinstein, H.J., Singh, S.L., Van Uitert, G.: J. Appl. Phys. **39** (1968) 4030.

69Akm   Akmanov, A.G., Akhmanov, S.A., Zhdanov, B.V., Kovrigin, A.I., Podsotskaya, N.K., Khokhlov, R.V.: Pisma Zh. Eksp. Teor. Fiz. **10** (1969) 244; JETP Lett. (English Transl.) **10** (1969) 154.
69Amm   Amman, E.O., Oshman, M.K., Foster, J.D., Yarborough, J.M.: Appl. Phys. Lett. **15** (1969) 131.
69Des   Deserno, V., Nath, G.: Phys. Lett. A **30** (1969) 483.
69Far   Faries, D.W., Gehring, K.A., Richards, P.L., Shen, Y.R.: Phys. Rev. **180** (1969) 363.
69Yaj   Yajima, T., Inoue, K.: Phys. Lett. A **26** (1968) 281; IEEE J. Quantum Electron. **5** (1969) 140.

70Che   Chesler, R.B., Karr, M.A., Geusic, J.E.: Proc. IEEE **58** (1970) 1899.
70Izr   Izrailenko, A.I., Kovrigin, A.I., Nikles, P.V.: Pisma Zh. Exp. Teor. Fiz. **12** (1970) 475; JETP Lett. (English Transl.) **12** (1970) 331.
70Nat   Nath, G., Nehmanesch, H., Gsänger, M.: Appl. Phys. Lett. **17** (1970) 286.
70Wal   Wallace, R.W.: Appl. Phys. Lett. **17** (1970) 497.

71Boy   Boyd, G.D., Gandrud, W.B., Buechler, E.: Appl. Phys. Lett. **18** (1971) 446.
71Cam   Campillo, A.J., Tang, C.L.: Appl. Phys. Lett. **19** (1971) 36.
71Dew   Dewey, C.F., Hocker, L.O.: Appl. Phys. Lett. **18** (1971) 58.
71Her   Herbst, R.L., Byer, R.L.: Appl. Phys. Lett. **19** (1971) 527.
71Oka   Okada, M., Ieiri, S.: Jpn. J. Appl. Phys. **10** (1971) 808.

71Yaj   Yajima, T., Takeuchi, N.: Jpn. J. Appl. Phys. **10** (1971) 907.
71Yar   Yarborough, J.M., Massey, G.A.: Appl. Phys. Lett. **18** (1971) 438.

72Boy   Boyd, G.D., Bridges, T.J., Patel, C.K.N., Buehler, E.: Appl. Phys. Lett. **21** (1972) 553.
72Cam   Campillo, A.J.: IEEE J. Quantum Electron. **8** (1972) 809.
72Dav   Davydov, A.A., Kulevsky, L.A., Prokhorov, A.M., Savelev, A.D., Smirnov, V.V.: Pisma Zh. Eksp. Teor. Fiz. **15** (1972) 725; JETP Lett. (English Transl.) **15** (1972) 513.
72Dmi   Dmitriev, V.G., Kushnir, V.R., Rustamov, S.R., Fomichev, A.A.: Kvantovaya Elektron. **2**(8) (1972) 111; Sov. J. Quantum Electron. (English Transl.) **2** (1972) 188.
72Gab   Gabel, C., Hercher, M.: IEEE J. Quantum Electron. **8** (1972) 850.
72Han   Hanna, D.C., Luther-Davies, B., Rutt, H.N., Smith, R.C.: Appl. Phys. Lett. **20** (1972) 34.
72Her   Herbst, R.L., Byer, R.L.: Appl. Phys. Lett. **21** (1972) 189.
72Kun   Kung, A.H., Young, J.F., Bjorklund, G.C., Harris, S.E.: Phys. Rev. Lett. **29** (1972) 985.
72Luc   Lucy, R.F.: Appl. Opt. **11** (1972) 1329.
72Mel   Meltzer, D.W., Goldberg, L.S.: Opt. Commun. **5** (1972) 209.
72Rab   Rabson, T.A., Ruiz, H.J., Shah, P.L., Tittel, F.K.: Appl. Phys. Lett. **20** (1972) 282.
72Tak   Takeuchi, N., Matsumoto, N., Yajima, T., Kishida, S.: Jpn. J. Appl. Phys. **11** (1972) 268.
72Tse   Tseng, D.Y.: Appl. Phys. Lett. **21** (1972) 382.

73Alc   Alcock, A.S., Walker, A.C.: Appl. Phys. Lett. **23** (1973) 468.
73Bai   Baird, K.M., Smith, D.S., Berger, W.E.: Opt. Commun. **7** (1973) 107.
73Dav   Davydov, A.A., Kulevsky, L.A., Prokhorov, A.M., Savelev, A.D., Smirnov, V.V., Shirkov, A.V.: Opt. Commun. **9** (1973) 234.
73Dmi   Dmitriev, V.G., Krasnyanskaya, V.N., Koldobskaya, M.F., Rez, I.S., Shalaev, E.A., Shvom, E.M.: Kvantovaya Elektron. **2**(14) (1973) 64; Sov. J. Quantum Electron. (English Transl.) **3** (1973) 126.
73Dun   Dunning, F.B., Tittel, F.K., Stebbings, R.F.: Opt. Commun. **7** (1973) 181.
73Gur   Gurski, T.R.: Appl. Phys. Lett. **23** (1973) 273.
73Han   Hanna, D.C., Rampal, V.V., Smith, R.C.: Opt. Commun. **8** (1973) 151.
73Jai   Jain, R.K., Gustafson, T.K.: IEEE J. Quantum Electron. **9** (1973) 859.
73Kil   Kildal, H., Mikkelsen, J.C.: Opt. Commun. **9** (1973) 315.
73Mat   Matsumoto, N., Yajima, T.: Jpn. J. Appl. Phys. **12** (1973) 90.
73Zer   Zernike, F., Midwinter, J.E.: Applied nonlinear optics, New York: Wiley, 1973.

74Amm   Ammann, E.O., Decker, C.D., Falk, J.: IEEE J. Quantum Electron. **10** (1974) 463.
74Ant   Antonov, E.N., Koloshnikov, V.G., Nikogosyan, D.N.: Opt. Spektrosk. **36** (1974) 768; Opt. Spectrosc. (USSR) (English Transl.) **36** (1974) 446.
74Bye   Byer, R.L., Choy, M.M., Herbst, R.L., Chemla, D.S., Feigelson, R.S.: Appl. Phys. Lett. **24** (1974) 65.
74Dav   Davydov, B.L., Koreneva, L.G., Lavrovsky, E.A.: Radiotekh. Elektron. **19** (1974) 1313; Radio Eng. Electron. Phys. (English Transl.) **19**(6) (1974) 130.
74Gul   Gulshaw, W., Kannelaud, J., Peterson, J.E.: IEEE J. Quantum Electron. **10** (1974) 253.
74Gur   Gurski, T.R., Epps, H.W., Maran, S.P.: Nature (London) **249** (1974) 638.
74Han   Hanna, D.C., Rampal, V.V., Smith, R.C.: IEEE J. Quantum Electron. **10** (1974) 461.
74Her   Herbst, R.L., Fleming, R.N., Byer, R.L.: Appl. Phys. Lett. **25** (1974) 520.
74Kat1 Kato, K.: Appl. Phys. Lett. **25** (1974) 342.
74Kat2 Kato, K.: IEEE J. Quantum Electron. **10** (1974) 616.
74Kat3 Kato, K.: IEEE J. Quantum Electron. **10** (1974) 622.
74Kat4 Kato, K., Alcock, A.J., Richardson, M.C.: Opt. Commun. **11** (1974) 5.
74Kat5 Kato, K., Nakao, S.: Jpn. J. Appl. Phys. **13** (1974) 1681.

74Kil	Kildal, H., Mikkelsen, J.C.: Opt. Commun. **10** (1974) 306.
74Mas	Massey, G.A., Elliott, R.A.: IEEE J. Quantum Electron. **10** (1974) 899.
74Pin	Pine, A.S.: J. Opt. Soc. Am. **64** (1974) 1683.
74Vor	Voronin, E.S., Solomatin, V.S., Shuvalov, V.V.: Opto-electronics (London) **6** (1974) 189.
74Wei	Weiss, J.A., Goldberg, L.S.: Appl. Phys. Lett. **24** (1974) 389.
75And	Andreev, R.B., Volosov, V.D., Kuznetsova, L.I.: Kvantovaya Elektron. **2** (1975) 420; Sov. J. Quantum Electron. (English Transl.) **5** (1975) 242.
75Aru	Arutyunyan, E.A., Mkrtchyan, V.S.: Kvantovaya Elektron. **2** (1975) 812; Sov. J. Quantum Electron. (English Transl.) **5** (1975) 450.
75Att	Attwood, D.T., Pierce, E.L., Coleman, L.W.: Opt. Commun. **15** (1975) 10.
75Bad	Badikov, V.V., Pivovarov, O.N., Skokov, Yu.V., Skrebneva, O.V., Trotsenko, N.K.: Kvantovaya Elektron. **2** (1975) 618; Sov. J. Quantum Electron. (English Transl.) **5** (1975) 350.
75Bet	Beterov, I.M., Stroganov, V.I., Trunov, V.I., Yurshin, B.Ya.: Kvantovaya Elektron. **2** (1975) 2440; Sov. J. Quantum Electron. (English Transl.) **5** (1975) 1329.
75Bri	Bridges, T.J., Nguyen, V.T., Burkhardt, E.G., Patel, C.K.N.: Appl. Phys. Lett. **27** (1975) 600.
75Cam	Campillo, A.J., Hyer, R.C., Shapiro, S.L.: Opt. Lett. **4** (1975) 357.
75Dew	Dewey jr., C.F., Cook jr., W.R., Hodgson, R.T., Wynne, J.J.: Appl. Phys. Lett. **26** (1975) 714.
75Gol	Goldberg, L.: Appl. Opt. **14** (1975) 653.
75Kat1	Kato, K.: J. Appl. Phys. **46** (1975) 2721.
75Kat2	Kato, K.: Opt. Commun. **13** (1975) 93.
75Kat3	Kato, K.: Opt. Commun. **13** (1975) 361.
75Kus	Kushida, T., Tanaka, Y., Ojima, M., Nakazaki, Y.: Jpn. J. Appl. Phys. **14** (1975) 1097.
75Mal1	Malz, D., Bergmann, J., Heise, J.: Exp. Tech. Phys. **23** (1975) 379.
75Mal2	Malz, D., Bergmann, J., Heise, J.: Exp. Tech. Phys. **23** (1975) 495.
75Nik1	Nikogosyan, D.N.: Kvantovaya Elektron. **2** (1975) 2524; Sov. J. Quantum Electron. (English Transl.) **5** (1975) 1378.
75Nik2	Nikogosyan, D.N., Sukhorukov, A.P., Golovey, M.I.: Kvantovaya Elektron. **2** (1975) 609; Sov. J. Quantum Electron. (English Transl.) **5** (1975) 344.
75Vor	Voronin, E.S., Solomatin, V.S., Cherepov, N.I., Shuvalov, V.V., Badikov, V.V., Pivovarov, O.N.: Kvantovaya Elektron. **2** (1975) 1090; Sov. J. Quantum Electron. (English Transl.) **5** (1975) 597.
75Zhd	Zhdanov, B.V., Kalitin, V.V., Kovrigin, A.I., Pershin, S.M.: Pisma Zh. Tech. Fiz. **1** (1975) 847; Sov. Tech. Phys. Lett. (English Transl.) **1** (1975) 368.
76Abd	Abdullaev, G.B., Kulevsky, L.A., Nikles, P.V., Prokhorov, A.M., Savelev, A.D., Salaev, E.Yu., Smirnov, V.V.: Kvantovaya Elektron. **3** (1976) 163; Sov. J. Quantum Electron. (English Transl.) **6** (1976) 88.
76Att	Attwood, D.T., Bliss, E.S., Pierce, E.L., Coleman, L.W.: IEEE J. Quantum Electron. **12** (1976) 203.
76Ave	Avetisyan, Yu.O., Pogosyan, P.S.: Pisma Zh. Tech. Fiz. **2** (1976) 1144; Sov. Tech. Phys. Lett. (English Transl.) **2** (1976) 450.
76Che	Chen, T.S., White, W.P.: IEEE J. Quantum Electron. **12** (1976) 436.
76Dew	Dewey, H.J.: IEEE J. Quantum Electron. **12** (1976) 303.
76Dun	Dunning, F.B., Stickel jr., R.E.: Appl. Opt. **15** (1976) 3131.
76Fer	Ferrario, A., Garbi, M.: Opt. Commun. **17** (1976) 158.
76Fro	Frölich, D., Stein, L., Schröder, H.W., Welling, H.: Appl. Phys. **11** (1976) 97.
76Hoc	Hocker, L.O., Dewey jr., C.F.: Appl. Phys. **11** (1976) 137.
76Hon	Hon, D.T.: IEEE J. Quantum Electron. **12** (1976) 148.

76Kat1   Kato, K.: Opt. Commun. **19** (1976) 332.
76Kat2   Kato, K., Adhav, R.S.: IEEE J. Quantum Electron. **12** (1976) 443.
76Liu   Liu, Y.S., Jones, W.B., Chernoch, J.P.: Appl. Phys. Lett. **29** (1976) 32.
76Mac   Machewirth, V.P., Webb, R., Anafi, D.: Laser Focus **12**(5) (1976) 104.
76Mas1   Massey, G.A., Johnson, J.C.: IEEE J. Quantum Electron. **12** (1976) 721.
76Mas2   Massey, G.A., Johnson, J.C., Elliott, R.A.: IEEE J. Quantum Electron. **12** (1976) 143.
76Men   Menyuk, N., Iseler, G.W., Mooradian, A.: Appl. Phys. Lett. **29** (1976) 422.
76Moo   Moore, C.A., Goldberg, L.S.: Opt. Commun. **16** (1976) 21.
76Sey   Seymour, R.J., Zernike, F.: Appl. Phys. Lett. **29** (1976) 705.
76Str   Stroganov, V.I., Trunov, V.I., Chernenko, A.A., Izrailenko, A.N.: Kvantovaya Elektron. **3** (1976) 1122; Sov. J. Quantum Electron. (English Transl.) **6** (1976) 601.
76Wel   Wellegehausen, B., Friede, D., Vogt, H., Shahdin, S.: Appl. Phys. **11** (1976) 363.
76Wen   Wenzel, R.G., Arnold, G.P.: Appl. Opt. **15** (1976) 1322.
76Zac   Zacharias, H., Anders, A., Halpern, J.B., Welge, K.H.: Opt. Commun. **19** (1976) 116.

77Aba   Abakumov, G.A., Bagdasarov, Kh.S., Vetrov, V.V., Vorobev, S.A., Zakharov, V.P., Pikelni, V.F., Smirnov, A.P., Fadeev, V.V., Fedorov, E.A.: Kvantovaya Elektron. **4** (1977) 1152; Sov. J. Quantum Electron. (English Transl.) **7** (1977) 656.
77And1   Andreev, S.A., Matveev, I.N., Nekrasov, I.P., Pshenichnikov, S.M., Sopina, N.P.: Kvantovaya Elektron. **4** (1977) 657; Sov. J. Quantum Electron. (English Transl.) **7** (1977) 366.
77And2   Andreou, D.: Opt. Commun. **23** (1977) 37.
77Bli   Blit, S., Weaver, E.G., Dunning, F.B., Tittel, F.K.: Opt. Lett. **1** (1977) 58.
77Dan   Danelyus, R., Dikchyus, G., Kabelka, V., Piskarskas, A., Stabinis, A., Yasevichyuto, Ya.: Kvantovaya Elektron. **4** (1977) 2379; Sov. J. Quantum Electron. (English Transl.) **7** (1977) 1360.
77Dob   Dobrzhansky, G.F., Kulevsky, L.A., Polivanov, Yu.N., Sayakhov, R.Sh., Sukhodolsky, A.T.: Kvantovaya Elektron. **4** (1977) 1794; Sov. J. Quantum Electron. (English Transl.) **7** (1977) 1019.
77Dzh   Dzhotyan, G.P., Dyakov, Yu.E., Pershin, S.M., Kholodnykh, A.I.: Kvantovaya Elektron. **4** (1977) 1215; Sov. J. Quantum Electron. (English Transl.) **7** (1977) 685.
77Fer   Ferguson, A.I., Dunn, M.H.: Opt. Commun. **23** (1977) 177.
77Fuk   Fukuda, T., Uematsu, Y.: Izv. Akad. Nauk SSSR, Ser. Fiz. **41** (1977) 548; Bull. Acad. Sci. USSR Phys. Ser. (English Transl.) **41**(3) (1977) 73.
77Hir   Hirth, A., Vollrath, K., Allain, J.V.: Opt. Commun. **20** (1977) 347.
77Iva   Ivanova, Z.I., Kabelka, V., Magnitsky, S.A., Piskarskas, A., Smilgyavichyus, V., Rubinina, N.M., Tunkin, V.G.: Kvantovaya Elektron. **4** (1977) 2469; Sov. J. Quantum Electron. (English Transl.) **7** (1977) 1414.
77Jan   Jantz, W., Koidl, P.: Appl. Phys. Lett. **31** (1977) 99.
77Kat1   Kato, K.: Appl. Phys. Lett. **30** (1977) 583.
77Kat2   Kato, K.: IEEE J. Quantum Electron. **13** (1977) 544.
77Kry   Kryukov, P.G., Matveets, Yu.A., Nikogosyan, D.N., Sharkov, A.V., Gordeev, E.M., Fanchenko, S.D.: Kvantovaya Elektron. **4** (1977) 211; Sov. J. Quantum Electron. (English Transl.) **7** (1977) 127.
77Nik   Nikogosyan, D.N.: Kvantovaya Elektron. **4** (1977) 5; Sov. J. Quantum Electron. (English Transl.) **7** (1977) 1.
77Oud   Oudar, J.L., Hierle, R.: J. Appl. Phys. **48** (1977) 2699.
77Rei   Reintjes, J.J., Eckardt, R.C.: Appl. Phys. Lett. **30** (1977) 91.
77Sey   Seymour, R.L., Choy, M.M.: Opt. Commun. **20** (1977) 101.
77Sti   Stickel jr., R.E., Dunning, F.B.: Appl. Opt. **16** (1977) 2356.

78Bia	Bianchi, A., Ferrario, A., Musci, M.: Opt. Commun. **25** (1978) 256.
78Dan1	Danelyus, R., Dikchyus, G., Kabelka, V., Piskarskas, A., Stabinis, A., Yasevichyute, Ya.: Litov. Fiz. Sb. **18** (1978) 93; Sov. Phys.-Collect. (English Transl.) **18**(1) (1978) 21.
78Dan2	Danelyus, R., Kabelka, V., Piskarskas, A., Smilgyavichyus, A.: Pisma Zh. Tech. Fiz. **4** (1978) 765; Sov. Tech. Phys. Lett. (English Transl.) **4** (1978) 308.
78Dan3	Danelyus, R., Kabelka, V., Piskarskas, A., Smilgyavichyus, V.: Kvantovaya Elektron. **5** (1978) 679; Sov. J. Quantum Electron. (English Transl.) **8** (1978) 398.
78Dmi	Dmitriev, V.G., Kulevsky, L.A.: Parametric generators of light, in: Handbook of lasers, Vol. 2, Prokhorov, A.M. (ed.), Moscow: Sovetskoye Radio, 1978, pp. 319–348 (in Russian).
78Har	Hargis, P.: Laser Focus **14**(7) (1978) 18.
78Koe	Koenig, R., Rosenfeld, A., Tam, N., Mory, S.: Opt. Commun. **24** (1978) 190.
78Kog	Kogan, R.M., Pixton, R.M., Crow, T.G.: Opt. Eng. **17** (1978) 120.
78Kry	Kryukov, P.G., Matveets, Yu.A., Nikogosyan, D.N., Sharkov, A.V.: Kvantovaya Elektron. **5** (1978) 2348; Sov. J. Quantum Electron. (English Transl.) **8** (1978) 1319.
78Mas	Massey, G.A., Jones, M.D., Johnson, J.C.: IEEE J. Quantum Electron. **14** (1978) 527.
78Mat	Matveets, Yu.A., Nikogosyan, D.N., Kabelka, V., Piskarskas, A.: Kvantovaya Elektron. **5** (1978) 664; Sov. J. Quantum Electron. (English Transl.) **8** (1978) 386.
78Pai	Paisner, J.A., Spaeth, M.L., Gerstenberger, D.C., Ruderman, I.W.: Appl. Phys. Lett. **32** (1978) 476.
78Sei	Seilmeier, A., Spanner, K., Laubereau, A., Kaiser, W.: Opt. Commun. **24** (1978) 237.
78Sti1	Stickel jr., R.E., Blit, S., Hidebrandt, G.F., Dahl, E.D., Dunning, F.B., Tittel, F.K.: Appl. Opt. **17** (1978) 2270.
78Sti2	Stickel jr., R.E., Dunning, F.B.: Appl. Opt. **17** (1978) 981. Erratum: p. 2132.
78Sti3	Stickel jr., R.E., Dunning, F.B.: Appl. Opt. **17** (1978) 1313.
78Vor	Voronin, E.S., Solomatin, V.S., Shuvalov, V.V.: Kvantovaya Elektron. **5** (1978) 2031; Sov. J. Quantum Electron. (English Transl.) **8** (1978) 1145.
79And1	Andreev, R.B., Volosov, V.D., Gorshkov, V.S.: Opt. Spektrosk. **46** (1979) 376; Opt. Spectrosc. (USSR) (English Transl.) **46** (1979) 207.
79And2	Andreeva, N.P., Andreev, S.A., Matveev, I.N., Pshenichnikov, S.M., Ustinov, N.D.: Kvantovaya Elektron. **6** (1979) 357; Sov. J. Quantum Electron. (English Transl.) **9** (1979) 208.
79Ang	Angelov, D.A., Gurzadyan, G.G., Nikogosyan, D.N.: Kvantovaya Elektron. **6** (1979) 2267; Sov. J. Quantum Electron. (English Transl.) **9** (1979) 1334.
79Ave	Avetisyan, Yu.O., Makaryan, A.O., Movsesyan, K.M., Pogosyan, P.S.: Pisma Zh. Tech. Fiz. **5** (1979) 233; Sov. Tech. Phys. Lett. (English Transl.) **5** (1979) 93.
79Dud	Dudina, N.S., Kopylov, S.M., Mikhailov, L.K., Cherednichenko, O.B.: Kvantovaya Elektron. **6** (1979) 2478; Sov. J. Quantum Electron. (English transl.) **9** (1979) 1468.
79Hal	Halbout, J.M., Blit, S., Donaldson, W., Tang, C.L.: IEEE J. Quantum Electron. **15** (1979) 1176.
79Jaa	Jaanimagi, P.A., Richardson, M.C., Isenor, N.R.: Opt. Lett. **4** (1979) 45.
79Jon	Jones, M.D., Massey, G.A.: IEEE J. Quantum Electron. **15** (1979) 204.
79Kab	Kabelka, V., Kutka, A., Piskarskas, A., Smillgyavichyus, V., Yasevichyute, Ya.: Kvantovaya Elektron. **6** (1979) 1735; Sov. J. Quantum Electron. (English Transl.) **9** (1979) 1022.
79Kat	Kato, K.: IEEE J. Quantum Electron. **15** (1979) 410.
79Kur	Kurtz, S.K., Jerphagnon, J., Choy, M.M.: Nonlinear dielectric susceptibilities, in: Landolt-Börnstein, Group 3, Vol. 11, Berlin, Heidelberg: Springer-Verlag, 1979, p. 671.
79Liu	Liu, P.L.: Appl. Opt. **18** (1979) 3543.
79Mar	Mariella jr., R.P.: Opt. Commun. **29** (1979) 100.
79Men	Menyuk, N., Iseler, G.W.: Opt. Lett. **4** (1979) 55.

80And   Andreev, S.A., Andreeva, N.P., Badikov, V.V., Matveev, I.N., Pshenichnikov, S.M.: Kvantovaya Elektron. **7** (1980) 2003; Sov. J. Quantum Electron. (English Transl.) **10** (1980) 1157.
80Aru   Arutyunyan, A.G., Atanesyan, V.G., Petrosyan, K.B., Pokhsraryan, K.M.: Pisma Zh. Tekh. Fiz. **6** (1980) 277; Sov. Tech. Phys. Lett. (English Transl.) **6** (1980) 120.
80Bar1   Bareika, B., Dikchyus, G., Isianova, E.D., Piskarskas, A., Sirutkaitis, V.: Pisma Zh. Tech. Fiz. **6** (1980) 694; Sov. Tech. Phys. Lett. (English Transl.) **6** (1980) 301.
80Bar2   Bareika, B., Dikchyus, G., Piskarskas, A., Sirutkaitis, V.: Kvantovaya Elektron. **7** (1980) 2204; Sov. J. Quantum Electron. (English Transl.) **10** (1980) 1277.
80Bar3   Baryshev, S.A., Pryalkin, V.I., Kholodnykh, A.I.: Pisma Zh. Tech. Fiz. **6** (1980) 964; Sov. Tech. Phys. Lett. (English Transl.) **6** (1980) 415.
80Bas   Bastow, S.J., Dunn, M.H.: Opt. Commun. **35** (1980) 259.
80Fil   Fill, E.E.: Opt. Commun. **33** (1980) 321.
80Gus   Gusev, Yu.A., Kirpichnikov, A.V., Konoplin, S.N., Marennikov, S.I., Nikles, P.V., Polivanov, Yu.N., Prokhorov, A.M., Savelev, A.D., Sayakhov, R.Sh., Smirnov, V.V., Chebotaev, V.P.: Pisma Zh. Tekh. Fiz. **6** (1980) 1262; Sov. Tech. Phys. Lett. (English Transl.) **6** (1980) 541.
80Kat1   Kato, K.: IEEE J. Quantum Electron. **16** (1980) 810.
80Kat2   Kato, K.: IEEE J. Quantum Electron. **16** (1980) 1017.
80Kat3   Kato, K.: IEEE J. Quantum Electron. **16** (1980) 1288.
80Koe   Koeneke, A., Hirth, A.: Opt. Commun. **34** (1980) 245.
80Lot   Loth, C., Bruneau, D., Fabre, E.: Appl. Opt. **19** (1980) 1022.
80Mak   Makarian, A.H., Movsessian, K.M., Pogossian, P.G.: Opt. Commun. **35** (1980) 147.
80See   See, Y.C., Falk, J.: Appl. Phys. Lett. **36** (1980) 503.
80Sek   Seka, W., Jakobs, S.D., Rizzo, J.E., Boni, R., Craxton, R.S.: Opt. Commun. **34** (1980) 469.
80Sha   Shaw, E.D., Patel, C.K.N., Chichester, R.J.: Opt. Commun. **33** (1980) 221.
80Web   Webster, C.R., Wöste, L., Zare, R.N.: Opt. Commun. **35** (1980) 435.
80Wel   Welford, D., Sibbett, W., Taylor, J.R.: Opt. Commun. **35** (1980) 283.
80Yam   Yamashita, M., Sibbett, W., Welford, D., Bradley, D.J.: J. Appl. Phys. **51** (1980) 3559.

81Akm   Akmanov, A.G., Valshin, A.M., Yamaletdinov, A.G.: Kvantovaya Elektron. **8** (1981) 408; Sov. J. Quantum Electron. (English Transl.) **11** (1981) 247.
81And   Andreev, S.A., Andreeva, N.P., Matveev, I.N., Pshenichnikov, S.A.: Kvantovaya Elektron. **8** (1981) 1361; Sov. J. Quantum Electron. (English Transl.) **11** (1981) 821.
81Bye   Byer, R.L., Park, Y.K., Feigelson, R.S., Kway, W.L.: Appl. Phys. Lett. **39** (1981) 17.
81Dmi   Dmitriev, V.G., Konvisar, P.G., Lyushnya, I.B., Mikhailov, V.Yu., Rustamov, S.R., Stelmakh, M.F.: Kvantovaya Elektron. **8** (1981) 906; Sov. J. Quantum Electron. (English Transl.) **11** (1981) 545.
81Kaz   Kazakov, A.A., Shavkunov, S.V., Shalaev, E.A.: Kvantovaya Elektron. **8** (1981) 2259; Sov. J. Quantum Electron. (English Transl.) **11** (1981) 1381.
81Lyu   Lyutskanov, V.L., Savov, S.D., Saltiel, S.M., Stamenov, K.V., Tomov, I.V.: Opt. Commun. **37** (1981) 149.
81Wit   Witte, K.J., Fill, E., Brederlow, G., Baumhacker, H., Volk, R.: IEEE J. Quantum Electron. **17** (1981) 1809.

82Ata   Atabaev, Sh., Polivanov, Yu.N., Poluektov, S.N.: Kvantovaya Elektron. **9** (1982) 378; Sov. J. Quantum Electron. (English Transl.) **12** (1982) 212.
82Bar   Bareika, B., Birmontas, A., Dikchyus, G., Piskarskas, A., Sirutkaitis, V., Stabinis, A.: Kvantovaya Elektron. **9** (1982) 2534; Sov. J. Quantum Electron. (English Transl.) **12** (1982) 1654.
82Ber   Bergquist, J.C., Hemmati, H., Itano, W.M.: Opt. Commun. **43** (1982) 437.

82Dan	Danelyus, R., Piskarskas, A., Sirutkaitis, V.: Kvantovaya Elektron. **9** (1982) 2491; Sov. J. Quantum Electron. (English Transl.) **12** (1982) 1626.
82Gol	Golyaev, Yu.D., Grodsky, S.A., Dmitriev, V.G., Konnvisar, P.G., Lantratov, S.V., Mikhailov, V.Yu., Rustamov, S.R.: Kvantovaya Elektron. **9** (1982) 2093; Sov. J. Quantum Electron. (English Transl.) **12** (1982) 1360.
82Ibr	Ibragimov, E.F., Redkorechev, V.I., Sukhorukov, A.P., Usmanov, T.: Kvantovaya Elektron. **9** (1982) 1131; Sov. J. Quantum Electron. (English Transl.) **12** (1982) 714.
82Kat	Kato, K.: IEEE J. Quantum Electron. **18** (1982) 451.
82Lin	Linford, G.J., Johnson, B.C., Hildum, J.S., Martin, W.E., Snyder, K., Boyd, R.D., Smith, W.L., Vercimak, C.L., Eimerl, D., Hunt, J.T.: Appl. Opt. **21** (1982) 3633.
82Liu	Liu, E., Dunning, F.B., Tittel, F.K.: Appl. Opt. **21** (1982) 3415.
82Tan	Tanaka, Y., Kuroda, H., Shionoya, S.: Opt. Commun. **41** (1982) 434.
83Bar	Bareika, B., Dikchyus, G., Piskarskas, A., Sirutkaitis, V., Yasevichyute, Ya.: Kvantovaya Elektron. **10** (1983) 2318; Sov. J. Quantum Electron. (English Transl.) **13** (1983) 1507.
83Bau	Baumert, J.C., Günter, P., Melchior, H.: Opt. Commun. **48** (1983) 215.
83Bre	Brederlow, G., Fill, E., Witte, K.J.: The high-power iodine laser, Springer Series Opt. Sci., Vol. 34, Berlin, Heidelberg: Springer-Verlag, 1983.
83Cou	Couillaud, B., Bloomfield, L.A., Hansh, T.W.: Opt. Lett. **8** (1983) 259.
83Dan	Danelyus, R., Piskarskas, A., Sirutkaitis, V., Stabinis, A., Yasevichyute, Ya.: Parametric generators of light and picosecond spectroscopy, Vilnius: Mokslas, 1983 (in Russian).
83Els	Elsaesser, T., Seilmeier, A., Kaiser, W.: Opt. Commun. **44** (1983) 293.
83Fil	Fill, E., Wildenauer, J.: Opt. Commun. **47** (1983) 412.
83Gul	Gulamov, A.A., Ibragimov, E.A., Redkorechev, V.I., Usmanov, T.: Kvantovaya Elektron. **10** (1983) 1305; Sov. J. Quantum Electron. (English Transl.) **13** (1983) 844.
83Hem1	Hemmati, H., Bergquist, J.C.: Opt. Commun. **47** (1983) 157.
83Hem2	Hemmati, H., Bergquist, J.C., Itano, W.M.: Opt. Lett. **8** (1983) 73.
83Joh	Johnson, A.M., Simpson, W.M.: Opt. Lett. **8** (1983) 554.
83Kac	Kaczmarek, F., Jendrzejczak, A.: Opt. Quantum Electron. **15** (1983) 187.
83Kaz	Kazakov, A.A., Konovalov, V.A., Shavkunov, S.V., Shalaev, E.A.: Kvantovaya Elektron. **10** (1983) 1603; Sov. J. Quantum Electron. (English Transl.) **13** (1983) 1054.
83Kop	Kopylov, S.M., Mikhailov, L.K., Cherednichenko, O.B.: Kvantovaya Elektron. **10** (1983) 625; Sov. J. Quantum Electron. (English Transl.) **13** (1983) 375.
83Maj	Majewsky, W.A.: Opt. Commun. **45** (1983) 201.
83Man	Mannik, L., Brown, S.K.: Opt. Commun. **47** (1983) 62.
83Mur	Muravev, A.A., Rubinov, A.N.: Pisma Zh. Eksp. Teor. Fiz. **37** (1983) 597; JETP Lett. (English Transl.) **37** (1983) 713.
83Oni	Onischukov, G.I., Fomichev, A.A., Kholodnykh, A.I.: Kvantovaya Elektron. **10** (1983) 1525; Sov. J. Quantum Electron. (English Transl.) **13** (1983) 1001.
83Pet	Petrosyan, K.B., Pogosyan, A.L., Pokhsraryan, K.M.: Izv. Akad. Nauk SSSR Ser. Fiz. **47** (1983) 1619; Bull. Acad. Sci. USSR Phys. Ser. (English Transl.) **47**(8) (1983) 155.
83Tak	Takagi, Y., Sumitani, M., Nakashima, N., O'Connor, D.V., Yoshihara, K.: Appl. Phys. Lett. **42** (1983) 489.
84Akh	Akhmanov, S.A., Begishev, I.E., Gulamov, A.A., Erofeev, E.A., Zhdanov, B.V., Kuznetsov, V.I., Rashkovich, L.N., Usmanov, T.B.: Kvantovaya Elektron. **11** (1984) 1701; Sov. J. Quantum Electron. (English Transl.) **14** (1984) 1145.
84And	Andreev, Yu.M., Voevodin, V.G., Gribenyukov, A.I., Zyryanov, O.Ya., Ippolitov, I.I., Morozov, A.N., Soskin, A.V., Khmelnitsky, G.S.: Kvantovaya Elektron. **11** (1984) 1511; Sov. J. Quantum Electron. (English Transl.) **14** (1984) 1021.

84Ash   Ashmarin, I.I., Bykovsky, Yu.A., Ukraintsev, V.A., Chistyakov, A.A., Shishonkov, L.V.: Kvantovaya Elektron. **11** (1984) 1847; Sov. J. Quantum Electron. (English Transl.) **14** (1984) 1237.
84Cot   Cotter, D., White, K.I.: Opt. Commun. **49** (1984) 205.
84Don   Donaldson, W.R., Tang, C.L.: Appl. Phys. Lett. **44** (1984) 25.
84Els   Elsaesser, T., Seilmeier, A., Kaiser, W., Koidl, P., Brandt, G.: Appl. Phys. Lett. **44** (1984) 383.
84Fan   Fan, Y.X., Eckardt, R.C., Byer, R.L., Route, R.K., Feigelson, R.S.: Appl. Phys. Lett. **45** (1984) 313.
84Foo   Foot, C.J., Baird, P.E.G., Boshier, M.G., Stacey, D.N., Woodgate, G.K.: Opt. Commun. **50** (1984) 199.
84Jer   Jerphagnon, J., Kurtz, S.K., Oudar, J.L.: Nonlinear dielectric susceptibilities, in: Landolt-Börnstein, Group 3, Vol. 18, Berlin, Heidelberg: Springer-Verlag, 1984, p. 456.
84Kat   Kato, K.: IEEE J. Quantum Electron. **20** (1984) 698.
84Liu   Liu, Y.S., Dentz, D., Belt, R.: Opt. Lett. **9** (1984) 76.
84Rud   Ruddock, I.S., Illingworth, R., Reekie, L.: Opt. Quantum Electron. **16** (1984) 87.
84She   Shen, Y.R.: The principles of nonlinear optics, New York: Wiley, 1984.
84Van   Van Stryland, E.W., Williams, W.E., Soileau, M.J., Smirl, A.L.: IEEE J. Quantum Electron. **20** (1984) 434.

85Ale   Aleksandrovsky, A.L., Akhmanov, S.A., Dyakov, V.A., Zheludev, N.I., Pryalkin, V.I.: Kvantovaya Elektron. **12** (1985) 1333; Sov. J. Quantum Electron. (English Transl.) **15** (1985) 885.
85And   Andreev, Yu.M., Vedernikova, T.V., Betin, A.A., Voevodin, V.G., Gribenyukov, A.I., Zyryanov, O.Ya., Ippolitov, I.I., Masychev, V.I., Mitropol'skii, O.V., Novikov, V.P., Novikov, M.A., Sosnin, A.V.: Kvantovaya Elektron. **12** (1985) 1535; Sov. J. Quantum Electron. (English Transl.) **15** (1985) 1014.
85Bau   Baumert, J.C., Hoffnagle, J., Günter, P.: Appl. Opt. **24** (1985) 1299.
85Bel   Belt, R.F., Gashurov, G., Liu, Y.S.: Laser Focus/Electro-Optics **21**(10) (1985) 110.
85Ber   Berg, M., Harris, C.B., Kenny, T.W., Richards, P.L.: Appl. Phys. Lett. **47** (1985) 206.
85Bog   Bogdanov, S.F., Konvisar, P.G., Rustamov, S.R.: Kvantovaya Elektron. **12** (1985) 2143; Sov. J. Quantum Electron. (English Transl.) **15** (1985) 1409.
85Bru   Bruneau, D., Tournade, A.M., Fabre, E.: Appl. Opt. **24** (1985) 3740.
85Cor   Corkum, P.B., Ho, P.P., Alfano, R.R., Manassah, J.T.: Opt. Lett. **10** (1985) 624.
85Eck   Eckardt, R.C., Fan, Y.X., Byer, R.L., Route, R.K., Feigelson, R.S., Van der Laan, J.: Appl. Phys. Lett. **47** (1985) 786.
85Els   Elsaesser, T., Lobentanzer, H., Seilmeier, A.: Opt. Commun. **52** (1985) 355.
85Kat   Kato, K.: IEEE J. Quantum Electron. **21** (1985) 119.
85Per   Perkins, P.E., Fahlen, T.S.: IEEE J. Quantum Electron. **21** (1985) 1636.
85Pok   Pokhsraryan, K.M.: Opt. Commun. **55** (1985) 439.
85Ree   Reekie, L., Ruddock, I.S., Illingworth, R.: Opt. Quantum Electron. **17** (1985) 169.
85Ros1   Rosker, M.J., Cheng, K., Tang, C.L.: IEEE J. Quantum Electron. **21** (1985) 1600.
85Ros2   Rosker, M.J., Tang, C.L.: J. Opt. Soc. Am. B **2** (1985) 691.
85Rya   Ryabov, S.G., Toropkin, G.N., Usoltsev, I.F.: Instruments of quantum electronics, Moscow: Radio i Svyaz, 1985 (in Russian).
85Tak   Takagi, Y., Sumitani, M., Nakashima, N., Yoshihara, K.: IEEE J. Quantum Electron. **21** (1985) 193.
85Vod   Vodopyanov, K.L., Voevodin, V.G., Gribenyukov, A.I., Kulevsky, L.A.: Izv. Akad. Nauk SSSR Ser. Fiz. **49** (1985) 569; Bull. Acad. Sci. USSR Phys. Ser. (English Transl.) **49**(3) (1985) 146.

86Bar	Bareika, B.F., Begishev, I.A., Burdulis, Sh.A., Gulamov, A.A., Erofeev, E.A., Piskarskas, A.S., Sirutkaitis, V.A., Usmanov, T.: Pisma Zh. Tekh. Fiz. **12** (1986) 186; Sov. Tech. Phys. Lett. (English Transl.) **12** (1986) 78.
86Bet	Bethune, D.S., Luntz, A.C.: Appl. Phys. B **40** (1986) 107.
86Bue	Buesener, H., Renn, A., Brieger, M., Von Moers, F., Hese, A.: Appl. Phys. B **39** (1986) 77.
86Che	Chen, C., Fan, Y.X., Eckardt, R.C., Byer, R.L.: Proc. SPIE (Int. Soc. Opt. Eng.) **681** (1986) 12.
86Dri	Driscoll, T.A., Hoffman, H.J., Stone, R.E., Perkins, P.E.: J. Opt. Soc. Am. B **3** (1986) 683.
86Eck	Eckardt, R.C., Fan, Y.X., Byer, R.L., Marquardt, C.L., Storm, M.E., Esterowitz, L.: Appl. Phys. Lett. **49** (1986) 608.
86Gar	Garmash, V.M., Ermakov, G.A., Pavlova, N.I., Tarasov, A.V.: Pisma Zh. Tekh. Fiz. **12** (1986) 1222; Sov. Tech. Phys. Lett. (English Transl.) **12** (1986) 505.
86Kat	Kato, K.: IEEE J. Quantum Electron. **22** (1986) 1013.
86Lav	Lavrovskaya, O.I., Pavlova, N.I., Tarasov, A.V.: Kristallografiya **31** (1986) 1145; Sov. Phys. Crystallogr. (English Transl.) **31** (1986) 678.
86Led	Ledoux, I., Zyss, J., Migus, A., Etchepare, J., Grillon, G., Antonetti, A.: Appl. Phys. Lett. **48** (1986) 1564.
86Miy	Miyazaki, K., Sakai, H., Sato, T.: Opt. Lett. **11** (1986) 797.
86Nik	Nikogosyan, D.N., Gurzadyan, G.G.: Kvantovaya Elektron. **13** (1986) 2519; Sov. J. Quantum Electron. (English Transl.) **16** (1986) 1633.
86Pis	Piskarskas, A., Smilgyavichyus, V., Umbrasas, A., Yodishyus, N.: Kvantovaya Elektron. **13** (1986) 1281; Sov. J. Quantum Electron. (English Transl.) **16** (1986) 841.
86Xin	Xinan, X., Shuzhong, Y., Fuyun, L.: Chin. J. Lasers **13** (1986) 892.
87Adh	Adhav, R.S., Adhav, S.R., Pelaprat, J.M.: Laser Focus/Electro-Optics **23**(9) (1987) 88.
87And1	Andreev, Yu.M., Baranov, V.Yu., Voevodin, V.G., Geiko, P.P., Gribenyukov, A.I., Izyumov, S.V., Kozochkin, S.M., Pismenny, V.D., Satov, Yu.A., Streltsov, A.P.: Kvantovaya Elektron. **14** (1987) 2252; Sov. J. Quantum Electron. (English Transl.) **17** (1987) 1435.
87And2	Andreev, Yu.M., Belykh, A.D., Voevodin, V.G., Geiko, P.P., Gribenyukov, A.I., Gurashvili, V.A., Izyumov, S.V.: Kvantovaya Elektron. **14** (1987) 782; Sov. J. Quantum Electron. **17** (1987) 490.
87And3	Andreev, Yu.M., Voevodin, V.G., Geiko, P.P., Gribenyukov, A.I., Dyadkin, A.P., Pigulsky, S.V., Starodubtsev, A.I.: Kvantovaya Elektron. **14** (1987) 784; Sov. J. Quantum Electron. (English Transl.) **17** (1987) 491.
87And4	Andreev, Yu.M., Voevodin, V.G., Geiko, P.P., Gribenyukov, A.I., Zuev, V.V., Solodukhin, A.S., Trushin, S.A., Churakov, V.V., Shubin, S.F.: Kvantovaya Elektron. **14** (1987) 2137; Sov. J. Quantum Electron. (English Transl.) **17** (1987) 1362.
87Aru	Arutyunyan, A.G., Buniatyan, G.R., Melkonyan, A.A., Mesropyan, A.V., Paityan, G.A., in: Nonlinear optical interactions, Yerevan: Yerevan University Press, 1987, p. 135–144.
87Beg	Begishev, I.A., Gulamov, A.A., Erofeev, E.A., Usmanov, T.: Pisma Zh. Tekh. Fiz. **13** (1987) 305; Sov. Tech. Phys. Lett. (English Transl.) **13** (1987) 125.
87Che	Chemla, D.S., Zyss, J. (eds): Nonlinear optical properties of organic molecules and crystals, Vols. 1, 2, New York: Academic, 1987.
87Dmi	Dmitriev, V., Tarasov, L.: Optique non lineaire appliquee, Moscow: Mir, 1987.
87Gla	Glab, W.L., Hessler, J.P.: Appl. Opt. **26** (1987) 3181.
87Ion	Ionushauskas, G., Piskarskas, A., Sirutkaitis, V., Yuozapavichyus, A.: Kvantovaya Elektron. **14** (1987) 2044; Sov. J. Quantum Electron. (English Transl.) **17** (1987) 1303.
87Led	Ledoux, I., Badan, J., Zyss, J., Migus, A., Hulin, D., Etchepare, J., Grillon, G., Antonetti, A.: J. Opt. Soc. Am. B **4** (1987) 987.
87Len	Lenfellner, H.: Opt. Lett. **12** (1987) 184.

87Moo1   Moody, S.E., Eggleston, J.M., Seamans, J.F.: IEEE J. Quantum Electron. **23** (1987) 335.
87Moo2   Moore, D.S., Schmidt, S.C.: Opt. Lett. **12** (1987) 480.
87Nik   Nikogosyan, D.N., Gurzadyan, G.G.: Kvantovaya Elektron. **14** (1987) 1529; Sov. J. Quantum Electron. (English Transl.) **17** (1987) 970.
87Nor   Norman, P.A., Bloor, D., Obhi, J.S., Karaulov, S.A., Hursthouse, M.B., Kolinsky, P.V., Jones, R.J., Hall, S.R.: J. Opt. Soc. Am. B **4** (1987) 1013.
87Pas   Pastel, R.L.: Appl. Opt. **26** (1987) 1574.
87Per   Perkins, P.E., Fahlen, T.S.: J. Opt. Soc. Am. B **4** (1987) 1066.
87Vod   Vodopyanov, K.L., Voevodin, V.G., Gribenyukov, A.I., Kulevsky, L.A.: Kvantovaya Elektron. **14** (1987) 1815; Sov. J. Quantum Electron. (English Transl.) **17** (1987) 1159.

88Akh   Akhmanov, S.A., Vysloukh, V.A., Chirkin, A.S.: Optics of femtosecond laser pulses, Moscow: Nauka, 1988; English version: Akhmanov, S.A., Vysloukh, V.A., Chirkin, A.S.: Optics of femtosecond laser pulses, New York: AIP, 1992.
88Amm   Ammann, E.O., Gush jr., S.: Appl. Phys. Lett. **52** (1988) 1374.
88Beg   Begishev, I.A., Ganeev, R.A., Gulamov, A.A., Erofeev, E.A., Kamalov, Sh.R., Usmanov, T., Khadzhaev, A.D.: Kvantovaya Elektron. **15** (1988) 353; Sov. J. Quantum Electron. (English Transl.) **18** (1988) 224.
88Bro   Bromley, L.J., Guy, A., Hanna, D.C.: Opt. Commun. **67** (1988) 316.
88Che   Cheng, L.K., Bosenberg, W.R., Tang, C.L.: Appl. Phys. Lett. **53** (1988) 175.
88Ede   Edelstein, D.C., Wachman, E.S., Cheng, L.K., Bosenberg, W.R., Tang, C.L.: Appl. Phys. Lett. **52** (1988) 2211.
88Fan   Fan, Y.X., Eckardt, R.C., Byer, R.L., Nolting, J., Wallenstein, R.: Appl. Phys. Lett. **53** (1988) 2014.
88Gar   Garayanz, N.P., Petrosyan, K.B., Pokhsraryan, K.M.: Izv. Akad. Nauk Arm. SSR Ser. Fiz. **23** (1988) 109; Sov. J. Contemp. Phys. Armen. Acad. Sci. (English Transl.) **23**(2) (1988).
88Gus   Gustafson, T.L.: Opt. Commun. **67** (1988) 53.
88Jos   Josse, D., Hierle, R., Ledoux, I., Zyss, J.: Appl. Phys. Lett. **53** (1988) 2251.
88Kom   Komine, H.: Opt. Lett. **13** (1988) 643.
88Koz   Kozlovsky, W.J., Nabors, C.D., Byer, R.L.: IEEE J. Quantum Electron. **24** (1988) 913.
88Lag   Lago, A., Wallenstein, R., Chen, C., Fan, Y.X., Byer, R.L.: Opt. Lett. **13** (1988) 221.
88Lok   Lokai, P., Burghardt, B., Mückenheim, W.: Appl. Phys. B **45** (1988) 245.
88Mue   Mückenheim, W., Lokai, P., Burghardt, B., Basting, D.: Appl. Phys. B **45** (1988) 259.
88Spe   Spears, K.G., Zhu, X., Yang, X., Wang, L.: Opt. Commun. **66** (1988) 167.
88Van   Vanherzeele, H., Bierlein, J.D., Zumsteg, F.Z.: Appl. Opt. **27** (1988) 3314.

89Abd   Abdullaev, G.B., Allakhverdiev, K.R., Karasev, M.E., Konov, V.I., Kulewskii, L.A., Mustafaev, N.B., Pashinin, P.P., Prokhorov, A.M., Starodumov, Yu.M., Chapliev, N.I.: Kvantovaya Elektron. **16** (1989) 757; Sov. J. Quantum Electron. (English Transl.) **19** (1989) 494.
89Aru   Arutyunyan, A.G., Gurzadyan, G.G., Ispiryan, R.K.: Kvantovaya Elektron. **16** (1989) 2493; Sov. J. Quantum Electron. (English Transl.) **19** (1989) 1602.
89Auy   Auyeung, R.C.Y., Zielke, D.M., Feldman, B.J.: Appl. Phys. B **48** (1989) 293.
89Bia   Biaggio, I., Looser, H., Günter, P.: Ferroelectrics **94** (1989) 157.
89Bos   Bosenberg, W.R., Pelouch, W.S., Tang, C.L.: Appl. Phys. Lett. **55** (1989) 1952.
89Cou   Coutts, D.W., Ainsworth, M.D., Piper, J.A.: IEEE J. Quantum Electron. **25** (1989) 1985.
89Dix   Dixon, G.J., Tanner, C.E., Wieman, C.E.: Opt. Lett. **14** (1989) 731.
89Ebr   Ebrahimzadeh, M., Dunn, M.H., Akerboom, F.: Opt. Lett. **14** (1989) 560.

89Fan   Fan, Y.X., Eckardt, R.C., Byer, R.L., Chen, C., Jiang, A.D.: IEEE J. Quantum Electron. **25** (1989) 1196.
89Gol   Goldberg, L., Chun, M.K.: Appl. Phys. Lett. **55** (1989) 218.
89Koz   Kozlovsky, W.J., Nabors, C.D., Eckardt, R.C., Byer, R.L.: Opt. Lett. **14** (1989) 66.
89Nab   Nabors, C.D., Eckardt, R.C., Kozlovsky, W.J., Byer, R.L.: Opt. Lett. **14** (1989) 1134.
89Pis   Piskarskas, A., Smilgevicius, V., Umbrasas, A.: Opt. Commun. **73** (1989) 322.
89Ris   Risk, W.P., Pon, R., Lenth, W.: Appl. Phys. Lett. **54** (1989) 1625.
89Sto   Stolzenberger, R.A., Hsu, C.C., Peyghambarian, N., Reid, J.J.E., Morgan, R.A.: IEEE Photon. Technol. Lett. **1** (1989) 446.
89Wu   Wu, B., Chen, N., Chen, C., Deng, D., Xu, Z.: Opt. Lett. **14** (1989) 1080.
89Zim   Zimmermann, C., Kallenbach, R., Hänsch, T.W., Sandberg, J.: Opt. Commun. **71** (1989) 229.

90Ben   Benicewicz, P.K., McGraw, D.: Opt. Lett. **15** (1990) 165.
90Bha1 Bhar, G.C., Chatterjee, U., Das, S.: Jpn. J. Appl. Phys. **29** (1990) L1127.
90Bha2 Bhar, G.C., Das, S., Datta, P.K.: Phys. Status Solidi (a) **119** (1990) K173.
90Bos   Bosenberg, W.R., Tang, C.L.: Appl. Phys. Lett. **56** (1990) 1819.
90Bur   Burdulis, S., Grigonis, R., Piskarskas, A., Sinkevicius, G., Sirutkaitis, V., Fix, A., Nolting, J., Wallenstein, R.: Opt. Commun. **74** (1990) 398.
90Hem   Hemmerich, A., McIntyre, D.H., Zimmermann, C., Hänsch, T.W.: Opt. Lett. **15** (1990) 372.
90Hua   Huang, J.Y., Zhang, J.Y., Shen, Y.R., Chen, C., Wu, B.: Appl. Phys. Lett. **57** (1990) 1961.
90Kat   Kato, K.: IEEE J. Quantum Electron. **26** (1990) 1173.
90Koz   Kozlovsky, W.J., Lenth, W., Latta, E.E., Moser, A., Bona, G.L.: Appl. Phys. Lett. **56** (1990) 2291.
90Lae1 Laenen, R., Graener, G., Laubereau, A.: Opt. Commun. **77** (1990) 226.
90Lae2 Laenen, R., Graener, G., Laubereau, A.: Opt. Lett. **15** (1990) 971.
90Lin1 Lin, J.T., Montgomery, J.L.: Opt. Commun. **75** (1990) 315.
90Lin2 Lin, J.T., Montgomery, J.L., Kato, K.: Opt. Commun. **80** (1990) 159.
90Mue   Müschenborn, H.J., Theiss, W., Demtröder, W.: Appl. Phys. B **50** (1990) 365.
90Pis   Piskarskas, A., Smil'gyavichyus, V., Umbrasas, A.: Kvantovaya Elektron. **17** (1990) 777; Sov. J. Quantum Electron. (English Transl.) **20** (1990) 701.
90Suk   Sukowski, U., Seilmeier, A.: Appl. Phys. B **50** (1990) 541.
90Wac   Wachman, E.S., Edelstein, D.C., Tang, C.L.: Opt. Lett. **15** (1990) 136.
90Wu   Wu, L.S., Looser, H., Günter, P.: Appl. Phys. Lett. **56** (1990) 2163.

91Bak   Bakker, H.J., Kennis, J.T.M., Kop, H.J., Lagendijk, A.: Opt. Commun. **86** (1991) 58.
91Bay   Bayanov, I.M., Gordienko, V.M., Djidjoev, M.S., Dyakov, V.A., Magnitskii, S.A., Pryalkin, V.I., Tarasevitch, A.P.: Proc. SPIE (Int. Soc. Opt. Eng.) **1800** (1991) 2.
91Bha   Bhar, G.C., Chatterjee, U., Das, S.: Appl. Phys. Lett. **58** (1991) 231.
91Bor1 Borsutzky, A., Brünger, R., Huang, Ch., Wallenstein, R.: Appl. Phys. B **52** (1991) 55.
91Bor2 Borsutzky, A., Brünger, R., Wallenstein, R.: Appl. Phys. B **52** (1991) 380.
91Dmi   Dmitriev, V.G., Gurzadyan, G.G., Nikogosyan, D.N.: Handbook of nonlinear optical crystals, Berlin: Springer, 1991.
91Ebr1 Ebrahimzadeh, M., Hall, G.J., Ferguson, A.I.: Opt. Lett. **16** (1991) 1744.
91Ebr2 Ebrahimzadeh, M., Robertson, G., Dunn, M.H.: Opt. Lett. **16** (1991) 767.
91Els   Elsaesser, T., Nuss, M.C.: Opt. Lett. **16** (1991) 411.
91Fix   Fix, A., Schröder, T., Wallenstein, R.: Laser Optoelektron. **3** (1991) 106.
91Ger   Gerstenberger, D.C., Tye, G.E., Wallace, R.W.: Opt. Lett. **16** (1991) 992.
91He   He, H., Lu, Y., Dong, J., Zhao, Q.: Proc. SPIE (Int. Soc. Opt. Eng.) **1409** (1991) 18.
91Hua   Huang, J.Y., Shen, Y.R., Chen, C., Wu, B.: Appl. Phys. Lett. **58** (1991) 1579.

91Joo   Joosen, W., Bakker, H.J., Noordam, L.D., Muller, H.G., Van Linden van den Heuvell, H.B.: J. Opt. Soc. Am. B **8** (1991) 2087.
91Jun   Jundt, D.H., Fejer, M.M., Byer, R.L., Norwood, R.G., Bordui, P.F.: Opt. Lett. **16** (1991) 1856.
91Kal   Kallenbach, R., Schmidt-Kaler, F., Weitz, M., Zimmermann, C., Hänsch, T.W.: Opt. Commun. **81** (1991) 63.
91Kat   Kato, K.: IEEE J. Quantum Electron. **27** (1991) 1137.
91Lae   Laenen, R., Graener, G., Laubereau, A.: J. Opt. Soc. Am. B **8** (1991) 1085.
91Lin   Lin, S., Huang, J.Y., Ling, J., Chen, C., Shen, Y.R.: Appl. Phys. Lett. **59** (1991) 2805.
91Neb   Nebel, A., Beigang, R.: Opt. Lett. **16** (1991) 1729.
91Pla   Planken, P.C.M., Snoeks, E., Noordam, L.D., Muller, H.G., Van Linden van den Heuvell, H.B.: Opt. Commun. **85** (1991) 31.
91Pol   Polzik, E.S., Kimble, H.J.: Opt. Lett. **16** (1991) 1400.
91Rob   Robertson, G., Henderson, A., Dunn, M.H.: Opt. Lett. **16** (1991) 1584.
91Skr   Skripko, G.A., Bartoshevich, S.G., Mikhnyuk, I.V., Tarazevich, I.G.: Opt. Lett. **16** (1991) 1726.
91Sug   Sugiyama, K., Yoda, J., Sakurai, T.: Opt. Lett. **16** (1991) 449.
91Suh   Suhre, D.R.: Appl. Phys. B **52** (1991) 367.
91Vod   Vodopyanov, K.L., Kulevskii, L.A., Voevodin, V.G., Gribenyukov, A.I., Allakhverdiev, K.R., Kerimov, T.A.: Opt. Commun. **83** (1991) 322.
91Wac   Wachman, E.S., Pelouch, W.S., Tang, C.L.: J. Appl. Phys. **70** (1991) 1893.
91Wan   Wang, Y., Xu, Z., Deng, D., Zheng, W., Wu, B., Chen, C.: Appl. Phys. Lett. **59** (1991) 531.
91Wat   Watanabe, M., Hayasaka, K., Imajo, H., Umezu, J., Urabe, S.: Appl. Phys. B **53** (1991) 11.
91Xie   Xie, F., Wu, B., You, G., Chen, C.: Opt. Lett. **16** (1991) 1237.
91Yan   Yang, S.T., Pohalski, C.C., Gustafson, E.K., Byer, R.L., Feigelson, R.S., Raymakers, R.J., Route, R.K.: Opt. Lett. **16** (1991) 1493.
91Yod   Yodh, A.G., Tom, H.W.K., Aumiller, G.D., Miranda, R.S.: J. Opt. Soc. Am. B **8** (1991) 1663.
91Zha   Zhang, J.Y., Huang, J.Y., Shen, Y.R., Chen, C., Wu, B.: Appl. Phys. Lett. **58** (1991) 213.

92Akh   Akhmanov, S.A., Bayanov, I.M., Gordienko, V.M., Dyakov, V.A., Magnitskii, S.A., Pryalkin, V.I., Tarasevich, A.P: Inst. Phys. Conf. Ser. Sect. I **126** (1992) 67.
92Ant   Anthon, D.W., Sipes, D.L., Pier, T.J., Ressl, M.R.: IEEE J. Quantum Electron. **28** (1992) 1148.
92Bha   Bhar, G.C., Datta, P.K., Chatterjee, U.: J. Phys. D **25** (1992) 1042.
92Bos   Bosenberg, W.R., Guyer, D.R.: Appl. Phys. Lett. **61** (1992) 387.
92Bro   Brown, A.J.W., Bowers, M.S., Kangas, K.W., Fisher, C.H.: Opt. Lett. **17** (1992) 109.
92Can   Canarelli, P., Benko, Z., Curl, R., Tittel, F.K.: J. Opt. Soc. Am. B **9** (1992) 197.
92Cui   Cui, Y., Dunn, M.H., Norrie, C.J., Sibbett, W., Sinclair, B.D., Tang, Y., Terry, J.A.C.: Opt. Lett. **17** (1992) 646.
92Dub   Dubietis, A., Jonusauskas, G., Piskarskas, A.: Opt. Commun. **88** (1992) 437.
92Ebr1   Ebrahimzadeh, M., Hall, G.J., Ferguson, A.I.: Appl. Phys. Lett. **60** (1992) 1421.
92Ebr2   Ebrahimzadeh, M., Hall, G.J., Ferguson, A.I.: Opt. Lett. **17** (1992) 652.
92Ell   Ellingson, R.J., Tang, C.L.: Opt. Lett. **17** (1992) 343.
92Fu   Fu, Q., Mak, G., Van Driel, H.M.: Opt. Lett. **17** (1992) 1006.
92Hei   Heitmann, U., Kötteritzsch, M., Heitz, S., Hese, A.: Appl. Phys. B **55** (1992) 419.
92Hie   Hielscher, A.H., Miller, C.E., Bayard, D.C., Simon, U., Smolka, K.P., Curl, R.F., Tittel, F.K.: J. Opt. Soc. Am. B **9** (1992) 1962.
92Hof   Hofmann, Th., Mossavi, K., Tittel, F.K., Szabo G.: Opt. Lett. **17** (1992) 1691.

92Hua	Huang, F., Huang, L.: Appl. Phys. Lett. **61** (1992) 1769.
92Jan	Jani, M.G., Murray, J.T., Petrin, R.R., Powell, R.C., Loiacono, D.N., Loiacono, G.M.: Appl. Phys. Lett. **60** (1992) 2327.
92Jos	Josse, D., Dou, S.X., Zyss, J., Andreazza, P., Perigaud, A.: Appl. Phys. Lett. **61** (1992) 121.
92Kat	Kato, K., Masutani, M.: Opt. Lett. **17** (1992) 178.
92Kea	Kean, P.N., Dixon, G.J.: Opt. Lett. **17** (1992) 127.
92Kra	Krause, H-J., Daum, W.: Appl. Phys. Lett. **60** (1992) 2180.
92Mak	Mak, G., Fu, Q., Van Driel, H.M.: Appl. Phys. Lett. **60** (1992) 542.
92Mal	Malcolm, G.P.A., Ebrahimzadeh, M., Ferguson, A.I.: IEEE J. Quantum Electron. **28** (1992) 1172.
92Mar	Marshall, L.R., Hays, A.D., Kaz, A., Burnham, R.L.: IEEE J. Quantum Electron. **28** (1992) 1158.
92McC	McCarthy, M.J., Hanna, D.C.: Opt. Lett. **17** (1992) 402.
92Neb	Nebel, A., Beigang, R.: Opt. Commun. **94** (1992) 369.
92Ou	Ou, Z.Y., Pereira, S.F., Polzik, E.S., Kimble, H.J.: Opt. Lett. **17** (1992) 640.
92Pel	Pelouch, W.S., Powers, P.E., Tang, C.L.: Opt. Lett. **17** (1992) 1070.
92Rin	Ringling, J., Kittelmann, O., Noack, F.: Opt. Lett. **17** (1992) 1794.
92Ris	Risk, W.P., Kozlovsky, W.J.: Opt. Lett. **17** (1992) 707.
92Rob	Robertson, G., Henderson, A., Dunn, M.H.: Appl. Phys. Lett. **60** (1992) 271.
92See	Seelert, W., Kortz, P., Rytz, D., Zysset, B., Ellgehausen, D., Mizell, G.: Opt. Lett. **17** (1992) 1432.
92Tai	Taira, Y.: Jpn. J. Appl. Phys. **31** (1992) L682.
92Tan	Tang, Y., Cui, Y., Dunn, M.H.: Opt. Lett. **17** (1992) 192.
92Tom	Tomov, I.V., Anderson, T., Rentzepis, P.M.: Appl. Phys. Lett. **61** (1992) 1157.
92Wat	Watanabe, M., Hayasaka, K., Imajo, H., Urabe, S.: Opt. Lett. **17** (1992) 46.
92Wu	Wu, B., Xie, F., Chen, C., Deng, D., Xu, Z.: Opt. Commun. **88** (1992) 451.
92Zhu	Zhu, X.D., Deng, L.: Appl. Phys. Lett. **61** (1992) 1490.
92Zim	Zimmermann, C., Hänsch, T.W., Byer, R., O'Brien, S., Welch, D.: Appl. Phys. Lett. **61** (1992) 2741.
93Agn	Agnesi, A., Reali, G.C., Kubecek, V., Kumazaki, S., Takagi, Y., Yoshihara, K.: J. Opt. Soc. Am. B **10** (1993) 2211.
93Ash	Ashworth, S., Iaconis, C., Votava, O., Riedle, E.: Opt. Commun. **97** (1993) 109.
93Ban	Banfi, G.P., Danielius, R., Piskarskas, A., Di Trapani, P., Foggi, P., Righini, R.: Opt. Lett. **18** (1993) 1633.
93Bar1	de Barros, M.R.X., Becker, P.C.: Opt. Lett. **18** (1993) 631.
93Bar2	Barykin, A.A., Davydov, S.V., Dorokhov, V.P., Zakharov, V.P., Butuzov, V.V.: Kvantovaya Elektron. **20** (1993) 794; Quantum Electron. (English Transl.) **23** (1993) 688.
93Bha	Bhar, G.C., Datta, P.K., Rudra, A.M.: Appl. Phys. B **57** (1993) 431.
93Bol	Bolt, R.J., Van der Mooren, M.: Opt. Commun. **100** (1993) 399.
93Bos	Bosenberg, W.R., Guyer, D.R.: J. Opt. Soc. Am. B **10** (1993) 1716.
93Bou	Bourzeix, S., Plimmer, M.D., Nez, F., Julien, L., Biraben, F.: Opt. Commun. **99** (1993) 89.
93Bud	Budni, P.A., Knights, M.G., Chicklis, E.P., Schepler, K.L.: Opt. Lett. **18** (1993) 1068.
93But	Butterworth, S.D., McCarthy, M.J., Hanna, D.C.: Opt. Lett. **18** (1993) 1429.
93Chu	Chung, J., Siegman, A.E.: J. Opt. Soc. Am. B **10** (1993) 2201.
93Col1	Colville, F.G., Henderson, A.J., Padgett, M.J., Zhang, J., Dunn, M.H.: Opt. Lett. **18** (1993) 205.
93Col2	Colville, F.G., Padgett, M.J., Henderson, A.J., Zhang, J., Dunn, M.H.: Opt. Lett. **18** (1993) 1065.

93Cui   Cui, Y., Withers, D.E., Rae, C.F., Norrie, C. J., Tang, Y., Sinclair, B.D., Sibbett, W., Dunn, M.H.: Opt. Lett. **18** (1993) 122.

93Dah   Dahinten, T., Plödereder, U., Seilmeier, A., Vodopyanov, K.L., Allakhverdiev, K.R., Ibragimov, Z.A.: IEEE J. Quantum Electron. **29** (1993) 2245.

93Dan   Danielius, R., Piskarskas, A., Stabinis, A., Banfi, G.P., Di Trapani, P., Righini, R.: J. Opt. Soc. Am. B **10** (1993) 2222.

93Dim   Dimov, S.S.: Opt. Quantum Electron. **25** (1993) 545.

93Dou   Dou, S.X., Josse, D., Zyss, J.: J. Opt. Soc. Am. B **10** (1993) 1708.

93Ebr   Ebrahimzadeh, M., Hall, G.J., Ferguson, A.I.: Opt. Lett. **18** (1993) 278.

93Fie   Fiedler, K., Schiller, S., Paschotta, R., Kürz, P., Mlynek, J.: Opt. Lett. **18** (1993) 1786.

93Fix   Fix, A., Schröder, T., Wallenstein, R., Haub, J.G., Johnson, M.J., Orr, B.J.: J. Opt. Soc. Am. B **10** (1993) 1744.

93Ger   Gerstenberger, D.C., Wallace, R.W.: J. Opt. Soc.Am. B **10** (1993) 1681.

93Gol   Goldberg, L., Busse, L.E., Mehuys, D.: Appl. Phys. Lett. **63** (1993) 2327.

93Gra   Grässer, Ch., Wang, D., Beigang, R., Wallenstein, R.: J. Opt. Soc. Am. B **10** (1993) 2218.

93Hal1   Hall, G.J., Ebrahimzadeh, M., Robertson, A., Malcolm, G.P.A., Ferguson, A.I.: J. Opt. Soc. Am. B **10** (1993) 2168.

93Hal2   Hall, G.J., Ferguson, A.I.: Opt. Lett. **18** (1993) 1511.

93Ham   Hamm, P., Lauterwasser, C., Zinth, W.: Opt. Lett. **18** (1993) 1943.

93Hua   Huang, F., Huang, L., Yin, B.-I., Hua, Y.-N.: Appl. Phys. Lett. **62** (1993) 672.

93Kea   Kean, P.N., Standley, R.W., Dixon, G.J.: Appl. Phys. Lett. **63** (1993) 302.

93Kra   Krause, H-J., Daum, W.: Appl. Phys. B **56** (1993) 8.

93Lae   Laenen, R., Wolfrum, K., Seilmeier, A., Laubereau, A.: J. Opt. Soc. Am. B **10** (1993) 2151.

93Lee   Lee, D., Wong, N.C.: J. Opt. Soc. Am. B **10** (1993) 1659.

93Lot   Lotshaw, W.T., Unternahrer, J.R., Kukla, M.J., Miyake, C.I., Braun, F.D.: J. Opt. Soc. Am. B **10** (1993) 2191.

93Mar1   Marshall, L.R., Kaz, A.: J. Opt. Soc. Am. B **10** (1993) 1730.

93Mar2   Marshall, L.R., Kaz, A., Aytur, O.: Opt. Lett. **18** (1993) 817.

93McC1   McCarthy, M.J., Butterworth, S.D., Hanna, D.C.: Opt. Commun. **102** (1993) 297.

93McC2   McCarthy, M.J., Hanna, D.C.: J. Opt. Soc. Am. B **10** (1993) 2180.

93Neb   Nebel, A., Fallnich, C., Beigang, R., Wallenstein, R.: J. Opt. Soc. Am. B **10** (1993) 2195.

93Poi   Poirier, P., Hanson, F.: Opt. Lett. **18** (1993) 1925.

93Pow1   Powers, P.E., Ellingson, R.J., Pelouch, W.S., Tang, C.L.: J. Opt. Soc. Am. B **10** (1993) 2162.

93Pow2   Powers, P.E., Ramakrishna, S., Tang, C.L., Cheng, L.K.: Opt. Lett. **18** (1993) 1171.

93Rin   Ringling, J., Kittelmann, O., Noack, F., Korn, G., Squier, J.: Opt. Lett. **18** (1993) 2035.

93Rob   Robertson, G., Henderson, A., Dunn, M.H.: Appl. Phys. Lett. **62** (1993) 123.

93Sch   Schiller, S., Byer, R.L.: J. Opt. Soc. Am. B **10** (1993) 1696.

93Sei   Seifert, F., Petrov, V.: Opt. Commun. **99** (1993) 413.

93Sim1   Simon, U., Miller, C.E., Bradley, C.C., Hulet, R.G., Curl, R.F., Tittel, F.K.: Opt. Lett. **18** (1993) 1062.

93Sim2   Simon, U., Tittel, F.K., Goldberg, L.: Opt. Lett. **18** (1993) 1931.

93Vod1   Vodopyanov, K.L.: J. Opt. Soc. Am. B **10** (1993) 1723.

93Vod2   Vodopyanov, K.L., Andreev, Yu.A., Bhar, G.C.: Kvantovaya Elektron. **20** (1993) 879; Quantum Electron. (English Transl.) **23** (1993) 763.

93Wan1   Wang, W., Outsu, M.: Opt. Commun. **102** (1993) 304.

93Wan2   Wang, W., Ohtsu, M.: Opt. Lett. **18** (1993) 876.

93Wu   Wu, R.: Appl. Opt. **32** (1993) 971.

93Yan1   Yang, S.T., Eckardt, R.C., Byer, R.L.: J. Opt. Soc. Am. B **10** (1993) 1684.

93Yan2	Yang, S.T., Eckardt, R.C., Byer, R.L.: Opt. Lett. **18** (1993) 971.
93Zha1	Zhang, J.Y., Huang, J.Y., Shen, Y.R., Chen, C.: J. Opt. Soc. Am. B **10** (1993) 1758.
93Zha2	Zhang, N., Yuan, D.R., Tao, X.T., Xu, D., Shao, Z.S., Jiang, M.H., Liu, M.G.: Opt. Commun. **99** (1993) 247.
93Zho	Zhou, H., Zhang, J., Chen, T., Chen, C., Shen, Y.R.: Appl. Phys. Lett. **62** (1993) 1457.
94Bac	Backus, S., Asaki, M.T., Shi, C., Kapteyn, H.C., Murnane, M.M.: Opt. Lett. **19** (1994) 399.
94Che	Cheung, E.C., Koch, K., Moore, G.T.: Opt. Lett. **19** (1994) 631.
94Col1	Colville, F.G., Ebrahimzadeh, M., Sibbett, W., Dunn, M.H.: Appl. Phys. Lett. **64** (1994) 1765.
94Col2	Colville, F.G., Padgett, M.J., Dunn, M.H.: Appl. Phys. Lett. **64** (1994) 1490.
94Dri	Driscoll, T.J., Gale, G.M., Hache, F.: Opt. Commun. **110** (1994) 638.
94Glo	Gloster, L.A.W., Jiang, Z.X., King, T.A.: IEEE J. Quantum Electron. **30** (1994) 2961.
94Han	Hanson, F., Poirier, P.: Opt. Lett. **19** (1994) 1526.
94Hua	Huang, F., Huang, L.: IEEE J. Quantum Electron.. **30** (1994) 2601.
94Lin	Lincoln, J.R., Ferguson, A.I.: Opt. Lett. **19** (1994) 1213.
94Loh	Lohner, A., Kruck, P., Rühle, W.W.: Appl. Phys. B **59** (1994) 211.
94Mas	Mason, P.D., Jackson, D.J., Gorton, E.K.: Opt. Commun. **110** (1994) 163. Erratum: **114** (1995) 529.
94Nis	Nisoli, M., De Silvestri, S., Magni, V., Svelto, O., Danielius, R., Piskarskas, A., Valiulis, G., Varanavicius, A.: Opt. Lett. **19** (1994) 1973.
94Pas	Paschotta, R., Kürz, P., Henking, R., Schiller, S., Mlynek, J.: Opt. Lett. **19** (1994) 1325.
94Pet	Petrov, V., Seifert, F.: Opt. Lett. **19** (1994) 40.
94Pow	Powers, P.E., Tang, C.L., Cheng, L.K.: Opt. Lett. **19** (1994) 1439.
94Rob1	Robertson, A., Ferguson, A.I.: Opt. Lett. **19** (1994) 117.
94Rob2	Robertson, G., Padgett, M.J., Dunn, M.H.: Opt. Lett. **19** (1994) 1735.
94Sch	Schröder, T., Boller, K.J., Fix, A., Wallenstein, R.: Appl. Phys. B **58** (1994) 425.
94Sei1	Seifert, F., Petrov, V., Noack, F.: Opt. Lett. **19** (1994) 837.
94Sei2	Seifert, F., Petrov, V., Woerner, M.: Opt. Lett. **19** (1994) 2009.
94Sei3	Seifert, F., Ringling, J., Noack, F., Petrov, V., Kittelmann, O.: Opt. Lett. **19** (1994) 1538.
94Tai	Taira, T., Kobayashi, T.: IEEE J. Quantum Electron. **30** (1994) 800.
94Ter	Terry, J.A.C., Cui, Y., Yang, Y., Sibbett, W., Dunn, M.H.: J. Opt. Soc. Am. B **11** (1994) 758.
94Umb	Umbrasas, A., Diels, J.-C., Jacob, J., Piskarskas, A.: Opt. Lett. **19** (1994) 1753.
94Wat	Watanabe, M., Ohmukai, R., Hayasaka, K., Imajo, H., Urabe, S.: Opt. Lett. **19** (1994) 637.
94Won	Wong, S.K., Oliver, R., Schepler, K.L., Fenimore, D.L.: Opt. Lett. **19** (1994) 1433.
94Yan	Yang, S.T., Eckardt, R.C., Byer, R.L.: Opt. Lett. **19** (1994) 475.
94Zen	Zenzie, H.H., Moulton, P.F.: Opt. Lett. **19** (1994) 963.
95Ash	Ashworth, S.H., Joschko, M., Woerner, M., Riedle, E., Elsaesser, T.: Opt. Lett. **20** (1995) 2120.
95Bha	Bhar, G.C., Das, S., Vodopyanov, K.L.: Appl. Phys. B **61** (1995) 187–190.
95Bre	Breitenbach, G., Schiller, S., Mlynek, J.: J. Opt. Soc. Am. B **12** (1995) 2095.
95But	Butterworth, S.D., Girard, S., Hanna, D.C.: J. Opt. Soc. Am. B **12** (1995) 2158.
95Can	Canto-Said, E.J., McCann, M.P., Wigley, P.C., Dixon, G.J.: Opt. Lett. **20** (1995) 1268.
95Cha1	Chatterjee, U., Rudra, A.M., Bhar, G.C.: Appl. Phys. B **61** (1995) 489–491.
95Cha2	Chatterjee, U., Rudra, A.M., Bhar, G.C.: Opt. Commun. **118** (1995) 367.
95Cha3	Chatterjee, U., Rudra, A.M., Bhar, G.C., Sasaki, T.: Kvantovaya Elektron. **22** (1995) 271; Quantum Electron. (English Transl.) **25** (1995) 255.

95Che   Chen, L.P., Wang, Y., Liu, J.M.: J. Opt. Soc. Am. B **12** (1995) 2192.
95Dhi   Dhirani, A., Guyot-Sionnest, P.: Opt. Lett. **20** (1995) 1104.
95DiT   Di Trapani, P., Andreoni, A., Podenas, D., Danielius, R., Piskarskas, A.: Opt. Commun. **118** (1995) 338.
95Ebr1   Ebrahimzadeh, M., French, S., Miller, A.: J. Opt. Soc. Am. B **12** (1995) 2180.
95Ebr2   Ebrahimzadeh, M., French, S., Sibbett, W., Miller, A.: Appl. Phys. B **60** (1995) 443.
95Ebr3   Ebrahimzadeh, M., French, S., Sibbett, W., Miller, A.: Opt. Lett. **20** (1995) 166.
95Gal1   Gale, G.M., Cavallari, M., Driscoll, T.J., Hache, F.: Opt. Commun. **119** (1995) 159.
95Gal2   Gale, G.M., Cavallari, M., Driscoll, T.J., Hache, F.: Opt. Lett. **20** (1995) 1562.
95Gol   Goldberg, L., Kliner, D.A.V.: Opt. Lett. **20** (1995) 1640.
95Gra   Gragson, D.E., Alavi, D.S., Richmond, G.L.: Opt. Lett. **20** (1995) 1991.
95Hol   Holtom, G.R., Crowell, R.A., Xie, X.S.: J. Opt. Soc. Am. B **12** (1995) 1723.
95Hui   Huisken, F., Kaloudis, M., Marquez, J., Chuzavkov, Yu.L., Orlov, S.N., Polivanov, Yu.N., Smirnov, V.V.: Opt. Lett. **20** (1995) 2306.
95Joh   Johnson, B.C., Newell, V.J., Clark, J.B., McPhee, E.S.: J. Opt. Soc. Am. B **12** (1995) 2122.
95Kaf   Kafka, J.D., Watts, M.L., Pieterse, J.W.: J. Opt. Soc. Am. B **12** (1995) 2147.
95Kat   Kato, K.: IEEE J. Quantum Electron. **31** (1995) 169.
95Kho   Khodja, S., Josse, D., Samuel, I.D.W., Zyss, J.: Appl. Phys. Lett. **67** (1995) 3841.
95Kry   Krylov, V., Rebane, A., Kalintsev, A.G., Schwoerer, H., Wild, U.P.: Opt. Lett. **20** (1995) 198.
95Kun   Kung, A.H.: Opt. Lett. **20** (1995) 1107.
95Liu   Liu, X., Xu, Z., Wu, B., Chen, C.: Appl. Phys. Lett. **66** (1995) 1446.
95McC   McCahon, S.W., Anson, S.A., Jang, D.-J., Boggess, T.F.: Opt. Lett. **20** (1995) 2309.
95Mor1   Mori, Y., Kuroda, I., Nakajima, S., Sasaki, T., Nakai, S.: Appl. Phys. Lett. **67** (1995) 1818.
95Mor2   Morrison, G.R., Ebrahimzadeh, M., Rae, C.F., Dunn, M.H.: Opt. Commun. **118** (1995) 55.
95Nie   Nielsen, J.S.: Opt. Lett. **20** (1995) 840.
95Qiu   Qiu, T., Tillert, T., Maier, M.: Opt. Commun. **119** (1995) 149.
95Rau   Rauscher, C., Roth, T., Laenen, R., Laubereau, A.: Opt. Lett. **20** (1995) 2003.
95Rei   Reid, D.T., Ebrahimzadeh, M., Sibbett, W.: Opt. Lett. **20** (1995) 55.
95Sch   Scheidt, M., Beier, B., Knappe, R., Boller, K.-J., Wallenstein, R.: J. Opt. Soc. Am. B **12** (1995) 2087.
95Spe   Spence, D.E., Wielandy, S., Tang, C.L., Bosshard, C., Günter, P.: Opt. Lett. **20** (1995) 680.
95Sri   Srinivasan, N., Kiriyama, H., Kimura, T., Ohmi, M., Yamanaka, M., Izawa, Y., Nakai, S., Yamanaka, C.: Opt. Lett. **20** (1995) 1265.
95Vod1   Vodopyanov, K.L., Voevodin, V.G.: Opt. Commun. **114** (1995) 333.
95Vod2   Vodopyanov, K.L., Voevodin, V.G.: Opt. Commun. **117** (1995) 277.
95Wal   Walker, D.R., Flood, C.J., Van Driel, H.M.: Opt. Lett. **20** (1995) 145.
95Zar   Zarrabi, J.H., Gavrilovic, P., Singh, S.: Appl. Phys. Lett. **67** (1995) 2439.
95Zho   Zhou, W.-L., Mori, Y., Nakai, S., Nakano, K., Niikura, S., Craig, B.: Appl. Phys. Lett. **66** (1995) 2463.
95Zim   Zimmermann, C., Vuletic, V., Hemmerich, A., Hansch, T.W.: Appl. Phys. Lett. **66** (1995) 2318.

96Aka   Akagawa, K., Wada, S., Nakamura, A., Tashiro, H.: Appl. Opt. **35** (1996) 2570.
96Apo   Apollonov, V.V., Gribenyukov, A.I., Korotkova, V.V., Suzdaltsev, A.G., Shakir, Yu.A.: Quantum Electron. **26** (1996) 469.
96Bha   Bhar, G.C., Rudra, A.M., Chaudhary, A.K., Sasaki, T., Mori, Y.: Appl. Phys. B **63** (1996) 141–144.

96Fre   French, S., Ebrahimzadeh, M., Miller, A.: Opt. Commun. **128** (1996) 166.
96Lin   Lin, S., Suzuki, T.: Opt. Lett.: **21** (1996) 579.
96Mom   Momose, T., Wakabayashi, T., Shida, T.: J. Opt. Soc. Am. B **13** (1996) 1706.
96Qia   Qian, L., Benjamin, S.D., Smith, P.W.E.: Opt. Commun. **127** (1996) 73.
96Sch   Schade, W., Blanke, T., Willer, U., Rempel, C.: Appl. Phys. B **63** (1996) 99.
96Sha   Shama, L.B., Daido, H., Kato, Y., Nakai, S., Zhang, T., Mori, Y., Sasaki, T.: Appl. Phys. Lett. **69** (1996) 3812.
96Spe   Spence, D.E., Wielandy, S., Tang, C.L., Bosshard, C., Günter, P.: Appl. Phys. Lett. **68** (1996) 452.
96Suh   Suhre, D.R., Taylor, L.H.: Appl. Phys. B **63** (1996) 225–228.
96Sut   Sutherland, R.L.: Handbook of nonlinear optics, New York: Marcel Dekker, 1996.
96Ume   Umemura, N., Kato, K.: Appl. Opt. **35** (1996) 5332.
96Yap   Yap, Y.K., Inagaki, M., Nakajima, S., Mori, Y., Sasaki, T.: Opt. Lett. **21** (1996) 1348.
96Zho   Zhou, W.-L., Mori, Y., Sasaki, T., Nakai, S.: Opt. Commun. **123** (1996) 583.

97Ber   Berkeland, D.J., Cruz, F.C., Bergquist, J.C.: Appl. Opt. **36** (1997) 4159.
97Cer   Cerullo, G., Nisoli, M., de Silvestri, S.: Appl. Phys. Lett. **71** (1997) 3616.
97Cha   Chandra, S., Allik, T.H., Catella, G., Utano, R., Hutchinson, J.A.: Appl. Phys. Lett. **71** (1997) 584.
97Coo   Cook, G.: Appl. Opt. **36** (1997) 2511.
97Kat   Kato, K.: Appl. Opt. **36** (1997) 2506.
97Kar   Kartaloglu, T., Koprulu, K.G., Aytur, O.: Opt. Lett. **22** (1997) 280.
97Kel   Kellner, T., Heine, F., Huber, G.: Appl. Phys. B **65** (1997) 789.
97Kom   Komatsu, R., Sugawara, T., Sassa, K., Sarukura, N., Liu, Z., Izumida, S., Segawa, Y., Uda, S., Fukuda, T., Yamanouchi, K.: Appl. Phys. Lett. **70** (1997) 3492.
97Lod   Lodahl, P., Sorensen, J.L., Polzik, E.K.: Appl. Phys. B **64** (1997) 383.
97Oie   Oien, A.L., McKinnie, I.T., Jain, P., Russel, N.A., Warrington, D.M.: Opt. Lett. **22** (1997) 859.
97Raf   Raffy, J., Debuisschert, T., Pocholle, J.P.: Opt. Lett. **22** (1997) 1589.
97Rei   Reid, D.T., McGowan, C., Sleat, W., Ebrahimzadeh, M., Sibbett, W.: Opt. Lett. **22** (1997) 525.
97Sch   Scheidt, M., Beier, B., Boller, K.J., Wallenstein, R.: Opt. Lett. **22** (1997) 1287.
97Sri   Srinivasan-Rao, T., Babzien, M., Sakai, F., Mori, Y., Sasaki, T.: Appl. Phys. Lett. **71** (1997) 1927.
97Sto   Stoll, K., Zondy, J.J., Acef, O.: Opt. Lett. **22** (1997) 1302.
97Tan   Tang, Y., Rae, C.F., Rahlff, C., Dunn, M.H.: J. Opt. Soc. Am. B **14** (1997) 3442.
97Ume   Umemura, N., Kato, K.: Appl. Opt. **36** (1997) 6794.
97Wan   Wang, T., Dunn, M.H., Rae, C.F.: Opt. Lett. **22** (1997) 763.
97Wu   Wu, Y.C., Fu, P.Z., Wang, J.X., Xu, Z., Zhang, L., Kong, Y., Cheng, C.: Opt. Lett. **22** (1997) 1840.

98Abe   Abedin, K.S., Haidar, S., Konno, Y., Takyu, C., Ito, H.: Appl. Opt: **37** (1998) 1642.
98Cho   Chou, H.P., Slater, R.C., Wang, Y.: Appl. Phys. B **66** (1998) 555.
98Dou   Douillet, A., Zondy, J.T.: Opt. Lett. **23** (1998) 1259.
98Edw   Edwards, T.J., Turnbull, G.A., Dunn, M.H., Ebrahimzadeh, M., Colville, F.G.: Appl. Phys. Lett. **72** (1998) 1527.
98Ehr   Ehret, S., Schneider, H.: Appl. Phys. B **66** (1998) 27.
98Gol   Golubovic, B., Reed, M.K.: Opt. Lett. **23** (1998) 1760.
98Lae   Laenen, R., Simeonidis, K., Laubereau, A.: J. Opt. Soc. Am. B **15** (1998) 1213.
98Mat   Matsubara, K., Tanaka, U., Imajo, H., Hayasaka, K., Ohmukai, R., Watanabe, M., Urabe, S.: Appl. Phys. B **67** (1998) 1.

98Pet1   Petrov, V., Rempel, C., Stolberg, K.P., Schade, W.: Appl. Opt: **37** (1998) 4925.
98Pet2   Petrov, V., Rotermund, F., Noack, F.: Electron. Lett. **34** (1998) 1748.
98Pet3   Petrov, V., Rotermund, F., Noack, F., Komatsu, R., Sugawara, T., Uda, D.: J. Appl. Phys. **84** (1998) 5887.
98Rot   Rotermund, F., Petrov, V.: Opt. Lett. **23** (1998) 1040.
98Ruf   Ruffing, B., Nebel, A., Wallenstein, R.: Appl. Phys. B **67** (1998) 537.
98Shi   Shirakawa, A., Kobayashi, T.: Appl. Phys. Lett. **72** (1998) 147.
98Ste   Steinbach, D., Hügel, W., Wegener, M.: J. Opt. Soc. Am. B **15** (1998) 1231.
98Tsu   Tsunekane, M., Kimura, S., Kimura, M., Taguchi, N., Inaba, H.: Appl. Phys. Lett. **72** (1998) 3414.
98Yap   Yap, I.K., Haramura, S., Taguchi, A., Mori, Y., Sasaki, T.: Opt. Commun. **145** (1998) 101.
98Zha   Zhang, J.Y., Huang, J.Y., Wang, H., Wong, K.S., Wong, G.K.: J. Opt. Soc. Am. B **15** (1998) 200.

99Dmi   Dmitriev, V.G., Gurzadyan, G.G., Nikogosyan, D.N.: Handbook of nonlinear optical crystals, third, revised edition, Berlin: Springer, 1999.
99Hai   Haidar, S., Nakamura, K., Niwa, E., Masumoto, K., Ito, H.: Appl. Opt. **38** (1999) 1798.
99Kou   Kouta, H., Kuwano, Y.: Opt. Lett. **24** (1999) 1230.

00Bha   Bhar, G.C., Kumbhakar, P., Chatterjee, U., Rudra, A.M., Nagahori, A.: Opt. Commun. **176** (2000) 199.
00Bye   Byer, R.L.: IEEE J. Sel. Top. Quantum Electron. **6** (2000) 911.
00Cha   Charra, F, Gurzadyan, G.G.: Nonlinear dielectric susceptibilities, Landolt-Börnstein, New series III, Vol. 30B, Berlin, Heidelberg: Springer, 2000, 486 pp. (ISBN 3-540-65567-0).
00Kag   Kagebayashi, Y., Deki, K., Morimoto, Y., Miyazawa, S., Sasaki, T.: Jpn. J. Appl. Phys. **39** (2000) L1224.
00Kon   Kondratyuk, N.V., Shagov, A.A., Demidchik, K.L., Yurkin, A.M., Kokh, A.E.: Kvantovaya Elektron. **30** (2000) 253; Quantum Electron. (English Transl.) **30** (2000) 253.
00Sas   Sasaki, T., Mori, Y., Yoshimura, M., Yap, Y.K., Kamimura, T.: Mater. Sci. Eng. **30** (2000) 1.

01Koj   Kojima, T., Konno, S., Fujikawa, S., Yasui, K.: Electr. Eng. Jpn. **137** (2001) 18.

4.2 Frequency conversion in gases and liquids

C.R. VIDAL

4.2.1 Fundamentals of nonlinear optics in gases and liquids

This chapter covers the properties of a nonlinear medium having spherical symmetry like gases and liquids. They therefore clearly differ from the properties of most solids (see Chap. 4.1).

Lasers have become so powerful these days that one can easily generate various kinds of optical overtones

$$\omega_\text{o} = \sum_{i,q} n_i \cdot \omega_i \pm k_q \cdot \omega_{\text{res},q} > 0 \,, \tag{4.2.1}$$

where n_i and k_q are some integer (including $n_i = 0$ or $k_q = 0$), using a suitable nonlinear medium with eigenfrequencies $\omega_{\text{res},q}$ and an incident laser frequency ω_i (conservation of energy).

In case of frequency conversion in gases one generally has $k_q = 0$ and deals with sum or difference frequency mixing

$$\omega_\text{s} = \omega_i \pm \sum_j \omega_j > 0 \tag{4.2.2}$$

which may be enhanced by exploiting suitable resonances of the atomic or molecular gas.

In case of stimulated scattering one generally has $n_i = 1$. Then $\omega_{\text{res},q}$ is a suitable manifold of atomic or molecular (rotational or vibrational) resonances of the gaseous or liquid scattering medium numbered by the index q. Like in classical spectroscopy the plus sign stands for Stokes processes, whereas the minus sign is responsible for Anti-Stokes processes.

4.2.1.1 Linear and nonlinear susceptibilities

Linear and nonlinear susceptibilities are discussed in [87Vid].

The complex linear susceptibility is given by

$$\chi^{(1)} = \bar{\chi}^{(1)} + i\,\tilde{\chi}^{(1)} = \frac{1}{\hbar} \sum_a \frac{|\mu_{ag}|^2}{(\Omega_{ag} - \omega)} \tag{4.2.3}$$

with the complex transition frequency

$$\Omega_{ag} = \omega_{ag} - i\,\Gamma_{ag} \tag{4.2.4}$$

and the dipole moment matrix elements μ_{ag} between the states $|a\rangle$ and $|g\rangle$. The nonlinear polarization is

$$\boldsymbol{P}^{\text{NL}} = \sum_{n=2}^{\infty} \boldsymbol{P}^{(n)} \,. \tag{4.2.5}$$

The definition of the electric field amplitude is given by

$$\boldsymbol{E}(r,t) = \frac{1}{2} \sum_j e_j \hat{E}(r,\omega_j) \exp(\mathrm{i}\, k_j r - \mathrm{i}\,\omega_j t) \;+\; c.c.\,, \tag{4.2.6}$$

resulting in the definition of the total polarization

$$\boldsymbol{P}(r,t) = \frac{1}{2} \sum_j e_j P(r,\omega_j) \exp(-\mathrm{i}\,\omega_j t) \;+\; c.c.\,, \tag{4.2.7}$$

where the nth-order polarization is given by

$$\boldsymbol{P}^{(n)}_{\alpha_\mathrm{s}}(r,\omega_\mathrm{s}) = \frac{n!N}{2^{n-1}} \epsilon_0 \sum_{\alpha_1 \ldots \alpha_n} \chi^{(n)}_{\alpha_\mathrm{s}\alpha_1\ldots\alpha_n}(-\omega_\mathrm{s};\omega_1\ldots\omega_n) \, \boldsymbol{E}_{\alpha_1}(r,\omega_1)\ldots\boldsymbol{E}_{\alpha_n}(r,\omega_n)\,. \tag{4.2.8}$$

The α_s are the unit vectors of the spatial coordinates, which may be cartesian, cylindrical, or spherical. The polarization can be expressed in terms of the density matrix [71Han]

$$\langle \boldsymbol{P}(t) \rangle = N\,\mathrm{Tr}\,[\boldsymbol{\rho}(t)\,\boldsymbol{\mu}] = N \sum_{mn} \rho_{mn}(t)\,\mu_{mn}\,, \tag{4.2.9}$$

whose elements are given by $\mathrm{i}\hbar \dot{\rho}_{mn} = [H,\rho]_{mn}$, where the Hamiltonian $H = H^0 + H'$ contains $H' = -\boldsymbol{\mu}\boldsymbol{E}(t)$. From a perturbation approach one obtains

$$\chi^{(n)}_{\alpha_1,\alpha_2\ldots\alpha_n}(-\omega_\mathrm{s};\omega_1\ldots\omega_n) =$$
$$\frac{1}{n!\,\hbar^n} \sum_{gb_1\ldots b_n} \rho(g)\, \frac{\langle g|e_\mathrm{s}\boldsymbol{\mu}|b_1\rangle\langle b_1|e_1\boldsymbol{\mu}|b_2\rangle\ldots\langle b_n|e_n\boldsymbol{\mu}|g\rangle}{(\Omega_{b_1g}-\omega_1-\cdots-\omega_n)(\Omega_{b_2g}-\omega_2-\cdots-\omega_n)\ldots\ldots(\Omega_{b_ng}-\omega_n)}\,. \tag{4.2.10}$$

4.2.1.2 Third-order nonlinear susceptibilities

These processes are responsible for the lowest-order frequency conversion in gases such as sum or difference frequency mixing, stimulated scattering processes and photorefraction. For the *degenerate* case the dominant terms in a system of spherical symmetry are [71Han]:

$$\chi^{(3)}(-3\omega;\omega,\omega,\omega) = \chi^{(3)}_\mathrm{T}(3\omega) = \hbar^{-3} \sum_{abc} \frac{\langle g|e_\mathrm{s}\boldsymbol{\mu}|a\rangle\langle a|e_1\boldsymbol{\mu}|b\rangle\langle b|e_2\boldsymbol{\mu}|c\rangle\langle c|e_3\boldsymbol{\mu}|g\rangle}{(\Omega_{ag}-\omega)(\Omega_{bg}-2\omega)(\Omega_{cg}-3\omega)}\,, \tag{4.2.11}$$

where the index T stands for the third harmonic generation.

For the *nondegenerate* case we have the general third-order nonlinear susceptibility [62Arm, 71Han]

$$\chi^{(3)}_{\alpha_1,\alpha_2,\alpha_3,\alpha_\mathrm{s}}(-\omega_\mathrm{s};\omega_1,\omega_2,\omega_3) =$$
$$\frac{1}{6\,\hbar^3} \sum_{gabc} \rho(g)\, \frac{\langle g|e_\mathrm{s}\boldsymbol{\mu}|a\rangle\langle a|e_1\boldsymbol{\mu}|b\rangle\langle b|e_2\boldsymbol{\mu}|c\rangle\langle c|e_3\boldsymbol{\mu}|g\rangle}{(\Omega_{ag}-\omega_1-\omega_2-\omega_3)(\Omega_{bg}-\omega_2-\omega_3)(\Omega_{cg}-\omega_3)} \tag{4.2.12}$$

obeying the conservation of energy

$$\omega_\mathrm{s} = \omega_1 + \omega_2 + \omega_3\,. \tag{4.2.13}$$

4.2.1.3 Fundamental equations of nonlinear optics

Maxwell's equations in SI units [62Jac, 87Vid] are given by (1.1.4)–(1.1.7) and the material equations (1.1.8) and (1.1.9), see Chap. 1.1.

With the following three approximations

1. magnetization $\boldsymbol{M} = 0$: $\mu_0 \boldsymbol{H} = \boldsymbol{B} \rightarrow \mu = 1$,
2. source-free medium: $\rho = 0$,
3. currentless medium: $\boldsymbol{j} = 0$

we get the simplified Maxwell equations

$$\nabla \times \boldsymbol{E} = -\mu_0 \frac{\partial \boldsymbol{H}}{\partial t} \tag{4.2.14}$$

$$\nabla \times \boldsymbol{H} = \epsilon_0 \frac{\partial \boldsymbol{E}}{\partial t} + \frac{\partial \boldsymbol{P}}{\partial t} \tag{4.2.15}$$

resulting in the wave equation

$$\Delta \boldsymbol{E} - \frac{1}{c^2} \frac{\partial^2 \boldsymbol{E}}{\partial t^2} = \frac{1}{\epsilon_0 c^2} \frac{\partial^2 \boldsymbol{P}}{\partial t^2} \tag{4.2.16}$$

with the polarization $\boldsymbol{P} = \boldsymbol{P}^{\mathrm{L}} + \boldsymbol{P}^{\mathrm{NL}}$. This gives the driven wave equation

$$\Delta \boldsymbol{E} - \frac{n^2}{c^2} \frac{\partial^2 \boldsymbol{E}}{\partial t^2} - \mathrm{i} \frac{\epsilon_0}{c^2} \frac{\partial^2 \boldsymbol{E}}{\partial t^2} = \frac{1}{\epsilon_0 c^2} \frac{\partial^2 \boldsymbol{P}^{\mathrm{NL}}}{\partial t^2} . \tag{4.2.17}$$

With the plane-wave approximation $\hat{E}(r,\omega) = \hat{E}(z,\omega)$ and the slow-amplitude approximation

$$\frac{\partial \hat{E}_j}{\partial t} \ll \omega \hat{E}_j , \quad \frac{\partial \hat{E}_j}{\partial z} \ll k \hat{E}_j , \tag{4.2.18}$$

we get the fundamental equations of nonlinear optics

$$\frac{\mathrm{d} \hat{E}_j}{\mathrm{d} z} = \mathrm{i} \frac{\omega_j}{2 \epsilon_0 c n_j} P_j^{\mathrm{NL}} \exp(-\mathrm{i} k_j z) - \frac{\kappa_j}{2} \hat{E}_j , \tag{4.2.19}$$

where κ is the absorption coefficient and where the total derivative is given by the partial derivatives

$$\frac{\mathrm{d} \hat{E}_j}{\mathrm{d} z} = \frac{\partial \hat{E}_j}{\partial z} + \frac{n_j}{c} \frac{\partial \hat{E}_j}{\partial t} , \tag{4.2.20}$$

and $\hat{E}_j$ is a slowly-varying-envelope function in space and time.

4.2.1.4 Small-signal limit

In this case the only nonlinear polarization for a medium of density N is given by

$$P_{\mathrm{s}}^{(3)}(\omega_{\mathrm{s}}) = \frac{3}{2} \epsilon_0 N \chi_{\mathrm{T}}^{(3)}(-\omega_{\mathrm{s}}; \omega_1, \omega_2, \omega_3) E_1 E_2 E_3 . \tag{4.2.21}$$

Within the plane-wave approximation one obtains

$$\frac{\mathrm{d} \hat{E}_{\mathrm{s}}}{\mathrm{d} z} = \mathrm{i} \frac{3 \pi \omega_{\mathrm{s}}}{c n_{\mathrm{s}}} N \chi_{\mathrm{T}}^{(3)} E_{10} E_{20} E_{30} \exp \left\{ \left(-\frac{\kappa_1 + \kappa_2 + \kappa_3}{2} - \mathrm{i} \Delta k \right) z \right\} , \tag{4.2.22}$$

where the wave-vector mismatch is given by the conservation of momenta

$$\Delta k = k_s - k_1 - k_2 - k_3 \,. \tag{4.2.23}$$

The wave vector k_j of the jth wave is given by the refractive index n_j

$$k_j = \frac{\omega_j n_j}{c} \,. \tag{4.2.24}$$

With the optical depth $\tau_j = \kappa_j L = \sigma_j^{(1)}(\omega_j) N L$ and the length L of the nonlinear medium we have

$$\hat{E}_s(L) = \mathrm{i}\, \frac{3\pi\omega_s}{c n_s} N L\, \epsilon_0 \chi_T^{(3)}\, E_{10} E_{20} E_{30}\, \frac{\exp\left(-\frac{\tau_s}{2}\right)}{\frac{\tau_s - \tau_0}{2} - \mathrm{i}\Delta k L} \left\{ \exp\left[\frac{\tau_s - \tau_0}{2} - \mathrm{i}\Delta k L\right] - 1 \right\}, \tag{4.2.25}$$

where the total optical depth $\tau_0 = \tau_1 + \tau_2 + \tau_3$. With the intensity

$$\Phi_j = \frac{\epsilon_0 n_j c}{2} |E_j|^2 \tag{4.2.26}$$

the intensity conversion is given by

$$\frac{\Phi_s}{n_s} = \left[\frac{6\pi\omega_s}{c^2 n_s} N L \chi_T^{(3)}(-\omega_s; \omega_1, \omega_2, \omega_3)\right]^2 \frac{\Phi_{10}\Phi_{20}\Phi_{30}}{n_1 n_2 n_3} F(\Delta k L, \tau_0, \tau_s) \,, \tag{4.2.27}$$

containing the general phase-matching factor

$$F(\Delta k L, \tau_0, \tau_s) = \frac{\exp(-\tau_0) + \exp(-\tau_s) - 2\exp\left(-\frac{\tau_0 + \tau_s}{2}\right) \cos(\Delta k L)}{\left(\frac{\tau_s - \tau_0}{2}\right)^2 + (\Delta k L)^2} < 1 \,. \tag{4.2.28}$$

4.2.1.5 Phase-matching condition

Maximum conversion efficiency is achieved for conservation of momenta k_j where $\Delta k = 0$

$$\omega_1 n_1 + \omega_2 n_2 + \omega_3 n_3 = \omega_s n_s \,. \tag{4.2.29}$$

In case of the third harmonic generation this gives $n_1 = n_s$. Frequency mixing in a two-component system results in

$$\frac{N_a}{N_b} = \frac{\omega_s \bar{\chi}_b^{(1)}(\omega_s) - \sum_{j=1}^{3} \omega_j \bar{\chi}_b^{(1)}(\omega_j)}{\sum_{j=1}^{3} \omega_j \bar{\chi}_a^{(1)}(\omega_j) - \omega_s \bar{\chi}_a^{(1)}(\omega_s)} \,. \tag{4.2.30}$$

For the third harmonic generation in a two-component system we have:

$$\frac{N_a}{N_b} = \frac{\bar{\chi}_b^{(1)}(3\omega) - \bar{\chi}_b^{(1)}(\omega)}{\bar{\chi}_a^{(1)}(\omega) - \bar{\chi}_a^{(1)}(3\omega)} \,. \tag{4.2.31}$$

The frequency mixing in a one-component system is given by:

$$\omega_s \bar{\chi}^{(1)}(\omega_s) = \sum_{j=1}^{3} \omega_j \bar{\chi}^{(1)}(\omega_j) \,. \tag{4.2.32}$$

4.2.2 Frequency conversion in gases

The following conditions have to be met for large conversion efficiencies:

1. a large nonlinear susceptibility $\chi_T^{(3)}$ which may be enhanced by a proper two-photon resonance,
2. large column densities with a proper phase matching,
3. small optical depths for the incident and generated waves to avoid reabsorption.

4.2.2.1 Metal-vapor inert gas mixtures

Metal-vapor inert gas mixtures are generally generated in concentric heat pipes because for efficient frequency mixing the phase matching can be accurately and independently adjusted through the partial pressures in the heat pipe [71Vid, 87Vid, 96Vid].

Tables of the multi-wave mixing experiments in different gaseous nonlinear media are arranged according to the elements, Table 4.2.1. For every element the wavelength is given together with the method of generation. The method of generation is indicated where $(\omega_1 + \omega_1)_{\text{Res}} + \omega_2$, for example, indicates a two-photon resonance of ω_1 in the particular atomic or molecular medium and the additional wave ω_2 can make the resulting radiation tunable.

4.2.2.2 Mixtures of different metal vapors

The modified concentric heat pipe [71Vid, 87Vid, 96Vid] is used for phase matching with small partial pressures avoiding strong homogeneous broadening.

In Table 4.2.2 mixtures of different metal vapors are listed.

4.2.2.3 Mixtures of gaseous media

For $\lambda > 106$ nm one prefers gas cells with lithium fluoride windows [87Vid]. For $\lambda < 106$ nm one should use either pulsed nozzle beams [87Bet] without windows or gas cells with a fast shutter [85Bon].

In Table 4.2.3 mixtures of gaseous media are given.

Table 4.2.1. Metal-vapor inert gas mixtures.

Vapor	Wavelength [nm]	Method	Ref.
Na	354.7	$3 \cdot \omega_1$	[75Blo1, 75Blo2, 76Oha]
Na	330.5	$(\omega_1 + \omega_1)_{\text{Res}} + \omega_2$	[74Blo]
Na	268 (cw)	$(\omega_1 + \omega_2)_{\text{Res}} + \omega_3$	[84Bol]
Na	231	$(\omega_1 + \omega_2)_{\text{Res}} + \omega_3$	[76Bjo]
Na	151.4	$7 \cdot \omega_1$	[77Gro2, 79Mit]
Na	117.7	$9 \cdot \omega_1$	[77Gro1, 79Mit]
Rb	354.7	$3 \cdot \omega_1$	[71You, 75Blo1, 76Pue, 76Oha]
Cs	213.4	$(\omega_1 + \omega_1)_{\text{Res}} + \omega_1$	[74Leu, 75War]
Be	121–123	$(\omega_1 + \omega_1)_{\text{Res}} + \omega_2$	[79Mah]
Mg	173.5	$2 \cdot \omega_1 + \omega_1 + \omega_1$	[85Hut]
Mg	140–160	$(\omega_1 + \omega_1)_{\text{Res}} + \omega_2$	[76Wal, 80Jun, 96Ste]
Mg	143.6 (cw)	$(\omega_1 + \omega_1)_{\text{Res}} + \omega_1$	[83Tim]
Mg	115, 121.2, 127	$(\omega_1 + \omega_1)_{\text{Res}} + \omega_2$	[81Car, 85Car]
Mg	121–129	$(\omega_1 + \omega_1)_{\text{Res}} + \omega_2$	[78McK]
Mg$^+$	123.6	$(\omega_1 + \omega_1)_{\text{Res}} + \omega_2$	[85Leb]
Ca	200	$3 \cdot \omega_1$	[76Fer]
Ca$^+$	127.8	$(\omega_1 + \omega_1)_{\text{Res}} + \omega_2$	[75Sor]
Zn	106–140	$(\omega_1 + \omega_1)_{\text{Res}} + \omega_2$	[82Jam]
Cd	118.2, 152, 177.3	$(\omega_1 + \omega_1)_{\text{Res}} + \omega_2$; $3 \cdot \omega_1$	[72Kun]
Cd	128.7–135.3	$(\omega_1 + \omega_1)_{\text{Res}} + \omega_2$	[84Miy]
Cd	145.3–171.1	$(\omega_1 + \omega_1)_{\text{Res}} - \omega_2$	[86Miy]
Sr	155.3, 166.7, 169.7, 173.5, 183.5 (cw)	$(\omega_1 + \omega_2)_{\text{Res}} + \omega_3$	[90Nol1]
Sr	184.0, 185.7, 192.0, 195.8, 208.8, 217.9 (cw)	$(\omega_1 + \omega_2)_{\text{Res}} + \omega_3$	[90Nol1]
Sr	171.2, 168.3, 169.7, 190 (cw)	$(\omega_1 + \omega_2)_{\text{Res}} + \omega_3$	[77Bjo, 78Fre]
		$(\omega_1 + \omega_1)_{\text{Res}} + \omega_2$	[90Nol2]
Ca	153.0, 159.5, 161.3, 163.3, 167.0 (cw)	$(\omega_1 + \omega_2)_{\text{Res}} + \omega_3$	[90Nol1]
Ca	169.7, 170.9, 172.3, 173.1, 176.9 (cw)	$(\omega_1 + \omega_2)_{\text{Res}} + \omega_3$	[90Nol1]
Zn	134.5–141.6 (cw)	$(\omega_1 + \omega_2)_{\text{Res}} + \omega_3$	[90Nol1]
Cd	138.1–140.3	$(\omega_1 + \omega_1)_{\text{Res}} + \omega_2$	[88Sch]
Sr	177.8–195.7	$(\omega_1 + \omega_1)_{\text{Res}} + \omega_2$	[74Hod, 75Sor, 76Sor]
Sr	165–166	$(\omega_1 + \omega_2)_{\text{Res}} + \omega_3$	[78Eco]
Sr	192.3	$(\omega_1 + \omega_1)_{\text{Res}} + \omega_1$	[80Pue, 81Egg]
Sr	171.2	$(\omega_1 + \omega_1)_{\text{Res}} + \omega_2$	[80Eco]
Ba	190–200	$(\omega_1 + \omega_1)_{\text{Res}} + \omega_2$	[80Hei]
Hg	109–196	$(\omega_1 + \omega_1)_{\text{Res}} \pm \omega_2$	[83Hil2]
Hg	184.9, 143.5, 140.1, 130.7, 125.9, 125.0	$(\omega_1 + \omega_1)_{\text{Res}} \pm \omega_2$	[81Bok]
Hg	125.1, 183.3, 208.5	$(\omega_1 + \omega_1)_{\text{Res}} \pm \omega_2$	[81Tom]
Hg	124.7–125.5, 122.8–123.5, 117.4–122	$(\omega_1 + \omega_1)_{\text{Res}} + \omega_2$	[82Mah, 82Tom]
Hg	120.3	$(\omega_1 + \omega_1)_{\text{Res}} + \omega_2$	[76Hsu]
Hg	87.5–105, 99.1–126.8	$(\omega_1 + \omega_1)_{\text{Res}} + \omega_2$	[85Her, 83Hil2]
Hg	89.6	$(\omega_1 + \omega_1)_{\text{Res}} + \omega_1$	[78Sla]
Hg	132–185	$(\omega_1 + \omega_1)_{\text{Res}} + \omega_2$	[86Hil2]
Tl	195.1	$(\omega_1 + \omega_1)_{\text{Res}} + \omega_1$	[75Wan]
Eu, Yb	185.5, 194	$(\omega_1 + \omega_1)_{\text{Res}} + \omega_2$	[75Sor, 76Sor]

Table 4.2.2. Mixtures of different metal vapors.

Mixture	Wavelength [nm]	Method	Ref.
Na + K	2–25 μm	$(\omega_1 - \omega_2)_{\text{Res}} - \omega_3$	[74Wyn]
Na + Mg	354.7	$3 \cdot \omega_1$	[75Blo2]

Table 4.2.3. Mixtures of gaseous media.

Gas	Wavelength [nm]	Method	Ref.
He, Ne, Ar, Kr, Xe	231.4	$3 \cdot \omega_1$	[67New, 69War]
He	53.2	$5 \cdot \omega_1$	[76Rei, 77Rei, 77She, 78Rei2, 78Rei1]
He	38	$7 \cdot \omega_1$	[77Rei, 78Rei2]
He	82.8, 50, 35.5	$3 \cdot \omega_1$; $5 \cdot \omega_1$; $7 \cdot \omega_1$	[83Bok]
He, Ne	106.4	$(\omega_1 + \omega_1)_{Res} + \omega_2$	[78Rei2]
He, Ne	88.7	$3 \cdot \omega_1$	[78Rei2, 78Rei1]
He, Ne	76, 70.9, 62.6, 59.1	$4 \cdot \omega_1 \pm \omega_2$	[77She, 78Rei2]
He, Xe	49.7, 35.5	$5 \cdot \omega_1$; $7 \cdot \omega_1$	[83Bok]
Ne	53.2, 118.2	$5 \cdot \omega_1$; $(\omega_1 + \omega_1)_{Res} + \omega_2$	[76Rei, 77She, 78Rei2]
Ne	72.05–73.58, 74.3–74.36	$3 \cdot \omega_1$	[84Hil]
Ar	120.4	$(\omega_1 + \omega_1)_{Res} + 3 \cdot \omega_1$	[80Din]
Ar	85.7–87.0, 97.4–104.8	$3 \cdot \omega_1$	[83Hil3, 83Mar]
Ar	106.7	$(\omega_1 + \omega_1)_{Res} + \omega_1$	[82Mil]
Ar	102.6–102.8	$3 \cdot \omega_1$	[79Rei, 80Rei]
Ar, Kr	61.6, 53.2	$5 \cdot \omega_1$	[78Rei2, 81Rei]
Ar	57	$(\omega_1 + \omega_1)_{Res} + \omega_1$	[76Hut]
Kr	131.2, 92.3, 92.8, 94.2	$(\omega_1 + \omega_1)_{Res} \pm \omega_2$	[85Bon]
Kr	121.6	$3 \cdot \omega_1$	[81Hil, 80Lan, 81Bat]
Kr	123.6	$(\omega_1 + \omega_1)_{Res} + \omega_1$	[82Mil]
Kr	112.4, 120.3–123.6	$(\omega_1 + \omega_1)_{Res} + \omega_2$	[79Cot1, 79Cot2]
Kr	131.2	$(\omega_1 + \omega_1)_{Res} - \omega_2$	[85Bon]
Kr	72.5–83.5; 127–180	$(\omega_1 + \omega_1)_{Res} \pm \omega_2$	[87Hil]
Kr	121–200	$(\omega_1 + \omega_1)_{Res} - \omega_2$	[90Mar]
Kr, Xe	71–92	$(\omega_1 + \omega_1)_{Res} + \omega_2$	[89Miy]
Kr, Xe	110–210	$(\omega_1 + \omega_1)_{Res} \pm \omega_2$	[82Hil]
Xe	84.6–109.5, 155–220	$(\omega_1 + \omega_1)_{Res} \pm \omega_2$	[82Hil, 82Hag, 83Hil1]
Xe	155	$(\omega_1 + \omega_1)_{Res} - \omega_2$	[83Hut]
Xe	147	$(\omega_1 + \omega_1)_{Res} + \omega_1$	[82Mil]
Xe	140.3–146.9	$3 \cdot \omega_1$; $(\omega_1 + \omega_1)_{Res} + \omega_2$	[81Hil, 83Val]
Xe	74.8, 75, 75.2	$(\omega_1 + \omega_1)_{Res} + \omega_1$	[82Mui]
Xe	125.4, 125.9, 126.1	$(\omega_1 + \omega_1)_{Res} - \omega_2$	[82Mui]
Xe	101.5, 101.8, 13.0	$(\omega_1 + \omega_1)_{Res} + \omega_2$	[82Mui]
Xe	163.1–194.6	$(\omega_1 + \omega_1)_{Res} \pm \omega_2$	[74Kun]
Xe	118.2	$3 \cdot \omega_1$	[83Kun, 76Kun, 82Gan, 83Bok]
Ar, Kr, Xe, CO, N_2	72; 90.4–102.5	$3 \cdot \omega_1$	[87Pag]
Ne, Ar, Kr, Xe, Hg	60–200	$(\omega_1 + \omega_1)_{Res} \pm \omega_2$	[86Hil1]
Ar, Xe, CO	74.2, 80.4, 95.1, 98.2, 100.1, 116.5, 117.8, 123.6	$\omega_1 + \omega_1 + \omega_2$; $3 \cdot \omega_1$	[89Cro]

References for 4.2

62Arm   Armstrong, J.A., Bloembergen, N., Ducuing, J., Pershan, P.S.: Phys. Rev. **127** (1962) 1918.
62Jac   Jackson, D.: Classical electrodynamics, New York: Wiley, 1962.

67New   New, G.H.C., Ward, J.F.: Phys. Rev. Lett. **19** (1967) 556.

69War   Ward, J.F., New, G.H.C.: Phys. Rev. **185** (1969) 57.

71Han   Hanna, D.C., Yuratich, M.A., Cotter, D.: Nonlinear optics of free atoms and molecules, Springer Series Opt. Sci., Vol. 17, Heidelberg: Springer-Verlag, 1979.
71Vid   Vidal, C.R., Haller, F.B.: Rev. Sci. Instrum. **42** (1971) 1779.
71You   Young, J.F., Bjorklund, G.C., Kung, A.H., Miles, R.B., Harris, S.E.: Phys. Rev. Lett. **27** (1971) 1551.

72Kun   Kung, A.H., Young, J.F., Bjorklund, G.C., Harris, S.E.: Phys. Rev. Lett. **29** (1972) 985.

74Blo   Bloom, D.M., Yardley, J.T., Young, J.F., Harris, S.E.: Appl. Phys. Lett. **24** (1974) 427.
74Hod   Hodgson, R.T., Sorokin, P.P., Wynne, J.J.: Phys. Rev. Lett. **32** (1974) 343.
74Kun   Kung, A.H.: Appl. Phys. Lett. **25** (1974) 653.
74Leu   Leung, K.M., Ward, J.F., Orr, B.J.: Phys. Rev. A **9** (1974) 2440.
74Wyn   Wynne, J.J., Sorokin, P.P., Lankard, J.R., in: Laser spectroscopy, Brewer, R.G., Mooradian, A. (eds.), New York: Plenum, 1974, p. 13–111.

75Bjo   Bjorklund, G.C.: IEEE J. Quantum Electron. **11** (1975) 287.
75Blo1   Bloom, D.M., Bekkers, G.W., Young, J.F., Harris, S.E.: Appl. Phys. Lett. **26** (1975) 687.
75Blo2   Bloom, D.M., Young, J.F., Harris, S.E.: Appl. Phys. Lett. **27** (1975) 390.
75Mat   Matsuoka, M., Nakatsuka, H., Okada, J.: Phys. Rev. A **12** (1975) 1062.
75Sor   Sorokin, P.P., Armstrong, J.A., Dreyfus, R.W., Hodgson, R.T., Lankard, J.R., Manganaro, L.H., Wynne, J.J., in: Laser spectroscopy, Haroche, S., Pebay-Peyroula, J.C., Haensch, T.W., Harris, S.E. (eds.), Heidelberg: Springer-Verlag, 1975, p. 46.
75Wan   Wang, C.C., Davis, L.I.: Phys. Rev. Lett. **35** (1975) 650.
75War   Ward, J.F., Smith, A.V: Phys. Rev. Lett. **35** (1975) 653.

76Bjo   Bjorklund, G.C., Bjorkholm, J.E., Liao, P.F., Storz, R.H.: Appl. Phys. Lett. **29** (1976) 729.
76Fer   Ferguson, A.I., Arthurs, E.G.: Phys. Lett. A **58** (1976) 298.
76Hsu   Hsu, K.S., Kung, A.H., Zych, L.J., Young, J.F., Harris, S.E.: IEEE J. Quantum Electron. **12** (1976) 60.
76Hut   Hutchinson, M.H.R., Ling, C.C., Bradley, D.J.: Opt. Commun. **18** (1976) 203.
76Kun   Kung, A.H., Young, J.F., Harris, S.E.: Appl. Phys. Lett. **22** (1973) 301; erratum: Appl. Phys. Lett. **28** (1976) 239.
76Oha   Ohashi, Y., Ishibashi, Y., Kobayashi, T., Inaba, H.: Jpn. J. Appl. Phys. **15** (1976) 1817.
76Pue   Puell, K.H., Spanner, K., Falkenstein, W., Kaiser, W., Vidal, C.R.: Phys. Rev. A **14** (1976) 2240.
76Rei   Reintjes, J., Eckardt, R.C., She, C.Y., Karangelen, N.E., Elton, R.C., Andrews, R.A.: Phys. Rev. Lett. **37** (1976) 1540.
76Roy   Royt, T.R., Lee, C.H., Faust, W.L.: Opt. Commun. **18** (1976) 108.

76Sor   Sorokin, P.P., Wynne, J.J., Armstrong, J.A., Hodgson, R.T.: Ann. N. Y. Acad. Sci. **267** (1976) 30.
76Tay   Taylor, J.R.: Opt. Commun. **18** (1976) 504.
76Wal   Wallace, S.C., Zdasiuk, G.: Appl. Phys. Lett. **28** (1976) 449.

77Bjo   Bjorklund, G.C., Bjorkholm, J.E., Freeman, R.R., Liao, P.F.: Appl. Phys. Lett. **31** (1977) 330.
77Dra   Drabovich, I.N., Mitev, V.M., Pavlov, L.I., tamenov, K.V.: Opt. Commun. **20** (1977) 350.
77Gro1   Grozeva, M.G., Metchkov, D.I., Mitev, V.M., Pavlov, L.I., Stamenov, K.V.: Opt. Commun. **23** (1977) 77.
77Gro2   Grozeva, M.G., Metchkov, D.I., Mitev, V.M., Pavlov, L.I., Stamenov, K.V.: Phys. Lett. A **64** (1977) 41.
77Met   Metchkov, D.I., Mitev, V.M., Pavlov, L.I., Stamenov, K.V.: Opt. Commun. **21** (1977) 391.
77Rei   Reintjes, J., She, C.Y., Eckardt, R.C., Karangelen, N.E., Andrews, R.A., Elton, R.C.: Appl. Phys. Lett. **30** (1977) 480.
77Roy   Royt, T.R., Lee, C.H.: Appl. Phys. Lett. **30** (1977) 332.
77She   She, C.Y., Reintjes, J.: Appl. Phys. Lett. **31** (1977) 95.

78Doi   Doitcheva, V.L., Mitev, V.M., Pavlov, L.I., Stamenov, K.V.: Opt. Quantum Electron. **10** (1978) 131.
78Eco   Economou, N.P., Freeman, R.R., Bjorklund, G.C.: Opt. Lett. **3** (1978) 209.
78Fre   Freeman, R.R., Bjorklund, G.C., Economou, N.P., Liao, P.F.: Appl. Phys. Lett. **33** (1978) 739.
78Mah   Mahon, R., McIlrath, T.J., Koopman, D.W.: Appl. Phys. Lett. **33** (1978) 305.
78McK   McKee, T.J., Stoicheff, B.P., Wallace, S.C.: Opt. Lett. **3** (1978) 207.
78Mit   Mitev, V.M., Pavlov, L.I., Stamenov, K.V.: J. Phys. B **11** (1978) 819.
78Rei1   Reintjes, J., She, C.Y.: Opt. Commun. **27** (1978) 469.
78Rei2   Reintjes, J., She, C.Y., Eckardt, R.C.: IEEE J. Quantum Electron. **14** (1978) 581.
78Sla   Slabko, V.V., Popov, A.K., Lukinykh, V.F.: Appl. Phys. **15** (1978) 239.

79Cot1   Cotter, D.: Opt. Commun. **31** (1979) 397.
79Cot2   Cotter, D.: Opt. Lett. **4** (1979) 134.
79Mah   Mahon, R., McIlrath, T.J., Tomkins, F.S., Kelleher, D.E.: Opt. Lett. **4** (1979) 360.
79Mit   Mitev, V.M., Pavlov, L.I., Stamenov, K.V.: Opt. Quantum Electron. **11** (1979) 229.
79Rei   Reintjes, J.: Opt. Lett. **4** (1979) 242.

80Din   Dinev, S.G., Marazov, O.R., Stamenov, K.V., Tomov, I.V.: Opt. Quantum Electron. **12** (1980) 183.
80Eco   Economou, N.P., Freeman, R.R., Heritage, J.P., Liao, P.F.: Appl. Phys. Lett. **36** (1980) 21.
80Egg   Egger, H., Hawkins, R.T., Bokor, J., Pummer, H., Rothschild, R., Rhodes, C.K.: Opt. Lett. **5** (1980) 282.
80Hei   Heinrich, J., Behmenburg, W.: Appl. Phys. **23** (1980) 333.
80Jun   Junginger, H., Puell, H.B., Scheingraber, H., Vidal, C.R.: IEEE J. Quantum Electron. **16** (1980) 1132.
80Lan   Langer, H., Puell, H., Roehr, H.: Opt. Commun. **34** (1980) 137.
80Mah   Mahon, R., Mui Yiu, Y.: Opt. Lett. **5** (1980) 279.
80Pue   Puell, H., Scheingraber, H., Vidal, C.R.: Phys. Rev. A **22** (1980) 1165.

80Rei   Reintjes, J.: Opt. Lett. **5** (1980) 342.
80Wal   Wallenstein, R.: Opt. Commun. **33** (1980) 119.

81Bat   Batishche, S.A., Burakov, V.S., Kostenich, Yu.V., Mostovnikov, A.V., Naumenkov, P.A., Tarasenko, N.V., Gladushchak, V.I., Moshkalev, S.A., Razdobarin, G.T., Semenov, V.V., Shreider, E.Ya.: Opt. Commun. **38** (1981) 71.
81Bok   Bokor, J., Freeman, R.R., Panock, R.L., White, J.C.: Opt. Lett. **6** (1981) 182.
81Car   Caro, R.G., Costela, A., Webb, C.E.: Opt. Lett. **6** (1981) 464.
81Egg   Egger, H., Srinivasan, T., Hohla, K., Scheingraber, H., Vidal, C.R., Pummer, H., Rhodes, C.K.: Appl. Phys. Lett. **39** (1981) 37.
81Hil   Hilbig, R., Wallenstein, R.: IEEE J. Quantum Electron. **17** (1981) 1566.
81Rei   Reintjes, J., Tankersley, L.L., Christensen, R.: Opt. Commun. **39** (1981) 334.
81Tom   Tomkins, F.S., Mahon, R.: Opt. Lett. **6** (1981) 179.
81Zap   Zapka, W., Cotter, D., Brackmann, U.: Opt. Commun. **36** (1981) 79.

82Gan   Ganeev, R.A., Kulagin, I.A., Usmanov, T., Khudalberganov, S.T.: Sov. J. Quantum Electron. (English Transl.) **12** (1982) 1637.
82Hag   Hager, J., Wallace, S.C.: Chem. Phys. Lett. **90** (1982) 472.
82Hil   Hilbig, R., Wallenstein, R.: Appl. Opt. **21** (1982) 913.
82Jam   Jamroz, W., La Rocque, P.E., Stoicheff, B.P.: Opt. Lett. **7** (1982) 617.
82Mah   Mahon, R., Tomkins, F.S.: IEEE J. Quantum Electron. **18** (1982) 913.
82Mil   Miller, J.C., Compton, R.N.: Phys. Rev. A **25** (1982) 2056.
82Mui   Mui Yiu, Y., Bonin, K.D., McIlrath, T.J.: Opt. Lett. **7** (1982) 268.
82Pum   Pummer, H., Srinivasan, T., Egger, H., Boyer, K., Luk, T.S., Rhodes, C.K.: Opt. Lett. **7** (1982) 93.
82Tom   Tomkins, F.S., Mahon, R.: Opt. Lett. **7** (1982) 304.

83Bok   Bokor, J., Bucksbaum, P.H., Freeman, R.R.: Opt. Lett. **8** (1983) 217.
83Dim   Dimov, S.S., Pavlov, L.I., Stamenov, K.V., Heller, Yu.I., Popov, A.K.: Appl. Phys. B **30** (1983) 35.
83Hil1   Hilbig, R., Wallenstein, R.: IEEE J. Quantum Electron. **19** (1983) 194.
83Hil2   Hilbig, R., Wallenstein, R.: IEEE J. Quantum Electron. **19** (1983) 1759.
83Hil3   Hilbig, R., Wallenstein, R.: Opt. Commun. **44** (1983) 283.
83Hut   Hutchinson, M.H.R., Thomas, K.J.: IEEE J. Quantum Electron. **19** (1983) 1823.
83Kun   Kung, A.H.: Opt. Lett. **8** (1983) 24.
83Mar   Marinero, E.E., Rettner, C.T., Zare, R.N.: Chem. Phys. Lett. **95** (1983) 486.
83Tim   Timmermann, A., Wallenstein, R.: Opt. Lett. **8** (1983) 517.
83Val   Vallée, F., DeRougement, F., Lukasik, J.: IEEE J. Quantum Electron. **19** (1983) 131.

84Bol   Bolotskikh, L.T., Yysotin, A.L., Tkhek-de, Im., Podavalova, O.P., Popov, A.K.: Appl. Phys. B **35** (1984) 249.
84Hil   Hilbig, R., Lago, A., Wallenstein, R.: Opt. Commun. **49** (1984) 297.
84Miy   Miyazaki, K., Sakai, H., Sato, T.: Opt. Lett. **9** (1984) 457.

85Bon   Bonin, K.D., McIlrath, T.J.: J. Opt. Soc. Am. B **2** (1985) 527.
85Car   Caro, R.G., Costela, A., Smith, N.P., Webb, C.E.: J. Phys. D **18** (1985) 1291.
85Her   Herman, P.R., Stoicheff, B.P.: Opt. Lett. **10** (1985) 502.
85Hut   Hutchinson, M.H.R., Manning, R.J.: Opt. Commun. **55** (1985) 55.
85Leb   Lebedev, V.V., Plyasulya, V.M., Troshin, B.I., Chebotaev, V.P.: Kvantovaya Elektron. (Moscow) **12** (1985) 866.

86Hil1	Hilbig, R., Hilber, G., Lago, A., Wallenstein, R.: Comments At. Mol. Phys. **18** (1986) 157.
86Hil2	Hilbig, R., Hilber, G., Wallenstein, R.: Appl. Phys. B **41** (1986) 225.
86Miy	Miyazaki, K., Sakai, H., Sato, T.: IEEE J. Quantum Electron. **22** (1986) 2266.
87Bet	Bethune, D.S., Rettner, C.T.: IEEE J. Quantum Electron. **23** (1987) 1348.
87Hil	Hilber, G., Lago, A., Wallenstein, R.: J. Opt. Soc. Am. B **4** (1987) 1753.
87Pag	Page, R.H., Larkin, R.J., Kung, A.H., Shen, Y.R., Lee, Y.T.: Rev. Sci. Instrum. **58** (1987) 1616.
87Vid	Vidal, C.R.: Four-wave frequency mixing in gases, in: Tunable lasers, Mollenauer, L.F., White, J.C. (eds.), Topics in applied physics, Vol. 59, New York: Springer-Verlag, 1987, p. 57–113.
88Sch	Schnitzer, A.M., Behmenburg, W.: Z. Phys. D **8** (1988) 141.
89Cro	Cromwell, E., Trickl, T., Lee, Y.T., Kung, A.H.: Rev. Sci. Instrum. **60** (1989) 2888.
89Miy	Miyazaki, K., Sakai, H., Sato, T.: Appl. Opt. **28** (1989) 699.
90Mar	Marangos, J.P., Shen, N., Ma, H., Hutchinson, M.H.R., Connerade, J.P.: J. Opt. Soc. Am. B **7** (1990) 1254.
90Nol1	Nolting, J., Kunze, H.J., Schuetz, I., Wallenstein, R.: Appl. Phys. B **50** (1990) 331.
90Nol2	Nolting, J., Wallenstein, R.: Opt. Commun. **79** (1990) 437.
96Ste	Steffes, B., Xinghua Li, Mellinger, A., Vidal, C.R.: Appl. Phys. B **62** (1996) 87.
96Vid	Vidal, C.R.: Vapor cells and heat pipes – experimental methods in the physical sciences, Vol. 29B: Atomic and molecular and optical physics, atoms and molecules, Dunning, F.B., Hulet, R. (eds.), New York: Academic, 1996, p. 67–83.

4.3 Stimulated scattering

A. LAUBEREAU

4.3.1 Introduction

The first example of stimulated scattering was incidentally discovered in 1962 as a "new laser line in the emission of a ruby laser" [62Woo]. The phenomenon occurred when the laser was equipped with a nitrobenzene cell for Q-switching operation. The emitted frequency component was identified as an amazingly intense Raman line [62Eck] due to stimulated Raman scattering predicted theoretically in 1931 [31Goe]. Hundreds of papers appeared since then on the novel phenomenon. Compared to the wealth of experimental evidence full quantitative information about the individual scattering processes however is rather scarce since many publications confine themselves to reported frequency shifts. A quantitative analysis is also often impeded by competing nonlinear effects and by the not too well known properties of the applied laser pulses. Three cases were investigated in detail: Stimulated Raman Scattering (SRS), Stimulated Brillouin Scattering (SBS), and stimulated Rayleigh scattering.

This chapter, Chap. 4.3, follows the discussions given by Maier and Kaiser [72Mai], by Maier [76Mai], and by Penzkofer et al. [79Pen]. Circular frequencies are denoted in the following by ω_i while the corresponding frequency values are represented by $\nu_i = \omega_i/2\pi$. The term "circular" is often omitted in context with the ω_i's.

4.3.1.1 Spontaneous scattering processes

Fluctuations of the molecular polarizability and of the number density of atoms or molecules give rise to various scattering processes when light passes a transparent medium. The scattering is characterized by the frequency ν_{sc} of the scattered light relative to the incident laser frequency ν_L, the linewidth $\delta\nu$, its polarization properties, and the scattering intensity. Here we introduce the scattering cross section $d\sigma/d\Omega$ relating the power P_{sc} of the light scattered into a solid angle $\Delta\Omega$ to the incident laser power P_L:

$$P_{sc} = N \frac{d\sigma}{d\Omega} \ell P_L \Delta\Omega \qquad (4.3.1)$$

with number density N of the (quasi-)particles generating the scattering. The interaction length is denoted by ℓ. $d\sigma/d\Omega$ is the differential cross section with respect to solid angle but integrated over the spectral lineshape. The spectrum of four scattering processes is depicted schematically in Fig. 4.3.1a. Two unshifted components are indicated: the narrow Rayleigh line scattered from nonpropagating entropy (temperature) fluctuations and the broader Rayleigh-wing line due to orientation fluctuations of anisotropic molecules. The lines are accompanied by the Brillouin doublet representing scattering from propagating isentropic density fluctuations. In the quantum-mechanical approach the Brillouin lines are related to the annihilation (frequency up-shifted anti-Stokes component) and creation (down-shifted Stokes line) of acoustic phonons with conservation of quantum energy and (pseudo-)momentum:

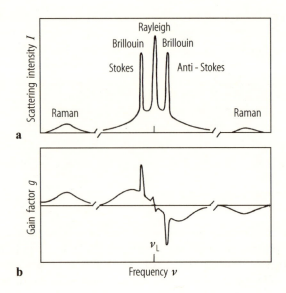

Fig. 4.3.1. (a) Schematic of the spectral intensity distribution of spontaneous light scattering in condensed matter with unshifted Rayleigh and Rayleigh-wing lines (quasi-elastic scattering) as well as Stokes- and anti-Stokes-shifted Brillouin and Raman lines (inelastic light scattering). (b) Frequency dependence of the corresponding gain factors of stimulated scattering (see text).

$$h\nu_L = h\nu_{sc} \pm h\nu_o, \tag{4.3.2}$$

$$\hbar k_L = \hbar k_{sc} \pm \hbar k_o \tag{4.3.3}$$

with Planck's constant h and wavevector k, $\hbar = h/2\pi$. Subscript "o" refers to the material excitation, i.e. acoustic phonons. The positive sign in (4.3.2) and (4.3.3) corresponds to the Stokes process (sc = S), while the negative sign applies for anti-Stokes scattering (sc = A). Due to the dispersion relation of acoustic phonons (phase velocity v of sound waves) the frequency shift is given by

$$\nu_o = \frac{vk_o}{2\pi} = 2v\frac{\nu_L n}{c}\sin\left(\frac{\theta}{2}\right). \tag{4.3.4}$$

Here c/n denotes the speed of light in the medium; θ is the scattering angle between wave vectors k_L and k_{sc} ($k_{sc} \cong k_L$, since $\nu_o \ll \nu_L$). Equation (4.3.4) refers to isotropic media, e.g. gases and liquids. For anisotropic solids three Brillouin doublets occur in the general case and (4.3.4) has to be modified according to the considered transverse or longitudinal acoustic phonon branch and the respective orientation-dependent sound velocity in the crystal. As a consequence of (4.3.3), $k_o \leq 2k_L$, so that only acoustic phonons close to the center of the first Brillouin zone are involved (note $k_L \sim 10^5$ cm^{-1}).

Figure 4.3.1a also schematically shows the Stokes and anti-Stokes line of Raman scattering off a molecular vibration or off an optical phonon branch, displaying a larger frequency shift. As before, only phonons of relatively small k_o are involved. Polyatomic molecules display a variety of such vibrational Raman lines. In gases many vibration-rotation Raman lines occur in addition and also rotational lines with small frequency shifts. In ionic crystals the relevant material excitation is of mixed phonon-photon character and termed polariton. Since the excited states of molecular vibrations and optical phonons are weakly populated, the anti-Stokes line intensity is also small compared to the corresponding Stokes line.

A further unshifted scattering component in liquids, the Mountain line [76Ber], is only mentioned here since it was not yet observed in stimulated scattering because of its weakness and broad width. Typical values for the frequency shift ν_o/c and the linewidth $\delta\nu/c$ (FWHM) in wavenumber units of the various processes are given in Tables 4.3.1–4.3.5. Some scattering cross sections for the Raman interaction are listed in Table 4.3.2. The scattered light intensity is small. Even for the large

4.3.1.2 Relationship between stimulated Stokes scattering and spontaneous scattering

The elementary interaction for Stokes scattering is illustrated in Fig. 4.3.2a (solid arrows). The process involves a transition from an initial to a final energy level of the medium (horizontal lines). The relationship between the stimulated and the spontaneous process is close and originates from the Boson character of photons, i.e. the analogy of the eigenmodes of the electromagnetic field with the harmonic oscillator, the transition probability of which increases with occupation number. As a result the rate of photons scattered into an eigenmode of the Stokes field (subscript "S") depends on the occupation number n_S of this mode. Under steady-state conditions we have:

$$\frac{d n_S}{d t} = \text{const.}\, n_L (1 + n_S) \,. \tag{4.3.5}$$

The first term in the bracket on the right-hand side of (4.3.5) represents spontaneous scattering depending linearly on incident photon number n_L or laser power, compare (4.3.1), as long as $n_S \ll 1$, i.e. a negligible number of scattered photons per mode of the radiation field is present. The second term on the right-hand side of (4.3.5) describes stimulated scattering that dominates for $n_S > 1$ and requires sufficiently high laser intensities. In this regime an avalanche build-up of scattered photons can occur.

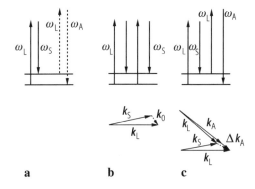

Fig. 4.3.2. (a) Schematic of the elementary scattering process of spontaneous scattering involving two energy levels (horizontal bars) of the medium with transition frequency ω_o; the Stokes (full arrows) and anti-Stokes (dashed arrows) processes are indicated. Corresponding diagrams for (b) stimulated Stokes scattering and (c) stimulated Stokes–anti-Stokes coupling in the stimulated scattering. Vertical arrows represent photons that are annihilated (upwards) or generated (downwards) in the interaction. The $\boldsymbol{k}$-vector geometries of the stimulated processes are depicted in the lower part of the figure (see text).

4.3.2 General properties of stimulated scattering

4.3.2.1 Exponential gain by stimulated Stokes scattering

Integration of (4.3.5) yields exponential growth of Stokes-scattered photons, $n_S \propto \exp(\text{const.}\, n_L\, t)$, or equivalently for forward scattering in the z-direction:

$$I_S(z) = I_S(0) \exp(g\, I_L\, z) \,. \tag{4.3.6}$$

Here we have replaced in the argument of the exponential the product "const. $n_L\, t$" by a more familiar term with laser intensity I_L, the gain factor g for stimulated Stokes scattering, and the interaction length z. Equation (4.3.6) indicates exponential amplification of an initial signal $I_S(0)$ that may be supplied by spontaneous scattering or by an additional input beam. The exponential growth of the scattered light is only limited by the energy conservation of (4.3.2), since for every scattered photon one incident laser photon has to be annihilated. The corresponding laser depletion leads to gain saturation not included in (4.3.6). Conversion efficiencies above 50 % have been observed for stimulated scattering in a number of cases. Equation (4.3.6) refers to steady state.

The gain factor g is an important material parameter for stimulated scattering. The dependence of g on the frequency shift of the scattering is indicated in Fig. 4.3.1b. Maximum gain occurs in the center of the down-shifted Brillouin and Raman lines (Stokes process). For stimulated Rayleigh scattering the peak gain occurs for a Stokes shift equal to half of the full width, $\delta\nu/2$, of the respective line. The negative gain values in Fig. 4.3.1b indicate loss via stimulated scattering on the anti-Stokes side.

Typical values of the peak gain factors are listed in Tables 4.3.2–4.3.5. Under steady-state conditions stimulated Brillouin scattering often represents the dominant interaction. In absorbing media additional mechanisms occur. The corresponding processes, stimulated thermal Brillouin and stimulated thermal Rayleigh scattering, are discussed below.

4.3.2.2 Experimental observation

Stimulated scattering was studied using the following three different experimental approaches:

1. generator setup,
2. oscillator setup,
3. stimulated amplification setup.

4.3.2.2.1 Generator setup

Here only an intense laser beam is directed into the sample. The kind of stimulated scattering is selected by the material and laser beam properties. As a general rule, a large gain of $g\, I_L\, z \cong 30$ is required under steady-state conditions for the traveling-wave situation with a single pass through the medium (length z), in order to observe the respective stimulated process. The scattering occurs in forward and/or backward direction because of a simple geometrical argument (maximum interaction length in these directions). The process builds up from an equivalent noise input $I_S(0)$, see (4.3.6), that can be estimated from zero point fluctuations of the electromagnetic field [79Pen]. The growth of the Stokes component is finally limited by the simultaneous decrease of incident laser radiation. The observations are difficult to analyze because of the competition of nonlinear interactions including optical self-focusing. The latter is often involved in liquid media. The observed frequency shift of the stimulated process may slightly deviate from the value known from spontaneous scattering (up to a few cm^{-1} in SRS) because of simultaneous self-phase modulation in the medium.

4.3.2.2.2 Oscillator setup

An optical resonator made up by mirrors or reflecting surfaces can provide feedback of the stimulated Stokes radiation so that the effective interaction length is increased by multiple passes through the medium. As a result the laser intensity requirements are lowered. The scattering angle is controlled by the cavity axis, so that off-axis emission is possible relative to the laser beam. The frequency-dependent feedback and the lower intensity level of the setup can be sufficient to select a specific stimulated scattering process. Among different Raman transitions only the one with largest gain factor g shows up in SRS in general.

4.3.2.2.3 Stimulated amplification setup

Two well defined beams representing the laser component and the incident Stokes radiation are directed into the scattering medium. Scattering angle and mechanism are determined by the direction and frequency shift of the incident Stokes beam. A second tunable laser is used for the latter in general. The pump intensity I_L is smaller by one or more orders of magnitude compared to the generator case, so that self-focusing and other competing effects including secondary scattering processes can be avoided. Quantitative information on the amplitude and/or frequency dependence of the gain factor $g(\nu_S)$ may be deduced from careful measurements of the amplification factor.

An example for the technique is Raman gain spectroscopy that is often applied in the low-intensity limit $g\,I_L\,z \ll 1$. An alternative is Raman loss spectroscopy of the transmitted laser component, since the production of Stokes photons corresponds to the annihilation of the same number of laser photons.

4.3.2.3 Four-wave interactions

4.3.2.3.1 Third-order nonlinear susceptibility

Stimulated Stokes scattering can be treated as a four-photon (or four-wave) interaction involving the third-order nonlinear susceptibility $\chi^{(3)}(-\omega_S; \omega_L, -\omega_L, \omega_S)$. The interaction is illustrated by the energy level scheme of Fig. 4.3.2b. The two waves are resonantly coupled via a difference frequency resonance, $\omega_L - \omega_S = \omega_o$, to the relevant material excitation. The latter is enhanced by the scattering thus increasing the coupling strength. The photons at frequencies ω_L and ω_S enter the process twice (see Fig. 4.3.2a). Stimulated amplification is provided in the resonant case by the imaginary part χ_3'' of $\chi^{(3)}$, while the real part leads to frequency modulation. The gain factor is related to the imaginary part by:

$$g \propto |\chi_3''|^2 \,. \tag{4.3.7}$$

Outside difference frequency resonances the real part of $\chi^{(3)}$ is also important for stimulated amplification. The general case of stimulated 4-photon amplification is treated in [79Pen]. The (fourth-rank) tensor character of $\chi^{(3)}$ is omitted here for brevity considering only parallel polarization of the light field components.

The corresponding wave-vector diagram is shown in the lower part of Fig. 4.3.2b. The general case with off-axis geometry is considered. The scattering couples to a material excitation with wave vector $\boldsymbol{k}_o$. The effective scattering angle is strongly influenced by interaction-length arguments. Because of the maximum interaction length, geometries with approximate forward and backward scattering are most important. In cases where the corresponding frequency shift ω_o vanishes, e.g. SBS, stimulated scattering exactly in forward direction is not possible. For backward scattering of short pulses, e.g. SRS of a picosecond laser, the interaction length ℓ may be governed by the duration t_p (FWHM of intensity envelope) of the incident laser pulse setting an upper limit of $\ell = t_p/2\,v_g$ (v_g: group velocity). In forward direction a less stringent limitation is set by group velocity dispersion between laser and Stokes pulses, $\ell = t_p\,\Delta(1/v_g)$. As a result SRS of picosecond pulses preferentially occurs in forward direction.

4.3.2.3.2 Stokes–anti-Stokes coupling

The stimulated Stokes scattering can be impeded by simultaneous anti-Stokes scattering, $\omega_A = \omega_L + \omega_o$. The anti-Stokes process is depicted in Fig. 4.3.2a (dashed arrows) and "consumes" material excitation, so that (4.3.6) is not applicable. The corresponding four-wave interaction via $\chi^{(3)}(-\omega_A; \omega_L, \omega_L, -\omega_S)$ is termed Stokes–anti-Stokes coupling and depicted in Fig. 4.3.2c. The significance of the process is determined by its wave vector mismatch Δk_A, depicted in the lower part of Fig. 4.3.2c, and the initial intensity ratio $I_A(0)/I_S(0)$ (I_A: anti-Stokes intensity). Δk_A is governed by the scattering angle and the color dispersion of the refractive index $n(\omega)$ of the medium since

$$k_i = n(\omega_i) \frac{\omega_i}{c}; \quad (i = A, L, S). \tag{4.3.8}$$

For a collinear geometry we simply have $\Delta k_A = k_A + k_S - 2k_L$. For $\Delta k_A = 0$ and $I_A/I_S = 1$, the inverse process of anti-Stokes scattering fully inhibits stimulated Stokes scattering. An example in this context is exact forward scattering in gases, where Δk_A is small, so that the observed weakness of SRS in exact forward direction is explained in this way. For a large mismatch, $|\Delta k_A| > 3\, g\, I_L$, on the other hand, the Stokes–anti-Stokes coupling is negligible. This condition is always fulfilled for backward scattering so that simultaneous anti-Stokes scattering cannot perturb the stimulated Stokes process notably. For $I_A \ll I_S$, the perturbation of Stokes scattering by anti-Stokes production is negligible, too. In this case the process of Fig. 4.3.2c is also called Coherent Anti-Stokes Raman Scattering, CARS, an important nonlinear spectroscopy (preferentially applied for phase-matching geometries, $\Delta k_A \cong 0$).

Outside Raman resonances the properties of Stokes–anti-Stokes coupling differ notably from the near-resonant case considered here.

4.3.2.3.3 Higher-order Stokes and anti-Stokes emission

For high conversion efficiency of the stimulated scattering the Stokes intensity I_S becomes comparable to the incident radiation I_L, and the material excitation is significant. As a consequence secondary processes show up, generating a cascade of higher-order Stokes and anti-Stokes lines with relative frequency shift ω_o and decreasing intensity levels. Two mechanisms are relevant here:

1. stimulated Stokes scattering where the intense first-order Stokes component serves as the pump radiation for generating the second-order line and so forth;
2. coherent Stokes or anti-Stokes scattering off the material excitation generated by the primary Stokes scattering producing new frequency-shifted lines. The mechanism is effected by wavevector mismatches of the individual processes.

The Stokes–anti-Stokes coupling discussed above is responsible for the generation of the first-order anti-Stokes component. Higher-order Stokes scattering limits the energy conversion efficiency of first-order Stokes production. The higher-order stimulated scattering should be distinguished from higher-order spontaneous scattering since only a fundamental material transition is involved in the former case.

4.3.2.4 Transient stimulated scattering

The build-up of a material excitation in stimulated scattering involves the response time T_2 (dephasing time) of the medium. When the pulse duration t_p of the incident laser is comparable to or smaller than T_2, the interaction becomes less efficient and the actual gain of the stimulated Stokes process is smaller than in the steady state. Equation (4.3.6) for the stationary case is not valid for $t_p/T_2 < 10$. The smaller transient gain for a given input situation may be overcome experimentally by increased pump intensities. For details the reader is referred to the literature [78Lau]. Here only three remarks are given:

1. For homogeneous broadening of the material transition ω_o involved in the stimulated scattering the relaxation time can be simply derived from the linewidth $\delta\nu$ (FWHM)

$$T_2 = (\pi\,\delta\nu)^{-1} = \frac{1}{\Gamma}\,. \tag{4.3.9}$$

For inhomogeneous broadening (4.3.9) may be also used to estimate an effective T_2^* from the line broadening that may be sufficient for a semi-quantitative discussion of the transient scattering. For the competition among different Raman transitions in transient SRS both gain factor g and dephasing time T_2 are relevant.

2. For frequency-modulated laser pulses the temporal behavior is not fully described by the duration t_p of the pulse envelope. Because of intensity fluctuations the effective duration of the pulse can be estimated to be $t_p^* \cong (2\,\delta\nu_L)^{-1} < t_p$ ($\delta\nu_L$: frequency width (FWHM) of the laser pulse). To ascertain steady-state conditions the condition

$$\frac{t_p^*}{T_2} > 10 \tag{4.3.10}$$

should be fulfilled.

3. Choice of a short t_p may allow to suppress stimulated scattering of transitions with longer T_2 that would have to occur in a less favorable transient situation. An example is SRS in liquids in forward direction with picosecond pulses that is observed in spite of the larger stationary gain factor of SBS. Here the different interaction lengths of forward (SRS) and backward scattering (SBS) also play a role.

4.3.3 Individual scattering processes

4.3.3.1 Stimulated Raman scattering (SRS)

The gain constant for stimulated amplification of the first Stokes component (4.3.6) at resonance, $\omega_S = \omega_L - \omega_o$ is given by

$$g_S = \frac{4\pi^2\,N\,(\partial\alpha/\partial q)^2\,\omega_S}{n_L\,n_S\,c^2\,m\,\omega_o\,\Gamma}\,. \tag{4.3.11}$$

Here N denotes the molecular number density. A highly polarized vibrational Raman line with halfwidth Γ (HWHM, isotropic scattering component) is considered. $(\partial\alpha/\partial q)$ is the isotropic part of the Raman polarizability (derivative of the molecular polarizability with respect to the vibrational coordinate q of transition ω_o). m represents the reduced mass of the molecular vibration. n_i ($i = $ L, S) is the refractive index at frequency ω_i. $(\partial\alpha/\partial q)$ is connected to the Raman scattering cross section by the relation:

$$\frac{d\sigma}{d\Omega} = \frac{(\partial\alpha/\partial q)^2\,\omega_S^4\,h\,n_S}{4\pi\,c^4\,m\,\omega_o\,n_L}\,. \tag{4.3.12}$$

The frequency dependence of the gain factor is given by:

$$g(\omega_S) = \frac{g_S \, \Gamma^2}{(\omega_S - \omega_L + \omega_o)^2 + \Gamma^2} \; . \tag{4.3.13}$$

A Lorentzian lineshape is assumed in (4.3.13) that holds well in gases at sufficiently high pressure, weakly associated liquids and solids. SRS of notably depolarized Raman lines is discussed in [78Lau]. Frequency shifts observed for SRS in the generator setup are compiled in Table 4.3.1. A list of gain factors g_S and other parameters is presented in Table 4.3.2. The relaxation time T_2 in condensed matter is in the range 10^{-12} to 10^{-10} s.

Table 4.3.1. Frequency shifts (in wavenumber units) observed in stimulated Raman scattering of various materials.

(a) Liquids

Medium	Stokes shift ν_0/c [cm^{-1}]	Excitation wavelength [nm]	Reference
Acetic acid	2944		[84Kru]
Acetone	2925	527	[68Bre, 69Col]
Aniline	997	694	[66Eck]
Benzaldehyde	1001	694	[66Bar]
Benzene	992	527	[67Sha, 68Bre, 69Col, 70Alf]
Benzene	3064	694	[66Eck]
Benzene-d$_6$	944	694	[67Blo]
Benzonitrile	2229	694	[66Eck]
Bromobenzene	998, 1000	527, 694	[66Eck, 67Sha]
Bromopropane	2962	694	[66Bar]
2-Bromopropane	2920	694	[66Bar]
1-Bromopropane	2935	694	[66Bar]
Butyl-benzene (tert.)	1000	694	[66Bar]
Carbondisulfide	656	527	[67Sha, 68Bre, 69Col, 70Alf]
Carbontetrachloride	460	694	[66Eck]
Chlorobenzene	1002	527	[67Sha, 69Col]
Chloromethylbutane	2927	694	[66Bar]
Chloroform	663	694	[66Eck]
Cyclohexane	2825		[84Kru]
Cyclohexanone	2683	694	[66Eck]
1,3-Dibromobenzene	992	694	[66Bar]
1,2-Dichloroethane	2958		[84Kru]
Dichloromethane	2989		[75Lau]
2,2-Dichlorodiethylether	2938	527	[83Tel]
1,2-Diethylbenzene	2934	694	[66Bar]
1,2-Dimethylcyclohexane	2853, 2921	694	[66Bar]
1,4-Dimethylcyclohexane	2876	694	[66Bar]
Dimethylhexadiene	2910	694	[66Bar]
1,4-Dioxane	2967		[84Kru]
DMSO, dimethylsulfoxide	2911		[95Go]
Ethanol	2928	527	[69Col, 71Lin]
Ethyl-Benzene	1002	694	[66Eck]
1-Fluoro-2-chlorobenzene	1034	694	[66Bar]
Fluorobenzene	1009	694	[66Eck]
Fluoromethane	2970	694	[78Map]
Isopropanol	2882	527	[69Col]
Methanol	2835	527	[69Col, 70Alf]
Methanol-d$_4$	2200	527	[73Lau]
3-Methylbutadiene	1638	694	[66Eck]

(continued)

Table 4.3.1a continued.

Medium	Stokes shift ν_0/c [cm^{-1}]	Excitation wavelength [nm]	Reference
Nitrobenzene	1344	527, 694	[67Blo, 67Sha, 69Col]
2-Nitropropane	2945	694	[66Bar]
Nitrogen ($T = 77$ K)	2326	527	[74Lau]
1,3-Pentadiene	1655	694	[66Eck]
Piperidine	2933	694	[66Eck]
Pyridine	992	694	[66Eck]
Siliciumtetrachloride	425	527	[71Lau]
Styrene	1315, 1631, 3056	694	[66Eck]
Tintetrabromide	221	527	[78Lau]
Tintetrachloride	368	527	[78Lau]
Tetrachloroethane	2984	694	[66Eck]
Tetrachloroethylene	448, 2939	527	[69Col, 72Lau, 66Bar]
Tetrahydrofuran	2849	694	[66Eck]
Toluene	1004	527, 694	[67Blo, 67Sha]
Water	3450	527	[68Bre, 69Col, 69Rah]
m-Xylene	2933	694	[66Eck]
o-Xylene	2913	694	[66Eck]
p-Xylene	2998	694	[66Eck]

(b) Solids

Medium	Stokes shift ν_0/c [cm^{-1}]	Excitation wavelength [nm]	Reference
Al$_2$O$_3$	416	532	[97Kam2]
1-Bromonaphthalene	1363	694	[66Eck]
Calcite	1086	527	[69Col]
1-Chloronaphthalene	1368	694	[66Eck]
Diamond	1332	527	[71Lau]
2-Ethylnaphthalene	1382	694	[66Bar]
Gd$_2$(MoO$_4$)$_3$	960	532	[97Kam1]
LiHCOO · H$_2$0	104, 1372	694	[90Lai]
NaClO$_3$	936	532	[97Kam3]
Naphthalene	1380	694	[66Eck]
Polydiacetylene	1200		[94Yos]
Sulfur	216, 470	694	[66Eck]

(c) Gases

Medium	Stokes shift ν_0/c [cm^{-1}]	Excitation wavelength [nm]	Reference
Ammonia	3339		[72Car]
Barium vapor	11395	552	[83Sap, 87Glo]
Ethylene (55 atm)	1344	694	[70Mac]
Butane (90 atm)	2920	694	[70Mac]
Carbondioxide (20–50 atm)	1385	694	[70Mac, 78Map]
Carbonmonoxide	2145	694	[72Car]

(continued)

Table 4.3.1c continued.

Medium	Stokes shift ν_0/c [cm^{-1}]	Excitation wavelength [nm]	Reference
Cesium vapor	14597		[84Har]
Chlorine	556	694	[72Car, 78Map]
Deuterium	2991	694	[67Blo]
Hydrogen	4160	694	[67Blo, 75Cha]
Hydrogenbromide (20 atm)	2558	694	[70Mac, 78Map]
Hydrogenchloride (35 atm)	2883	694	[70Mac, 78Map]
Methane (10 atm)	2917		[70Mac]
Nitrogen (55–100 atm)	2330	694	[70Mac, 75Cha]
N$_2$O (50 atm)	774	694	[70Mac, 78Map]
NO	1877	694	[72Car]
Oxygen (50–100 atm)	1550	694	[70Mac]
SF$_6$ (15–20 atm)	1551	694	[70Mac]
SF$_6$ (18 atm)	775	694	[72Car]

Table 4.3.2. Gain factor and other parameters of stimulated Raman scattering.

(a) Liquids

Medium	Stokes shift ν_0/c [cm^{-1}]	Scattering coefficient $N \times d\sigma/d\Omega$ [10^7 m^{-1} sr^{-1}]	Linewidth $\delta\nu/c$ [cm^{-1}]	Gain factor g_s [10^{12} m/W]	Excitation wavelength [nm]	Ref.
Acetone	2925		17.4	12	530	[69Col]
Benzene	992		2.2	28	694	[72Mai]
Bromobenzene	1000	15	1.9	15	694	[72Mai]
Carbondisulfide	655	75	0.50	240	694	[72Mai]
Chlorobenzene	1002	15	1.6	19	694	[72Mai]
Ethanol	2928		17.4	51	530	[69Col]
Isopropanol	2882		26.7	9.2	530	[69Col]
Methanol	2834		18.7	23	530	[69Col]
Methanol	2944		26.5	18	530	[69Col]
Nitrogen	2326	2.9	0.067	170	694	[72Mai]
Nitrobenzene	1345	64	6.6	21	694	[72Mai]
Oxygen	1552	4.8	0.117	140	694	[72Mai]
Tetrachloroethylene	447			17	598	[76Mai]
Toluene	1003	11	1.9	12	694	[72Mai]
1,1,1-Trichloroethane	2939		5.2	51	530	[69Col]
Water	3450	430		1.4	530	[69Col]

(b) Solids

Medium	Stokes shift ν_0/c [cm^{-1}]	Linewidth $\delta\nu/c$ [cm^{-1}]	Gain factor g_S [10^{12} m/W]	Excitation wavelength [nm]	Ref.
Ba$_2$NaNb$_5$O$_{15}$	650		67	694	[72Mai]
Calcite	1086	1.1	1.4	530	[69Col]
CuAlS$_2$	314		21,000	514	[97Bai]
GaP	403		19,000	632	[97Bai]
^{6}LiNbO$_3$	256		180	694	[72Mai]
^{7}LiNbO$_3$	256		89	694	[72Mai]
^{6}LiTaO$_3$	600		43	694	[72Mai]
Quartz	467		0.15	527	[67Wig]

(c) Gases

Medium	Stokes shift ν_0/c [cm^{-1}]	Differential scattering cross section $d\sigma/d\Omega$ [10^{36} m^2 sr^{-1}]	Dephasing time T_2 [ps]	Gain factor g_S [10^{12} m/W]	Excitation wavelength [nm]	Ref.
H$_2$, Q(1)	4155	1.2	208	9.7	1064	[86Han]
H$_2$, Q(1)	4155	79	208	27.6	532	[86Han]
D$_2$, Q(2)	2987	2.0	150	3.7	1064	[86Han]
D$_2$, Q(2)	2987	8.0	150	10	532	[86Han]
Methane, Q	2917	7.0	16	3.3	1064	[86Han]
Methane, Q	2917	270	16	8.6	532	[86Han]

4.3.3.2 Stimulated Brillouin scattering (SBS) and stimulated thermal Brillouin scattering (STBS)

Stimulated Brillouin scattering was extensively studied in liquids, solids, and gases. In many substances it is the dominant process under stationary conditions and occurs generally in backward direction. The scattering originates from two coupling mechanisms between the electromagnetic field and the medium: electrostriction and absorption. In transparent media only electrostriction is relevant. In absorbing media the second contribution called Stimulated Thermal Brillouin Scattering (STBS) is caused by absorption-induced local temperature changes leading to propagating density waves. The frequency dependencies of the gain factors for the two mechanisms are different. The peak values of stimulated gain are given by:

$$g_B^e = \frac{(\partial\varepsilon/\partial\rho)_T^2 \, \omega_S^2 \, \rho_o}{2 \, c^3 \, n_S \, v \, \Gamma_B} \qquad (4.3.14)$$

for the electrostrictive contribution (superscript "e"), and by

$$g_B^a = \frac{\alpha \, (\partial\varepsilon/\partial\rho)_T \, \omega_S \, \beta_T}{4 \, c \, n_S \, C_p \, \Gamma_B} \qquad (4.3.15)$$

for STBS. Here $(\partial\varepsilon/\partial\rho)_T$ is the change of the relative dielectric constant with mass density ρ at constant temperature T. ρ_o is the equilibrium density value. v denotes the sound velocity at frequency $\omega_o = \omega_L - \omega_S$ (see (4.3.4) for backward scattering, $\theta = 180°$). The parameters in (4.3.15) are the absorption coefficient of the laser intensity α and the relative volume expansion coefficient β_T. The half-width $\Gamma_B = \pi\,\delta\nu$ of the corresponding spontaneous Brillouin line that displays an approximately quadratic frequency dependence also enters the expressions above. For liquids one can write:

$$\frac{\Gamma_B}{\omega_o^2} = \frac{\frac{4}{3}\eta_S + \Lambda\left(\frac{1}{C_V} - \frac{1}{C_p}\right) + \eta_V}{2\,\rho_o\,v^2}, \qquad (4.3.16)$$

where η_S and η_V, respectively, denote the shear and volume viscosity; the latter is to some extent frequency-dependent via relaxation phenomena. Λ is the thermal conductivity. C_V and C_p are the specific heat per unit mass at constant volume and pressure, respectively. The phonon lifetime τ of the involved acoustic phonons with circular frequency ω_o is related to the linewidth by $\tau = T_2/2 = 1/(2\,\Gamma_B)$. The peak gain value g_B^a increases proportional to α and is of same order of magnitude as g_B^e for $\alpha \approx 1$ cm^{-1}.

The total frequency-dependent gain factor for the (first-order) Stokes component of SBS including STBS is given by

$$g(\omega_S) = \frac{g_B^e\,\Gamma_B^2}{(\omega_S - \omega_L + \omega_o)^2 + \Gamma_B^2} - \frac{g_B^a\,2\,\Gamma_B\,(\omega_S - \omega_L + \omega_o)}{(\omega_S - \omega_L + \omega_o)^2 + \Gamma_B^2}. \qquad (4.3.17)$$

The maximum contribution of STBS is red-shifted relative to the Brillouin line and occurs at $\omega_S = \omega_L - \omega_o - \Gamma_B$. In the blue wing of the Brillouin Stokes line the mechanism produces stimulated loss. Equation (4.3.17) states that the Stokes shift observed in the stimulated Brillouin scattering of absorbing media in the generator or oscillator setup – occurring at the peak value of $g(\omega_S)$ – is modified compared to the spontaneous Brillouin line.

A list of frequency shifts observed in SBS of transparent media is presented in Table 4.3.3 where values for the Brillouin linewidth $\delta\nu$ and the gain parameters g_B^a/α and g_B^e are also compiled. The relaxation time $T_2\,(=1/\pi\,\delta\nu)$ in condensed matter is in the order of 10^{-9} s so that SBS is close to steady state for giant laser pulses with $t_p \approx 10^{-8}$ s (if self-focusing is avoided), but is of transient character in the subnanosecond time domain.

4.3.3.3 Stimulated Rayleigh scattering processes, SRLS, STRS, and SRWS

Three mechanisms can be distinguished:

1. Stimulated Rayleigh Line Scattering in transparent substances, SRLS, by electrostrictive coupling to non-propagating density changes,
2. Stimulated Thermal Rayleigh Scattering, STRS, by absorptive coupling similar to the STBS case, and
3. Stimulated Rayleigh Wing Scattering, SRWS, in liquids by orientational changes of anisotropic molecules.

The frequency shifts of the Stokes component of the first two cases are considerably smaller than for SBS. SRLS is difficult to observe because of the small gain factor and the relatively long relaxation time $T_2 \approx 10^{-8}$ s for backward scattering leading to transient scattering for nanosecond pulses.

Table 4.3.3. Stimulated Brillouin scattering in backward direction: frequency shift, linewidth, and gain factor.

Medium	Stokes shift ν_0/c [cm^{-1}]	Linewidth $\delta\nu$ [MHz]	Gain coefficient g_B^a/α (calculated) [cm^2 / MW]	Gain factor g_B^e (calculated) [cm / MW]	Gain factor g_B^e (measured) [cm / MW]	Ref.
Acetic acid	0.152					[67Wig]
Acetone	0.154	180	0.022	0.017	0.020	[70Poh]
Aniline	0.259					[67Wig]
Benzaldehyde	0.224					[67Wig]
Benzene	0.211		0.024	0.024	0.018	[68Den]
Bromobenzene	0.188					[67Wig]
Carbondisulfide	0.194	75	0.213	0.197	0.068	[97Jo]
Carbontetrachloride	0.146	650	0.0134	0.0084	0.006	[68Den, 70Poh]
Chloroform	0.148					[67Wig]
Cyclohexane	0.180			0.007	0.0068	[68Den]
p-Dichlorobenzene	0.184					[67Wig]
Ethanol	0.152		0.010	0.012		[72Mai]
Fluorinert FC 72					0.006	[97Yos]
Fluorinert FC 75					0.005	[97Yos]
Glass BSC-2	0.866					[67Wig]
Glass DF-3	0.638					[67Wig]
Glycerol	0.386					[67Wig]
n-Hexane		220		0.027	0.026	[68Den, 70Poh]
InSb						[95Lim]
Methanol	0.142		0.013	0.013	0.013	[68Den, 70Poh]
Methyliodide	0.166					[67Wig]
Nitrobenzene	0.228					[67Wig]
m-Nitrotoluene	0.229					[67Wig]
n-Nitrotoluene	0.217					[67Wig]
Octanol	0.194					[67Wig]
Pyridine	0.226					[67Wig]
Quartz	1.16				0.005	[89Agr]
Sulfurhexafluoride (20 atm)					0.0015	[93Fil, 97Jo]
Tetrabromomethane	0.173					[67Wig]
Toluene	0.193	480		0.013	0.013	[68Den, 70Poh]
Water	0.197		0.0008	0.0066	0.0048	[68Den, 77Rys, 94Yos]
p-Xylene	0.199					[67Wig]

1. and 2. The peak value of the stimulated gain for SRLS is given by:

$$g_{RL}^e = \frac{(\partial\varepsilon/\partial\rho)_T^2 \, \omega_S \, \rho_o \, (\gamma - 1)}{4 \, c^2 \, n_S^2 \, v^2} \, , \tag{4.3.18}$$

where $\gamma = C_p/C_V$. It is interesting to notice that g_{RL}^e does not depend on scattering angle ($\omega_S \cong \omega_L$). A finite optical absorption coefficient α of the medium gives rise to a second contribution with peak value:

$$g_{RL}^a = \frac{\alpha \, (\partial\varepsilon/\partial\rho)_T \, \omega_S \, \beta_T}{2 \, c \, n_S \, C_p \, \Gamma_{RL}} \, . \tag{4.3.19}$$

Table 4.3.4. Stimulated Rayleigh scattering: gain factors and linewidth values.

Medium	Gain factor g_{RL}^e [cm/MW]	Gain coefficient $g_{RL}^a(\text{max.})/\alpha$ [cm^2/MW]	Linewidth $\delta\nu$ [MHz]	Reference
Acetone	2	0.47	21	[70Rot]
Benzene	2.2	0.57	24	[70Rot]
Carbondisulfide	6	0.62	36	[70Rot]
Ethanol		0.38	18	[70Rot]
Methanol	8.4	0.32	20	[70Rot]
Tetrachloromethane	2.6×10^{-4}	0.82	17	[70Rot]
Water	0.02	0.019	27.5	[70Rot]

Here $\Gamma_{RL} = \pi\, \delta\nu$ is the halfwidth (HWHM) of the spontaneous Rayleigh line in circular frequency units that originates from the damping of entropy changes via thermal conductivity Λ:

$$\Gamma_{RL} = \frac{4\, k_L^2\, \Lambda\, \sin^2(\theta/2)}{\rho_o\, C_p} \,. \tag{4.3.20}$$

Via Γ_{RL} the gain constant g_{RL}^a strongly depends on scattering angle. The frequency dependencies of SRLS and STRS have opposite sign, leading to the total gain factor:

$$g(\omega_S) = \frac{(g_B^e - g_B^a)\, 2\, \Gamma_{RL}(\omega_L - \omega_S)}{(\omega_L - \omega_S)^2 + \Gamma_{RL}^2} \,. \tag{4.3.21}$$

For dominant coupling via electrostriction positive gain occurs on the Stokes side, $\omega_S < \omega_L$. Small absorption values, $\alpha > 10^{-3}$ cm^{-1}, can be sufficient for dominant STRS that produces gain on the anti-Stokes side. For zero frequency shift, $\omega_S = \omega_L$, the gain factor vanishes in the steady state, (4.3.21), but not in the transient case. Stimulated scattering in forward scattering is particularly delicate since $\Gamma_{RL} \to 0$, (4.3.20), and Stokes–anti-Stokes coupling has to be included. Values for g and $\delta\nu$ are listed in Table 4.3.4.

3. Stimulated Rayleigh wing scattering is connected with the overdamped rotational motion of liquid molecules in combination with an anisotropic polarizability tensor. The latter is also involved in the optical Kerr effect enhancing the nonlinear refractive index of the medium (optical self-focusing, self-phase modulation). The maximum steady-state gain factor for SRWS is given by:

$$g_{RW} = \frac{16\, \pi^2\, N\, \omega_S\, (\alpha_\parallel - \alpha_\perp)^2}{45\, k_B\, T_o\, c^2\, n_S^2} \,. \tag{4.3.22}$$

The difference of the molecular polarizability parallel and perpendicular to the (assumed) molecular axis of rotational symmetry is denoted by $\alpha_\parallel - \alpha_\perp$. k_B is the Boltzmann constant, T_o the sample temperature. The frequency dependence of the gain factor is analogous to the previous cases:

$$g(\omega_S) = \frac{g_{RW}\, 2\, \Gamma_{RW}\, (\omega_L - \omega_S)}{(\omega_L - \omega_S)^2 + \Gamma_{RW}^2} \,. \tag{4.3.23}$$

Maximum gain of SRWS occurs for $\omega_S = \omega_L - \Gamma_{RW}$. The halfwidth Γ_{RW} of the Rayleigh wing line may be taken from spontaneous scattering observations or from the reorientational time τ_{or}. The latter can be derived from spontaneous Raman spectroscopy, NMR, or time-resolved spectroscopy, e.g. transient optical Kerr effect observations: $\tau_{or} = T_{2,RW} = 1/\Gamma_{RW}$. An estimate of the halfwidth may be computed from shear viscosity and the size of the molecules using the Debye theory:

$$\Gamma_{RW} = \frac{3\, k_B\, T_o}{8\, \pi\, R\, \eta_S} \,. \tag{4.3.24}$$

Here R denotes an effective mean radius of the molecule. The proportionality $\Gamma_{RW} \propto \eta_S$ was demonstrated experimentally for numerous examples.

Equations (4.3.6), (4.3.22), and (4.3.23) hold for large scattering angles where Stokes–anti-Stokes coupling can be neglected. Close to forward direction simultaneous anti-Stokes scattering enhances the stimulated Stokes scattering, in contrast to SRS. Including the Stokes–anti-Stokes coupling maximum gain is predicted for an optimum scattering angle

$$\theta_{\text{opt}} = \left(\frac{2 g_{RW} I_L c}{n_L \omega_L} \right)^{\frac{1}{2}}. \tag{4.3.25}$$

The corresponding gain factor for stimulated amplification in the scattering direction θ_{opt} without frequency shift, $\omega_S = \omega_A = \omega_L$, amounts to:

$$g_{\text{opt}} = 2 g_{RW}, \tag{4.3.26}$$

where g_{RW} is given by (4.3.22). For more general cases the reader is referred to the literature, e.g. [72Mai]. Frequency shift and gain factor numbers are compiled in Table 4.3.5.

Table 4.3.5. Frequency shift and gain factor of stimulated Rayleigh wing scattering.

Medium	Frequency shift × c^{-1} [cm^{-1}]	Gain factor G_{RW} [10^{12} m/W]	Reference
Azoxybenzene	0.036		[68Fol]
Benzene		6	[72Mai]
Benzonitrol	0.198		[68Fol]
Benzoylchloride	0.184		[68Fol]
Benzylidenaniline	0.065		[68Fol]
Bromobenzene		14	[72Mai]
1-Bromonaphthalene	0.076		[68Fol]
Carbondisulfide		30	[72Mai]
Chlorobenzene		10	[72Mai]
Chloronaphthalene	0.100		[68Fol]
1,4-Dimethylnitrobenzene	0.090		[68Fol]
m-Dinitrobenzene	0.116		[68Fol]
2,4-Dinitrotoluene	0.098		[68Fol]
Naphthalene	0.5		[68Fol]
Nitroacetophenone	0.105		[68Fol]
o-Nitroaniline	0.107		[68Fol]
p-Nitroanisol	0.075		[68Fol]
Nitrobenzaldehyde	0.101		[68Fol]
Nitrobenzene	0.111	76	[72Mai]
o-Nitrophenol	0.078		[68Fol]
m-Nitrotoluene	0.097		[68Fol]
o-Nitrotoluene	0.133		[68Fol]
p-Nitrotoluene	0.145		[68Fol]
Styrene	0.4		[68Fol]
Toluene		20	[72Mai]

References for 4.3

31Goe   Göppert-Mayer, M.: Ann. Phys. (Leipzig) **9** (1931) 273.

62Eck   Eckhardt, G., Hellwarth, R.W., McClung, F.J., Schwarz, S.E., Weiner, D., Woodbury, E.J.: Phys. Rev. Lett. **9** (1962) 455.
62Woo   Woodbury, E.J., Ng, W.K.: Proc. IRE **50** (1962) 2367.

66Bar   Barret, J.J., Tobin, M.C.: J. Opt. Soc. Am. **56** (1966) 129.
66Eck   Eckhardt, G.: IEEE J. Quantum Electron. **2** (1966) 1.

67Blo   Bloembergen, N.: Am. J. Phys. **35** (1967) 989.
67Sha   Shapiro, S.L., Giordmaine, J.A., Wecht, K.W.: Phys. Rev. Lett. **19** (1967) 1093.
67Wig   Wiggins, T.A., Wick, R.V., Foltz, N.D., Cho, C.W., Rank, D.H.: J. Opt. Soc. Am. **57** (1967) 661.

68Bre   Bret, C.G., Weber, H.P.: IEEE J. Quantum Electron. **4** (1968) 807.
68Den   Denariez, M., Bret, G.: C. R. Acad Sci. (Paris) **265** (1968) 144.
68Fol   Foltz, N.D., Cho, C.W., Rank, D.H., Wiggins, T.A.: Phys. Rev. **165** (1968) 396.

69Col   Colles, M.J.: Opt. Commun. **1** (1969) 169.
69Rah   Rahn, O., Maier, M., Kaiser, W.: Opt. Commun. **1** (1969) 109.

70Alf   Alfano, R.R., Shapiro, S.L.: Phys. Rev. A **2** (1970) 2376.
70Mac   Mack, M.E., Carman, R.L., Reintjes, J., Bloembergen, N.: Appl. Phys. Lett. **16** (1970) 209.
70Poh   Pohl, D., Kaiser, W.: Phys. Rev. B **4** (1970) 31.
70Rot   Rother, W.: Z. Naturforsch. A **25** (1970) 1120.

71Lau   Laubereau, A., von der Linde, D., Kaiser, W.: Phys. Rev. Lett **27** (1971) 802.
71Lin   von der Linde, D., Laubereau, A., Kaiser, W.: Phys. Rev. Lett. **26** (1971) 954.

72Car   Carman, R.L., Mack, M.E.: Phys. Rev. A **5** (1972) 341.
72Lau   Laubereau, A., von der Linde, D., Kaiser, W.: Phys. Rev. Lett. **28** (1972) 1162.
72Mai   Maier, M. Kaiser, W., in: Laser Handbook, Arecchi, F.T., Schulz-Dubois, E.O. (eds.), Vol. 2, Amsterdam: North Holland, 1972, p. 1077.

73Lau   Laubereau, A., von der Linde, D., Kaiser, W.: Opt. Commun. **1** (1973) 173.

74Lau   Laubereau, A.: Chem. Phys. Lett. **27** (1974) 600.

75Cha   Chatelet, M., Ogsengorn, B.: Chem. Phys. Lett. **36** (1975) 73.
75Lau   Laubereau, A.: Unpublished data, 1975.

76Ber   Berne, B.J., Pecora, R.: Dynamic light scattering, New York: Wiley, 1976, Chap. 10.5.
76Mai   Maier, M.: Appl. Phys. **11** (1976) 209.

77Rys   Rysakov, V.M., Korotkov, V.I.: Sov. J. Quantum Electron. (English Transl.) **7** (1977) 83.

78Lau	Laubereau, A
78Map	Maple, J.R.,
79Pen	Penzkofer, A ...ctron. **6** (1979) 55.
83Sap	Sapondzhy... ron. (English Transl.) **13** (1983) 1062.
83Tel	Telle, H.R., Laubereau, A.: Chem. Phys. Lett. **94** (1983).
84Har	Harris, A.L., Berg, M., Brown, J.K., Harris, C.B., in: Ultrafast phenomena IV, Auston, D.H., Eisenthal, K.B. (eds.), Springer Ser. Chem. Phys., Vol. 38, Berlin, Heidelberg, New York, Tokyo: Springer-Verlag, 1984.
84Kru	Kruminsh, A.V., Nikogosyan, D.N., Oraevsky, A.A.: Sov. J. Quantum Electron. (English Transl.) **11** (1984) 1479.
86Han	Hanna, D.C., Pointer, D.J., Pratt, D.J.: IEEE J. Quantum Electron. **22** (1986) 332.
87Glo	Glownia, J.H., Misewich, J., Sorokin, P.P.: Opt. Lett. **12** (1987) 19.
89Agr	Agrawal, G.P.: Nonlinear fiber optics, Boston: Academic, 1989.
90Lai	Lai, K.K., Schüsslbauer, W., Amler, H., Bogner, U., Maier, M., Jordan, M., Jodl, H.J.: Phys. Rev. B **42** (1990) 5834.
93Fil	Filippo, A.A., Perrone, M.R.: Appl. Phys. B **57** (1993) 103.
94Yos	Yoshizawa, M., Hattori, Y., Kobayashi, T.: Phys. Rev. B **49** (1994) 13259.
95Go	Go, C.S., Lee, J.H., Chang, J.S.: Appl. Opt. **34** (1995) 2671.
95Lim	Limaye, R., Sen, P.K.: Phys. Rev. B **51** (1995) 1546.
97Bai	Bairamov, B.H., Aydinli, A., Bodnar, I.V., Rud, Yu.V., Nogoduyko, V.K., Toporov, V.V.: Inst. Phys. Conf. Ser. No. 155, Bristol: IOP, 1997, p. 993.
97Jo	Jo, M.S., Nam, C.H.: Appl. Opt. **36** (1997) 1149.
97Kam1	Kaminskii, A.A., Butashin, A.V., Eichler, H.-J., Grebe, D., MacDonald, R., Ueda, K., Nishioka, H., Odajima, W., Tateno, M., Song, J., Musha, M., Bagaev, S.N., Pavlyuk, A.A.: Opt. Mater. **7** (1997) 59.
97Kam2	Kaminskii, A.A., Eichler, H.-J., Grebe, D., MacDonald, R., Butashin, A.V.: Phys. Status Solidi (b) **199** (1997) R3.
97Kam3	Kaminskii, A.A., Hulliger, J., Eichler, H.-J., Findeisen, J., Butashin, A.V., MacDonald, R., Bagaev, S.N.: Phys. Status Solidi (b) **203** (1997) R9.
97Yos	Yoshida, H., Kmetik, V., Fujita, H., Nakatsuka, M., Yamanaka, T., Yoshida, K.: Appl. Opt. **36** (1997) 3739.

4.4 Phase conjugation

H.J. Eichler, A. Hermerschmidt, O. Mehl

4.4.1 Introduction

Phase conjugation is a nonlinear optical process which generates a light beam having the same wavefronts as an incoming light beam but opposite propagation direction, see Fig. 4.4.1. Therefore phase conjugation is also called wavefront reversal. A nonlinear optical device generating a phase-conjugated wave is called a *phase conjugator* or *Phase-Conjugate Mirror (PCM)*.

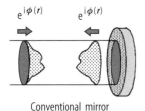
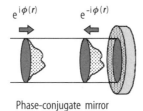

Fig. 4.4.1. Wavefront reflection at a conventional mirror and at a phase-conjugate mirror (PCM).

In Fig. 4.4.2 we consider the conjugation property of a PCM on a probe wave emanating from a point source. A diverging beam, after "reflection" from an ideal PCM, gives rise to a converging conjugate wave that precisely retraces the path of the incident probe wave, and therefore propagates in a time-reversed sense back to the same initial point source.

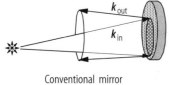
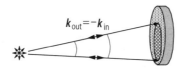

Fig. 4.4.2. Beam propagation after reflection at a conventional mirror and a PCM, both illuminated by a point source.

A phase conjugator reflects light, mostly laser beams, only if the incident power is high enough (*self-pumped phase conjugator*) or if the nonlinear material in the phase conjugator is pumped by additional laser beams, e.g. two additional beams in a *degenerate four-wave mixing* arrangement. In principle phase conjugation could be achieved also by a *deformable mirror* which is controlled by a wavefront sensor adapting the local mirror curvature to the incoming wavefront. Instead of a deformable mirror also a *2-dimensional phase modulator* could be used. However, deformable mirrors and other phase modulators up to now are more complicated set-ups with longer reaction periods than nonlinear optical phase conjugators to solve practical problems requiring phase conjugation.

4.4.2 Basic mathematical description

The incoming wave E_{in} is given by (4.4.1) with frequency f, where the amplitude E_0 and phase Φ are combined to the complex amplitude A. The complex conjugate is denoted by c.c.

$$E_{\text{in}}(x,y,z,t) = \frac{1}{2} E_0(x,y,z)\, e^{2\pi i (ft + \Phi(x,y,z))} + \text{c.c.} = A(x,y,z)\, e^{i\omega t} + \text{c.c.}, \qquad (4.4.1)$$

$$A(x,y,z) = \frac{1}{2} E_0(x,y,z)\, e^{2\pi i \Phi(x,y,z)}. \qquad (4.4.2)$$

The phase-conjugated wave exhibits the same wavefronts, however the sign of the phase Φ is inverted due to the inverted propagation direction. Thus, the phase-conjugated wave E_{pc} can be written as (4.4.3):

$$E_{\text{pc}}(x,y,z,t) = \frac{1}{2} E_0(x,y,z)\, e^{2\pi i (ft - \Phi(x,y,z))} + \text{c.c.} = A_{\text{pc}}(x,y,z)\, e^{i\omega t} + \text{c.c.}, \qquad (4.4.3)$$

$$A_{\text{pc}}(x,y,z) = \frac{1}{2} E_0(x,y,z)\, e^{-2\pi i \Phi(x,y,z)} = A^*(x,y,z). \qquad (4.4.4)$$

As can be seen A_{pc} equals the complex-conjugated A^*, which explains the term *phase conjugation*. From (4.4.1) and (4.4.3) we derive that the incident and phase-conjugated wave are also related to each other by

$$E_{\text{in}}(x,y,z,-t) = E_{\text{pc}}(x,y,z,t). \qquad (4.4.5)$$

Thus, the phase-conjugate wave E_{pc} propagates as if one would reverse the temporal evolution of the incident wave E_{in}. Therefore the term *"time-reversed replica"* is sometimes used to describe the phase-conjugate wave.

An ideal PCM also maintains the polarization state of an incident wave after phase conjugation. As an example, a probe wave that is Right-Handed Circularly Polarized (RHCP) will result in a RHCP-reflected wave after conjugation. This is in contrast to a conventional mirror, which reflects an incident RHCP field to yield a Left-Handed Circularly Polarized (LHCP) wave [82Pep].

One should realize that an ideal phase-conjugated wave exhibits the same frequency f as the incident wave and reveals the same polarization state. Often, real phase conjugators do not have these properties. However, if an PCM maintains the polarization state it is called a *"vector phase conjugator"*.

The nonlinear optical process which comes closest to yielding an ideal phase-conjugate wave is the backward-going, degenerate four-wave mixing interaction. Other classes of interaction (e.g. stimulated effects) result in nonideal conjugate waves due to frequency shifts, nonconjugated field polarization, etc. Although the application of stimulated effects, especially Stimulated Brillouin Scattering (SBS), yields to nonideal phase-conjugate mirrors they are used the most to solve practical problems requiring phase conjugation (e.g. compensation of phase distortions in high average power laser systems [99Eic]).

4.4.3 Phase conjugation by degenerate four-wave mixing

Four-wave mixing can be understood as a real-time holographic process, which facilitates phase conjugation. If the frequencies of the incoming wave, the two additionally required pump waves, and the phase-conjugated or reflected wave are equal the process is called *Degenerate Four-Wave Mixing (DFWM)*.

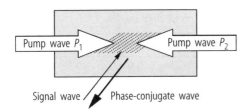

Fig. 4.4.3. Setup for phase conjugation by four-wave mixing.

In Fig. 4.4.3 the setup for phase conjugation by four-wave mixing is shown.

Interference of the incoming wave $E_{\mathrm{in}}(x,y,z,t)$ with the pump wave $P_1(x,y,z,t)$ results in a spatially periodic intensity pattern which modulates the absorption coefficient or refractive index of the optical material resulting in a dynamic or transient amplitude or phase grating. The other pump $P_2(x,y,z,t)$ is diffracted at this grating producing the phase-conjugated wave. This corresponds to the conventional holographic process where the read-out wave is replaced by the second pump wave counterpropagating to the first pump or reference wave.

Recording of a hologram is the first step in phase conjugation and leads to a transmission function t in the hologram plane (variables will not be noted furthermore to simplify the readability):

$$t \propto |P_1 + E_{\mathrm{in}}|^2 = \cdots = |P_1|^2 + P_1 E_{\mathrm{in}}^* + P_1^* E_{\mathrm{in}} + |E_{\mathrm{in}}|^2 \ . \tag{4.4.6}$$

During the read-out the phase-conjugate wave can be generated. Therefore, the hologram is illuminated with a second pump wave P_2, propagating in the opposite direction to P_1. This is in contrast to standard holography. Since P_2 precisely retraces the path of P_1 in the opposite propagation direction, P_2 equals P_1^*. This means, that the two pump beams should be phase-conjugated to each other, so that their spatial phases cancel and do not influence the phases of the reflected beam.

In the hologram plane we obtain a field strength distribution as follows:

$$P_2 t = P_1^* t \propto P_1^* |P_1|^2 + |P_1|^2 E_{\mathrm{in}}^* + (P_1^*)^2 E_{\mathrm{in}} + P_1^* |E_{\mathrm{in}}|^2 \ . \tag{4.4.7}$$

The second term $|P_1|^2 E_{\mathrm{in}}^*$ corresponds to the phase-conjugate wave of E_{in}. The other expressions lead to three additional waves which are not of interest here. They can be suppressed in thick nonlinear media in case of Bragg diffraction.

Common dynamic grating materials for phase conjugation are:

– photorefractive crystals ($LiNbO_3$, $BaTiO_3$, ...),
– liquid crystals (molecular reorientation effects),
– laser crystals (spatial hole-burning, excited-state absorption),
– saturable absorbers,
– absorbing gases and liquids (thermal gratings),
– semiconductors (Si, GaAs, ...).

The disadvantage of phase conjugation by four-wave mixing is the requirement of two additional pump waves for the nonlinear medium. However, this facilitates amplification of the phase-conjugate wave in the nonlinear medium at the same time. Vector phase conjugation is not achieved by this simple DFWM scheme, but requires polarization-dependant interactions.

4.4.4 Self-pumped phase conjugation

Self-pumped phase conjugation of continuous-wave laser beams in the lower power range (mW ... W) can be realized in Four-Wave Mixing (FWM) loop arrangements using photorefractive media, see Sect. 4.4.6 for detailed discussion.

Table 4.4.1. Brillouin gain coefficient g and phonon lifetime τ for different SBS media.

SBS medium	Brillouin gain coefficient g [cm/GW]	Phonon lifetime τ [ns]
SF_6 (20 bar)	25	15
Xe (50 bar)	90	33
C_2F_6 (30 bar)	60	10
CS_2	130	5.2
CCl_4	6	0.6
Acetone	20	2.1
Quartz	2.4	5

For pulsed lasers, self-pumped phase conjugation is achieved by stimulated scattering. For practical application, stimulated Brillouin scattering [72Kai] in

- gases (SF_6, Xe, C_2F_6, CH_4, N_2, ...) under high pressure,
- liquids (CS_2, CCl_4, acetone, freon, $GeCl_4$, methanol, ...), and
- solids (bulk quartz glass, glass fibers)

is used.

Table 4.4.1 shows the Brillouin gain coefficient g and the phonon lifetime τ for different gaseous, liquid, and solid-state SBS media.

A phase-conjugate mirror consists simply of a gas or liquid cell or a fiber piece. The incoming wave is focused into the material where an oppositely traveling wave is generated initially by spontaneous scattering. This wave interferes with the incoming wave and induces a sound wave or another type of phase grating reflecting the incoming beam similarly as a dielectric multilayer mirror. The induced density variations have the frequency of the initial sound wave, which is amplified therefore and reinforces the backscattering. A detailed discussion of stimulated Brillouin scattering is given in Sect. 4.3.3.2.

The amplification depends strongly on the extension of the interference area. Therefore the phase-conjugated backscattered part dominates, leading to an exponential rise of the reflected phase-conjugated signal. The wavefronts of the sound-wave grating match the wavefronts of the incoming beam. Any disturbance of the incident wavefront will result in a self-adapted mirror curvature with response times in the ns range.

For applications the "threshold", reflectivity, and conjugation fidelity are the most important parameters that characterize the performance of a Brillouin-scattering phase-conjugate mirror. A sharply defined threshold does not exist for the nonlinear SBS process. However, after exceeding a certain input energy a steep increase of reflectivity can be observed. Often this is called the energy threshold of the phase-conjugate medium. For long pulses as compared to the phonon lifetime (typically several ns) the SBS is expected to become stationary. In this case the energy threshold can be substituted by a power threshold. Well above this threshold, the reflectivity is not stationary but exhibits statistical fluctuations because SBS starts from noise.

It is important to emphasize that the power and not the intensity determines the "threshold" in case of strongly monochromatic input waves. Slight focusing leads to lower intensity, but also to a longer Rayleigh length and a larger interaction area. Stronger focusing reduces the interaction length, but results in stronger refractive-index modulation. Both effects compensate each other if the interaction length is not limited by the coherence length.

Practically, for most laser sources the coherence length is rather short. Here the interaction length should be short compared to the coherence length. This requires adequate focusing of the beam into the SBS medium. Focal length and scattering material have to be chosen suitable to achieve a high SBS reflectivity and a good reproduction of the wavefront. Side effects in the material like absorption, optical breakdown, or other scattering processes have to be avoided. Figure 4.4.4

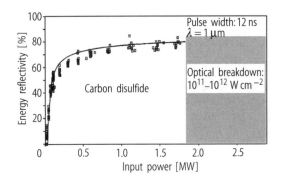

Fig. 4.4.4. Commonly used carbon disulfide (CS$_2$) shows an SBS threshold of about 18 kW (pulse peak power) under stationary conditions.

Table 4.4.2. SBS threshold, max. reflectivity, far-field fidelity, M^2-limit, and power limit for different fiber phase conjugators, coherence length 1.5 m. The reflectivity is corrected with respect to Fresnel and coupling losses.

Core diameter [µm]	SBS threshold [kW]	Maximum reflectivity [%]	Far-field fidelity [%]	M^2-limit	Power limit [kW]
200	17	80	93	63	160
100	6.4	80	91	31	40
50	2.0	88	70	16	10
25	0.3	86	–	8	2.5

shows the energy reflectivity of carbon disulfide as a function of the input power at 1 µm wavelength. Carbon disulphide shows one of the smallest power thresholds for liquids of about 18 kW. Applying gases as SBS media, the power thresholds are about one order of magnitude higher. A saturation of energy reflectivity close to 80 % is a typical value for liquid SBS media, although reflectivities up to 96 % had been demonstrated [91Cro].

For high-power input pulses bulk solid-state media like quartz are investigated as SBS media, too [97Yos]. To reduce the power threshold of SBS a waveguide geometry can be applied [95Jac]. The beam intensity inside the waveguide is high within a long interaction length resulting in low power thresholds. To avoid toxic liquids and gases under high pressure multimode quartz fibers can be used [97Eic]. The lower Brillouin gain of quartz glass compared to suitable SBS gases and liquids can be overcome using fibers with lengths of several meters resulting in SBS thresholds down to 200 W peak power [98Eic].

The power threshold P_{th} can be estimated from (4.4.8), where A_{eff} is the effective mode field area inside the fiber core, L_{eff} the effective interaction length, which depends on the coherence length, and g is the Brillouin gain coefficient; for quartz g is about 2.4 cm/GW [89Agr].

$$P_{th} = \frac{21 \, A_{eff}}{L_{eff} \, g} \, . \tag{4.4.8}$$

Table 4.4.2 shows the power threshold, the maximum energy reflectivity, the far-field fidelity, the M^2-limit (see below), and an approximated power limit of fiber phase conjugators with different core diameters. The used quartz-quartz fibers had a step-index geometry and a numerical aperture of 0.22. They were investigated with an Nd:YAG oscillator amplifier system generating pulses of 30 ns (FWHM) at 1.06 µm wavelength. Regarding applications it is important to couple also spatially aberrated beams into the fiber. The upper limit for the beam parameter product is due to the finite numerical aperture and the core diameter of the fiber. This can by expressed by a "times diffraction limit value" M^2, see Chap. 2.2 for further information about beam characterization. The upper power limit is approximated assuming a damage threshold above 500 MW/cm^2 for ns pulses.

An important feature of a fiber phase conjugator is the threshold behavior for different M^2-values of the incoming beam. In case of a fiber the SBS threshold is nearly independent of the incoming beam quality. This is caused by mode conversion inside the fiber resulting in homogeneous illumination and therefore in constant SBS reflectivity. In case of a Brillouin cell the reflectivity depends on the far-field distribution of the incoming beam. Here phase distortions result in amplitude fluctuations in the focal region. A comparison between a diffraction-limited beam ($M^2 = 1.0$) and a highly distorted beam ($M^2 = 10$) showed an increase of the SBS threshold of 300 % in case of the Brillouin cell. For the fiber phase conjugator no remarkable changes of the power threshold were observed [97Eic].

Practically, the reproduction of the initial wavefront is not perfect after phase conjugation. To characterize the deviation with respect to the reference wave the term *fidelity* F is introduced [77Zel]:

$$F = \frac{\left|\int E_{\text{in}} E_{\text{p}}^* \, d^2 r\right|^2}{\int |E_{\text{in}}|^2 d^2 r \cdot \int |E_{\text{p}}|^2 d^2 r} \,. \tag{4.4.9}$$

The fidelity equals unity in case of perfect wavefront reproduction and is smaller than unity for practical cases. To calculate the fidelity, the electric field distribution of the incident signal E_{in} and the not perfectly phase-conjugated wave E_{p} – the perfectly phase-conjugated wave is denoted E_{pc} in Sect. 4.4.2 – has to be known. The determination requires sophisticated measurement equipment. In contrary, the *far-field fidelity* can be measured with less effort and is therefore often used. The transmission through an aperture of the phase-conjugate signal is compared with the transmission of the input signal. The ratio is called far-field fidelity, because the aperture is placed in the focal plane of a focusing lens.

4.4.5 Applications of SBS phase conjugation

Phase conjugation generates a wave which retraces the incoming wave in a time-reversed way. Thereby it is possible to eliminate phase distortions in optical systems. For example, in a solid-state laser amplifier, the incoming beam is not only amplified but suffers also from phase distortions due to thermal refractive-index changes in the laser crystal. After passing this amplifier crystal, the beam is reflected by a phase conjugator and passes the crystal a second time. As the wavefronts are inverted with respect to the propagation direction, the refractive-index changes reduce the phase distortions and after the second passage, these distortions disappear so that the beam quality of the incoming wave is reproduced. In Fig. 4.4.5 a double-pass scheme with phase-conjugate mirror to compensate for phase distortions is shown.

Typically, phase conjugators are applied in *Master Oscillator Power Amplifier* (*MOPA*) setups, where a nearly diffraction-limited master oscillator beam is increased in power within an amplifier

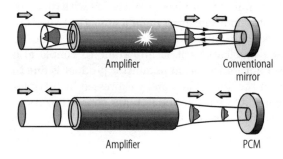

Fig. 4.4.5. Double-pass scheme with phase-conjugate mirror to compensate for phase distortions.

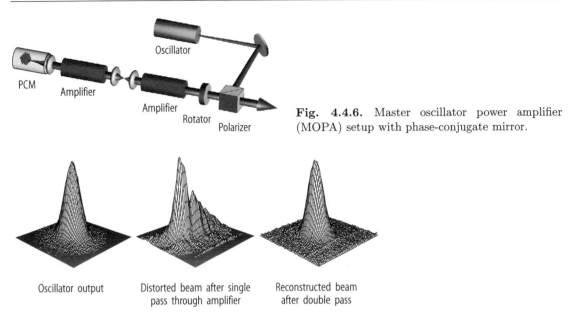

Fig. 4.4.6. Master oscillator power amplifier (MOPA) setup with phase-conjugate mirror.

Fig. 4.4.7. Far-field intensity distributions of the oscillator beam, the distorted beam after single-pass amplification, and the highly amplified beam after double-pass amplification with phase conjugation.

arrangement, see Fig. 4.4.6. After the first amplification pass the beam quality is reduced due to thermally induced phase distortions. The spatial-distorted beam enters the SBS mirror and becomes phase-conjugated. The initial beam quality of the master oscillator can be roughly reproduced after the second amplification pass. The amplified beam is extracted with an optical isolation, which consists in this case of a Faraday rotator and a polarizer.

Figure 4.4.6 shows a MOPA system producing up to 210 W average output power at 2 kHz average repetition rate (1.08 μm wavelength). The system is part of an advanced setup yielding up to 520 W average output power [99Eic]. The oscillator beam has a nearly diffraction-limited beam quality ($M^2 < 1.2$) which is already reduced in front of the first amplifier ($M^2 \cong 1.5$). This results from optical components between oscillator and amplifier which introduce phase distortions. After single-pass amplification the beam quality decreases to $M^2 \cong 5$ due to phase distortions introduced by both pumped amplifier rods at 6.5 kW pumping power for each amplifier. After phase conjugation and double-pass amplification the initial beam quality can be nearly reproduced ($M^2 < 1.9$). Differences between the initial and final beam quality are caused by a fidelity smaller than unity and diffraction at several apertures in the amplifier chain.

The performance of the phase-conjugate mirror can be illustrated by far-field intensity profiles recorded at different positions in the setup. In Fig. 4.4.7 the oscillator output beam exhibits a smooth Gaussian profile corresponding to the nearly diffraction-limited beam quality. After single-pass amplification the reduction of beam quality is confirmed by a strongly aberrated far-field profile. After phase conjugation and double-pass amplification the initial intensity distribution can be nearly reproduced. In this example the average power of the master oscillator beam (approx. 1 W) was increased to 130 W after double-pass amplification.

Presently, phase distortion elimination in double or multipass laser amplifiers is the most often application of phase conjugation. In addition phase conjugators are useful as mirrors in laser oscillators replacing one of the conventional mirrors. Again, the phase conjugator eliminates phase distortions in the laser medium induced by optical or discharge pumping. For recent advances and applications of SBS-phase-conjugation see [02Eic, 03Rie, 04Rie].

4.4.6 Photorefraction

The photorefractive effect belongs to the nonlinear optical effects with the highest sensitivity for operation at low optical intensity levels. Photorefractive phase conjugators are able to operate using intensities of only mW/cm^2. The price paid of the low intensity performance is diminished speed. The response times of recent photorefractive phase conjugators span in the range of milliseconds to several minutes.

The photorefractive effect describes light-induced refractive-index changes in the material when the incident light is spatially nonuniform [88Gue, 93Yeh, 95Nol, 96Sol]. The spatial nonuniformity distinguishes the photorefractive effect from other common nonlinear optical effects that occur under spatial uniform intensity. The maximum refractive-index change induced in a photorefractive material does not occur necessarily locally where the light intensity is a maximum. The nonlocal response occurs because electric charges move and are stored inside the material. In case of classical nonorganic bulk photorefractive materials, such as ferroelectric oxides (BaTiO$_3$, LiNbO$_3$, KNbO$_3$), sillenites (Bi$_{12}$SiO$_{20}$, Bi$_{12}$TiO$_{20}$, Bi$_{12}$GeO$_{20}$) or semi-insulating semiconductors (GaAs, InP, CdTe), electrons (or holes) are photoexcited from localized impurity centers or defect sites, which are energetically located deep in the band gap of the material, into the conduction (or valence) band.

The energy of the exciting photons is smaller than the band-gap energy. Free carriers excited in bright crystal regions move due to diffusion and drift into dark crystal regions where they are trapped by empty defect sites, see Fig. 4.4.8. As a consequence of separated and trapped electric charges the formation of space-charge electric fields occurs. These electric fields change the refractive index of the material by electrooptics effects, usually the Pockels effect.

Nonuniform illumination occurs when two coherent laser beams interfere in the crystal. The intersecting beams create a periodical interference pattern. The formation of a photorefractive index grating due to a sinusoidal intensity pattern is shown in Fig. 4.4.9. When diffusion is the main effect for the transport of the excited charge carriers (there is no external electric field applied on the crystal) the electric-field maxima are shifted by a quarter fringe spacing relative to the intensity maxima. This $\pi/2$ phase shift of the induced index grating plays a fundamental role in photorefractive non-linear optical wave mixing. It allows for an energy transfer between the two beams writing the grating in a process called two-wave mixing. One of the beams (called signal beam) is amplified at the expense of the other beam (called pump beam).

A phase-conjugate beam can be created by four-wave mixing processes. In this case the two-wave mixing arrangement is extended with a second pump beam which counterpropagates with respect to the first pump beam, see Fig. 4.4.3. In case of external pump beams, the phase-conjugation

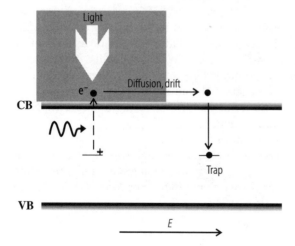

Fig. 4.4.8. Band transport model of photorefraction.

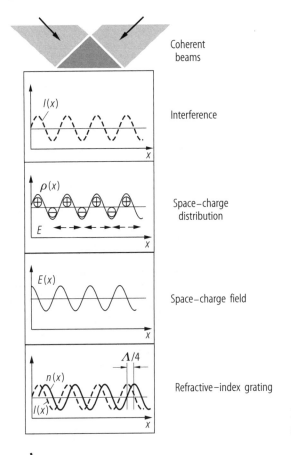

Fig. 4.4.9. Formation of a photorefractive index grating due to a sinusoidal intensity pattern.

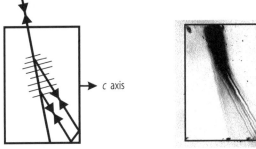

Fig. 4.4.10. Scheme of photorefractive total-internal-reflection phase conjugator (cat conjugator), left. The light propagation in the crystal can be seen due to scattering, right.

process may be highly efficient leading to large reflectivities well above 100 % in relation to the incoming power.

Self-pumped phase conjugators require only a single incident beam and because of their simplicity they are more advantageous for practical applications. The operation of photorefractive self-pumped phase conjugators is based on a non-linear optical process called beam fanning. When a single beam is incident on a photorefractive crystal, some light is scattered inside the crystal. This scattered light forms a set of gratings with the incident light and is amplified by two-wave mixing. This process was named fanning because a broad fan of scattered amplified light is generated emerging from the crystal.

Perhaps the most commonly used photorefractive self-pumped phase conjugator type is the so-called cat conjugator [82Fei]. In this case the first pump beam is generated from the incident beam by fanning, the second pump beam by backreflection on the crystal corner. Figure 4.4.10 shows a rhodium-doped barium titanate crystal which acts as a cat conjugator for an incident beam of 5 mW

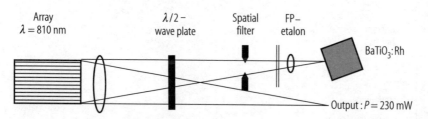

Fig. 4.4.11. Coherent diode laser array coupled by a phase-conjugating BaTiO$_3$:Rh crystal [98Lob].

optical power at 808 nm wavelength. The formed internal phase-conjugation loops can be observed in the lower right-hand corner of the crystal. Self-pumped phase-conjugate reflectivities as high as 60–80 % have been reported for visible and near-infrared wavelengths by numerous investigators using photorefractive crystals in various arrangements [85Gue, 86Pep, 95Mu, 94Wec, 97Huo].

The efficient operation of photorefractive phase conjugators at low and moderate power levels makes this type of device attractive especially for diode-laser applications. Free-running high-power diode laser arrays emit laser beams of poor spatial and spectral quality. Optical phase-conjugate feedback can increase both the spatial and the temporal coherence of the radiation. Figure 4.4.11 shows an external-cavity diode laser system comprising a photorefractive BaTiO$_3$ crystal as phase conjugator, a Fabry-Perot etalon, and a spatial filter forcing the laser diode array to operate in a single spatial and a single longitudinal mode [98Lob]. The coherence length of the phase-conjugate laser system has been increased by a factor of 70 and the output has become almost diffraction-limited. The output power is reduced from 440 mW to 230 mW.

References for 4.4

72Kai Kaiser, W., Maier M.: Stimulated Rayleigh, Brillouin and Raman Spectroscopy, Laser Handbook, Arecchi, F.T., Schulz-DuBois, E.O. (eds.), Amsterdam: North-Holland Publ. Co., 1972, pp. 1077–1150.

77Zel Zel'dovich, B.Ya., Shkunov, V.V.: Wavefront reproduction in stimulated Raman scattering; Sov. J. Quantum Electron. (English Transl.) **7** (5) (1977) 610–615.

82Fei Feinberg, J.: Self-pumped, continuous-wave phase conjugator using internal reflection; Opt. Lett. **7** (1982) 486.

82Pep Pepper, D.M.: Nonlinear optical phase conjugation; Opt. Eng. **21** (1982) 156–183.

85Gue Günter, P., Voit, E., Zha, M.Z., Albers, J.: Self-pulsation and optical chaos in self-pumped photorefractive $BaTiO_3$; Opt. Commun. **55** (1985) 210–214.

86Pep Pepper, D.M.: Hybrid phase conjuagtor/modulators using self-pumped 0°-cut and 45°-cut $BaTiO_3$ crystals; Appl. Phys. Lett. **49** (16) (1986) 1001–1003.

88Gue Günter, P., Huignard, J.-P.: Photorefractive materials and their applications I–II, Topics in Applied Physics, Vol. 61–62, Berlin: Springer-Verlag, 1988.

89Agr Agrawal, G.P.: Nonlinear fiber optics, Boston: Academic Press, 1989.

91Cro Crofts, G.J., Damzen, M.J., Lamb, R.A.: Experimental and theoretical investigation of two-cell stimulated-Brillouin-scattering systems; J. Opt. Soc. Am. B **8** (1991) 2282–2288.

93Yeh Yeh, P.: Introduction to photorefractive nonlinear optics, New York: John Wiley & Sons, Inc 1993.

94Wec Wechsler, B.A., Klein, M.B., Nelson, C.C., Schwartz, R.N.: Spectroscopic and photorefractive properties of infrared-sensitive rhodium-doped barium titanate; Opt. Lett. **19** (8) (1994) 536–538.

95Jac Jackel, S., et al.: Low threshold, high fidelity, phase conjugate mirrors based on CS_2 filled hollow waveguide structures; JNOPM **11** (1995) 89–97.

95Mu Mu, X., Shao, Z., Yue, X., Chen, J., Guan, Q., Wang, J.: High reflectivity self-pumped phase conjugation in an unusually cut Fe-doped $KTa_{1-x}Nb_xO_3$ crystal; Appl. Phys. Lett. **66** (1995) 1047.

95Nol Nolte, D.D.: Photorefractive effects and materials, Norwell: Kluwer Academic Publishers, 1995.

96Sol Solymar, L., Webb, D.J., Grunnet-Jepsen, A.: The physics and applications of photorefractive materials, Oxford: Clarendon Press, 1996.

97Eic Eichler, H.J., Haase, A., Kunde, J., Liu, B., Mehl, O.: Fiber phase-conjugator as reflecting mirror in a MOPA-arrangement, Solid State Lasers VI, San José (California); Proc. SPIE (Int. Soc. Opt. Eng.) **2986** (1997) 46–54.

97Eic Eichler, H.J., Kunde, J., Liu, B.: Quartz fibre phase conjugators with high fidelity and reflectivity; Opt. Commun. 139 (1997) 327–334.

97Huo   Huot, N., Jonathan, J.M.C., Rytz, D., Roosen, G.: Self-pumped phase conjugation in a ring cavity at 1.06 μm in cw and nanosecond regimes using photorefractive $BaTiO_3$:Rh; Opt. Commun. **140** (1997) 296.

97Yos   Yoshida et al.: SBS phase conjugation in a bulk fused-silica glass at high energy operation, CLEO 1997; OSA Tech. Dig. Ser. **11** (1997) 117–118.

98Eic   Eichler, H.J., Dehn, A., Haase, A., Liu, B., Mehl, O., Rücknagel, S.: High repetition rate continuously pumped solid state lasers with phase conjugation, Solid State Lasers VII, San José (California); Proc. SPIE (Int. Soc. Opt. Eng.) **3265** (1998) 200–210.

98Lob   Lobel, M., Petersen, P.M., Johansen, P.M.: Single-mode operation of a laser-diode array with frequency-selective phase-conjugate feedback; Opt. Lett. **23** (1998) 825.

99Eic   Eichler, H.J., Mehl, O.: Multi amplifier arrangements with phase conjugation for power-scaling of high-beam quality solid state lasers, Solid State Lasers VIII, San José (California): Proc. SPIE (Int. Soc. Opt. Eng.) **3613** (1999) 483–492.

02Eic   Eichler, H.J., Mocofanescu, A., Riesbeck, T., Risse, E., Bedau, D.: Stimulated Brillouin scattering in multimode fibers for optical phase conjugation; Opt. Commun. **209** (2002) 391–395.

03Rie   Riesbeck, T., Risse, E., Eichler, H.J.: Pulsed solid-state laser systems with high brightness by fiber phase conjugation; Proc. SPIE (Int. Soc. Opt. Eng.) **5120** (2003) 494–499.

04Rie   Riesbeck, T., Risse, E., Mehl, O., Eichler, H.J.: Multi-kilohertz pulsed laser systems with high beam quality by phase conjugation in liquids and fibers; in: Brignon, A., Huignard, J.-P. (eds.): Phase conjugate laser optics, Hoboken, New Jersey: John Wiley & Sons, 2004.

Index

$\alpha-\mathrm{HIO_3}$ 143, 183
$\alpha-$Iodic Acid   143
$\beta-\mathrm{BaB_2O_4}$ 142
π-pulse   36
$^6\mathrm{LiNbO_3}$ 227
$^6\mathrm{LiTaO_3}$ 227
$^7\mathrm{LiNbO_3}$ 227
1,1,1-Trichloroethane   226
1,2-Dichloroethane   224
1,2-Diethylbenzene   224
1,2-Dimethylcyclohexane   224
1,3-Dibromobenzene   224
1,3-Pentadiene   225
1,4-Dimethylcyclohexane   224
1,4-Dimethylnitrobenzene   231
1,4-Dioxane   224
1-Bromonaphthalene   225, 231
1-Bromopropane   224
1-Chloronaphthalene   225
1-Fluoro-2-chlorobenzene   224
2π-pulse   35
2,2-Dichlorodiethylether   224
2,4-Dinitrotoluene   231
2-Bromopropane   224
2-Ethylnaphthalene   225
2-Nitropropane   225
3-Methyl-4-Nitro-Pyridine-1-Oxide   142
3-Methylbutadiene   224
4-Hydroxy-3-Methoxy-Benzaldehyde (Vanillin)   142

Abbe's number   97
ABCD
  matrix   111
  transformation   120
Aberration   127
  lens   117
  spherical   117
    third-order   119
Absorber
  saturable   237
Absorbing
  gas   237
  liquid   237
Absorption
  excited-state   237

Acetic acid   224, 229
Acetone   224, 226, 229, 230, 238
Achromatic correction   97
Acoustic phonons   217
ADA   143, 165
Adiabatic
  equations   20
  pulse amplification   25
ADP   143, 160, 161, 164, 166, 168, 176, 180
$\mathrm{Ag_3AsS_3}$ 142, 166, 171, 173, 186
$\mathrm{Ag_3SbS_3}$ 142
$\mathrm{AgGaS_2}$ 142, 166, 171, 173, 186
$\mathrm{AgGaSe_2}$ 142, 166, 174, 186
Air
  thick lens in   113
Airy's disc   92
$\mathrm{Al_2O_3}$ 225
Ammonia   225
Ammonium Dihydrogen Arsenate   143
Ammonium Dihydrogen Phosphate   143
Amplification
  stimulated   221
Amplifier
  feed-back   3
Angle
  birefringence   145
  Brewster's   101
  divergence   82
  walk-off   145
Angular
  moment   56
  -spectrum representation   87
Aniline   224, 229
Annular aperture   92
Anti-Stokes
  emission, higher-order   222
  line   218
  Raman scattering, coherent   222
  scattering   222
Aperture
  annular   92
  circular   90
  length   151
  rectangular   89
Apodization
  Gaussian   114

Approach
  perturbation 206
Approximation
  dipole 12
  far-field, Fraunhofer 87
  Fraunhofer 88
    far-field 87
  Fresnel's 85, 86
  Fresnel–Kirchhoff 86
  plane-wave 152
  Rayleigh–Sommerfeld 86
  rotating-wave 14
Ar 211
Argon laser 166
Astigmatic
  beam
    general 121, 123
    simple 120, 121
  general 64
  simple 63
  system
    general 116
Astigmatism
  inner 64
  intrinsic 60, 65
Axis
  principal 58
  of the beam 57
Azoxybenzene 231

Ba 210
$Ba_2NaNb_5O_{15}$ 142, 227
Banana (Barium Sodium Niobate) 142, 158, 179, 183
Barium Sodium Niobate (Banana) 142, 158, 179, 183
Barium vapor 225
Basov 3
$BaTiO_3$ 237, 242, 244
BBO 142, 157–161, 163–166, 168, 172, 177, 181, 184
Be 210
Beam
  Bessel
    diffraction-free 79
    real 80
    vectorial 80
  characterization 53
  classification 61
  conversion 65
  diameter
    generalized 57
  diffraction-free 79
  elliptical Gaussian 57
  extraordinary 145
  fanning 243
  fluctuation matrix 67

Gauss
  –Hermite 81
  –Laguerre 83
  –Schell model 53
Gaussian 53, 81
  elliptical 57
  general astigmatic 64, 121, 123
  Hermite–Gaussian 53
    higher-order 123
  Laguerre–Gaussian 53
  ordinary 145
  partially coherent 54
  positional stability 67
  principal axis of 57
  propagation 120
    Gaussian 111
  ratio, effective 60
  pseudo-symmetric 64
  simple astigmatic 63, 120, 121
  stigmatic 62, 120, 121
  waist 82
  width 57
    long-term 69
Benzaldehyde 224, 229
Benzene 224, 226, 229–231
Benzene-d_6 224
Benzonitrile 224
Benzonitrol 231
Benzoylchloride 231
Benzylidenaniline 231
Bessel beam
  diffraction-free 79
  real 80
  vectorial 80
Beta-Barium Borate 142
$Bi_{12}GeO_{20}$ 242
$Bi_{12}SiO_{20}$ 242
$Bi_{12}TiO_{20}$ 242
Biaxial crystal 105, 145
Birefringence 106
  angle 145
Bisectrix 147
Black-body radiator 47
Bloch vector 33
Bolometer 48
Brewster angle 101
  -tilted plate 76
Brillouin
  doublet 218
  gain coefficient 238
  lines 217
  scattering
    stimulated 217, 227, 229, 236
    stimulated, thermal 220, 227
Broadening
  collision 28

Doppler 28
　homogeneous 29, 223
　inhomogeneous 30, 223
　line 26, 28
　pressure 28
　saturation 29
　types of 29
Bromobenzene 224, 226, 229, 231
Bromopropane 224
Butane 225
Butyl-benzene (tert.) 224

$C_{10}H_{11}N_3O_6$ 142
$C_{10}H_{13}N_3O_3$ 142
$C_{11}H_{14}N_2O_3$ 142
C_2F_6 238
$C_6H_6N_2O_3$ 142
$C_8H_8O_3$ 142
Ca 210
Ca^+ 210
Cadmium Germanium Arsenide 142
Cadmium Selenide 142
Calcite 225, 227
Calculus
　Jones 75
　Mueller 77
Calorimeter 50
Calorimetry
　photoacoustic 50
Carbondioxide 225
Carbondisulfide 224, 226, 229–231
Carbonmonoxide 225
Carbontetrachloride 224, 229
Cardinal
　plane 110
　point 110
CARS (coherent anti-Stokes Raman scattering) 222
Cat conjugator 243
CBO 142, 159, 160
CCl_4 238
Cd 210
CDA 142, 156, 180
$CdGeAs_2$ 142, 166, 167, 174
CdSe 142, 171, 174, 186
CdTe 242
Centered moment 55
Cesium Borate 142
Cesium Dideuterium Arsenate 142
Cesium Dihydrogen Arsenate 142
Cesium Lithium Borate 142
Cesium vapor 226
CH_4 238
Characterization
　beam 53
Chlorine 226
Chlorobenzene 224, 226, 231

Chloroform 224, 229
Chloromethylbutane 224
Chloronaphthalene 231
Circular aperture 90
Classification 104
　beam 61
CLBO 142, 157, 160, 161
CO 211
　laser 166
CO_2 laser radiation
　harmonic generation of 166
　up-conversion of 171
Coefficient
　gain 18, 229
　Miller 147
　nonlinear
　　effective 150
　scattering 226
　Seebeck 48
Coherence length 238
Coherent
　anti-Stokes Raman scattering (CARS) 222
　interaction 31
　partially 54
　pulses, resonant 34
Collins integral 126
Collision broadening 28
Coma 119
Complex
　notation 6
　q-parameter 120
　refractive index 8, 103
　susceptibility 10
Concave
　grating
　　Rowland 114
　mirror
　　spherical 114
Condition
　phase-matching 208
　resonance 5
　steady-state 5
　symmetry
　　Kleinman 150
　threshold 4
Conductivity
　electric 6
Conjugation fidelity 238
Conjugator
　cat 243
Conservation
　of momenta 208
Constant
　time 31
Construction
　Listing's 110

Continuous wave
  optical parametric oscillation 176
Continuum generation
  picosecond 186
    in crystals 186
Contribution
  electrostrictive 227
Conversion
  beam 65
  efficiency
    quantum 153
  factor for SI and CGS-esu systems 150
  frequency
    efficiency 151
    in gases 209
Correction
  achromatic 97
Coupling
  laser fiber 129
  Stokes–anti-Stokes 222
  waveguide 127, 128
Cross section 18
  differential 217
  Raman scattering 223
  scattering 217
    differential 227
    Raman 223
Cross-spectral density 53
Cryogenic radiometer 47, 50
Crystal
  biaxial 105, 145
  data of 108
  isotropic 105
  laser 237
  liquid 237
  negative uniaxial 105
  optical 105
  optics 104
  photonic, data of 108
  photorefractive 237
  picosecond continuum generation in 186
  positive uniaxial 105
  uniaxial 145
    negative 146
    positive 146
Cs 210
CS_2 238
CsB_3O_5 142
CsD_2AsO_4 142
CsH_2AsO_4 142
$CsLiB_6O_{10}$ 142
$CuAlS_2$ 227
Cubic 105
Current density 5
Curvature
  phase, generalized 59

Cyclohexane 224, 229
Cyclohexanone 224
Cylindrical
  vector wave 78
  wave 79

D_2, $Q(2)$ 227
DAN 142, 157
Data
  of crystals 108
  of gases 108
  of infrared materials 108
  of liquids 108
  of metals 108
  of negative-refractive-index materials 108
  of optical glass 108
  of photonic crystals 108
  of polymeric materials 108
  of semiconductors 108
  of solid state laser materials 108
DCDA 142, 156, 161
Decay
  time T_1 15
  time T_2 16
Degenerate four-wave mixing 235, 236
Degree
  of polarization 77
Delta formulation, Miller 147
Density
  cross-spectral 53
  current 5
  distribution
    power 55
    power, far-field 55
  inversion 12, 13, 17
  matrix 206
  of electric charges 5
Dephasing 227
Depth
  penetration 102
Detector
  pyroelectric 49
  quantum 49
  thermal 48
Determination
  of the ten second-order moments 66
Deuterium 226
DFWM (Degenerate Four-Wave Mixing) 236
Diameter
  beam, generalized 57
Diamond 225
Dichloromethane 224
Dielectric medium 96
Dielectrics
  homogeneous 6
  isotropic 6
  linear 6

Difference frequency generation   144, 153, 172
   generation of IR radiation by   172
   in the far IR region   175
Differential
   cross section   217
   scattering
      cross section   227
Diffraction   84
   efficiency   92
   figure
      Fresnel's   93
   -free
      beam   79
      Bessel beam   79
   length   151
   pattern
      Fraunhofer   89
   scalar theory of   85
   theory
      Rayleigh–Sommerfeld–Debye   89
      time-dependent   89
   vector theory of   85
Diffractive optics   94
Diffuse emitter   47
Dimethylhexadiene   224
Dimethylsulfoxide (DMSO)   224
Dipole
   approximation   12
   Hertz's   8
   moment   11, 15
   oscillating   8
Disc
   Airy's   92
Dispersion
   formula   97
   -spreading length   151
Displacement
   electric   5
Distance
   Rayleigh   82
Distribution
   power density   55
      far-field   55
   Wigner   53
Divergence angle   82
DKB5   143, 165
DKDA   143, 165
DKDP   143, 156, 159, 160, 162
DMSO (dimethylsulfoxide)   224
Doped media
   propagation in   10
Doppler broadening   28
Doublet
   Brillouin   218
Dye laser radiation
   second harmonic generation of   164

Effective
   beam propagation ratio   60
   length   151
   nonlinear coefficient   150
   nonlinearity   147
Efficiency
   conversion
      quantum   153
   diffraction   92
   frequency-conversion   151
Einstein coefficient
   of induced emission   21
   of spontaneous emission   21
Electric
   charges, density of   5
   conductivity   6
   displacement   5
   field   5
   polarization   6
   susceptibility   6
Electromagnetic field   5
Electrostrictive contribution   227
Ellipse
   variance   57
Ellipsoid
   index   105
Ellipsometry   104
Elliptical Gaussian beam   57
Emission
   anti-Stokes, higher-order   222
   induced
      Einstein coefficient   21
   spontaneous   4, 15, 28
      Einstein coefficient   21
   Stokes, higher-order   222
Emitter
   diffuse   47
Energy
   flux   7
   radiant   45, 46
   relaxation   15
Entropy relaxation   16
Equation
   adiabatic   20
   fundamental
      of nonlinear optics   207
   Helmholtz   74, 79
   Maxwell's   5
   Maxwell–Bloch   15, 16
   rate   20
   slowly varying envelope (SVE)   74, 80
   SVE (slowly varying envelope)   74, 80
   wave   73, 78
Ethanol   224, 226, 229, 230
Ethyl-Benzene   224
Ethylene   225

Eu 210
Excitance
  radiant 46
Excited-state absorption 237
External reflection 101
Extraordinary beam 145

Factor
  conversion
    for SI and CGS-esu systems 150
  gain 220, 226, 227
  slit 92
Far field 58
  approximation
    Fraunhofer 87
  fidelity 239
  Fraunhofer approximation 87
  power density distribution 55
Far IR region
  difference frequency generation in 175
Faraday rotator 76
Feed-back amplifier 3
Femtosecond optical parametric oscillation 184
Ferroelectric oxide 242
Feynman representation 32
Fiber
  laser coupling 129
  phase conjugator 240
Fidelity 240
  conjugation 238
  far-field 239
Field
  electric 5
  electromagnetic 5
  far 58
  magnetic 5
  near 56
Fifth harmonic generation 153, 156
  of Nd:YAG laser radiation 161
Fluctuation matrix
  beam 67
Fluorinert
  FC 72 229
  FC 75 229
Fluorobenzene 224
Fluoromethane 224
Formula
  dispersion 97
  Fresnel's 98, 99
  Sellmeier's 97
Four-level system 24
Four-wave
  interaction 221
  mixing, degenerate 235, 236
Fourier
  optics 94
  transform 88

Fourth harmonic generation 153, 156
  of Nd:YAG laser radiation 160
  of Ti:sapphire laser radiation 164
Fraunhofer
  approximation 88
    far-field 87
  diffraction pattern 89
  far-field approximation 87
Free-space propagation 61
Freon 238
Frequency
  conversion
    efficiency 151
    gases 205
    in gases 209
    liquids 205
  mixing 208
  Rabi 33
  spatial 88
Fresnel
  approximation 85, 86
  diffraction figure 93
  formulae 98, 99
  –Kirchhoff approximation 86
  number 88
Fundamental equations
  of nonlinear optics 207

GaAs 175, 186, 242
Gain
  coefficient 18, 229
    Brillouin 238
  factor 220, 226, 227
  small-signal 11
Gallium Selenide 142
GaP 175, 227
Gas 205
  absorbing 237
  data of 108
  -eous media, mixture of 209, 211
  frequency conversion in 205, 209
  laser radiation
    second harmonic generation of 166
  mixture, metal-vapor inert 209, 210
  Raman
    parameters 227
    scattering 225
  scattering
    Raman 225
GaSe 142, 167, 174, 186
Gauss
  –Hermite beam 81
  –Laguerre beam 83
  –Schell model 53
  beam 53
Gaussian
  apodization 114

beam 53, 81
  elliptical 57
  propagation 111
  imaging 108
  line shape 27
  matrix 111
$Gd_2(MoO_4)_3$ 225
$GeCl_4$ 238
General
  astigmatic
    beam 64, 121, 123
    system 116
  parabolic
    system 112
  ray-transfer matrix 117
Generalized
  beam diameter 57
  phase curvature 59
Generation
  continuum
    picosecond 186
    picosecond, in crystals 186
  difference-frequency 144, 153, 172
    generation of IR radiation by 172
    in the far IR region 175
  fifth harmonic 153, 156
    of Nd:YAG laser radiation 161
  fourth harmonic 153, 156
    of Nd:YAG laser radiation 160
    of Ti:sapphire laser radiation 164
  harmonic 156
    of CO_2 laser radiation 166
  of harmonics
    of high-power Nd:glass laser radiation 162
    of iodine laser radiation 162
    of Nd:YAG laser radiation 161
  second harmonic 144, 152, 156
    in "nonlinear regime" 154
    of dye laser radiation 164
    of gas laser radiation 166
    of Nd:YAG laser radiation 156, 157, 159
    of Nd:YAG laser radiation, intracavity 158
    of ruby laser radiation 163
    of semiconductor laser radiation 164
    of Ti:sapphire laser radiation 163
  sixth harmonic 156
    of Nd:YAG laser radiation 161
  sum-frequency 144, 153, 167
    of UV radiation 167–169
  third harmonic 152, 156, 208
    of Nd:YAG laser radiation 159
    of Ti:sapphire laser radiation 163
Geometric optical radiance 54
Geometrical optics 108
Glass
  BSC-2 229

DF-3 229
  optical 97
    data of 108
Glycerol 229
Goos–Hänchen shift 103
Gradient-index lens 113
Grating 92
  concave
    Rowland 114
  thermal 237
Group
  point 150

H_2, Q(1) 227
Half-wave plate 76
Harmonic generation 156
  of CO_2 laser radiation 166
  of high-power Nd:glass laser radiation 162
  of iodine laser radiation 162
  of Nd:YAG laser radiation 161
Harmonics
  of high-power Nd:glass laser radiation 162
  of iodine laser radiation 162
  of Nd:YAG laser radiation 161
He 211
He-Ne laser 166
Heat pipe 209
Helmholtz equation 74, 79
Hermite–Gaussian beam 53
  higher-order 123
Hertz's dipole 8
Hexagonal 105
Hg 210, 211
$HgGa_2S_4$ 143, 171
Higher order
  anti-Stokes emission 222
  Hermite–Gaussian beam 123
  Stokes emission 222
Hole-burning
  spatial 237
Holography 237
Homogeneous 6
  broadening 29, 223
  dielectrics 6
  system 16
Huygens' principle 8, 85
Hydrogen 226
Hydrogenbromide 226
Hydrogenchloride 226
Hyperbolic
  propagation law 60

Imaging
  Gaussian 108
Index
  ellipsoid 105
  surface 105, 146

Indicatrix
  optical 145
Induced emission
  Einstein coefficient 21
Induction
  magnetic 5
Infrared material, data of 108
Inhomogeneous
  broadening 30, 223
  system 16
Inner astigmatism 64
InP 242
Input
  noise 220
InSb 229
Integral
  Collins 126
Intensity 8
  radiant 46
  saturation 18
Interaction 14
  coherent 31
  four-wave 221
  Hamiltonian 14
  length
    nonlinear 152
    quasistatic 151
  nonlinear
    length 152
  quasistatic
    length 151
  three-wave 144
Internal reflection 101
Intracavity second harmonic generation of Nd:YAG laser radiation 158
Intrinsic astigmatism 60, 65
Inversion density 12, 13, 17
Iodine laser radiation
  harmonics of 162
IR
  far
    region, difference frequency generation in 175
  material, data of 108
  mid
    region, optical parametric oscillation in 186
  near
    region, cw optical parametric oscillation in 176
    region, femtosecond optical parametric oscillation in 184
    region, nanosecond optical parametric oscillation in 176
    region, picosecond optical parametric oscillation in 180
  radiation

generation by difference frequency generation 172
Irradiance 46
Isopropanol 224, 226
Isotropic 7
  crystal 105
  dielectrics 6

Jones
  calculus 75
  matrix 75
  vector 75

K 210
$KB_5O_8 \cdot 4D_2O$ 143
$KB_5O_8 \cdot 4H_2O$ 143
KB5 143, 161, 162, 165, 166, 169
KD_2AsO_4 143
KD_2PO_4 143
KDP 143, 156, 159–164, 166, 167, 176, 180, 186
KH_2PO_4 143
Kleinman symmetry conditions 150
$KNbO_3$ 143, 157, 158, 163–165, 179, 185, 242
Kr 211
Kramers–Kronig relation 97
KTA 143, 173, 179, 183, 185
$KTiOAsO_4$ 143
$KTiOPO_4$ 143
KTP 143, 156–159, 170, 173, 178, 179, 182–185

Laguerre
  –Gaussian beam 53
  polynomial 83
Lamb dip 26
Lambert's cosine law 47
Lambertian radiator 47
Laser
  argon 166
  CO 166
  crystal 237
  energy meter 50
  fiber coupling 129
  He-Ne 166
  NH_3 166
  oscillator 3
  theory, semiclassical 4
Law
  Planck's 21
  propagation, hyperbolic 60
LBO 143, 157–159, 161, 163, 169, 177, 178, 181, 182, 184
Length
  aperture 151
  coherence 238
  diffraction 151
  dispersion-spreading 151
  effective 151

interaction
  nonlinear 152
  quasistatic 151
nonlinear interaction 152
quasistatic interaction 151
Rayleigh 61
Lens
  aberration 117
  gradient-index 113
  shape factor of 117
  thermal 113
  thick 110
    in air 113
  thin 113
LFM 143, 156, 165
$Li_2B_4O_7$ 160, 161
LiB_3O_5 143
$LiCOOH\ H_2O$ 143
Lifetime
  upper-level, T_1 18
Light
  -induced refractive-index change 242
  partially polarized 77
  pressure 50
LiHCOO 225
$LiIO_3$ 143, 156, 158, 159, 161, 163, 165, 166, 170, 172, 175, 176, 180, 186
$LiNbO_3$ 143, 156, 158, 159, 161, 166, 170, 172, 175, 176, 180, 186, 237, 242
$LiNbO_3$:MgO 143, 159, 176, 180
Line
  anti-Stokes 218
  broadening 26, 28
  Raman
    vibration-rotation 218
    vibrational 218, 223
  scattering
    Rayleigh, stimulated 228
  shape 26
    Gaussian 27
    Lorentzian 27
    normalization of 27
    normalized 27
    spectral 17
  Stokes 218
  width 26
Linear 7
  dielectrics 6
  optics 73
  susceptibility 205
Linewidth 226, 227
Liquid 205
  absorbing 237
  crystal 237
  data of 108
  frequency conversion in 205

Raman
  parameters 226
  scattering 224
scattering
  Raman 224
Listing's construction 110
Lithium Fomate 143
Lithium Iodate 143
Lithium Niobate 143
Lithium Triborate 143
Long-term beam width 69
Lorentzian line shape 27

m-Dinitrobenzene 231
m-Nitrotoluene 229, 231
m-Xylene 225
Magnetic
  field 5
  induction 5
  polarization 6
  susceptibility 6
Maiman 3
MAP 142, 157
Master oscillator power amplifier 240
Matching
  mode 128
  phase 144
    condition 208
    noncollinear 146
Material
  infrared, data of 108
  negative-refractive-index, data of 108
  optical 95
  photorefractive 242
  polymeric, data of 108
  solid-state-laser, data of 108
Matrix
  ABCD 111
  density 206
  fluctuation
    beam 67
  Gaussian 111
  Jones 75
  ray-transfer 111
    general 117
  system 55
  variance 56
Maxwell
  –Bloch equations 15, 16
  equations 5
Medium
  dielectric 96
Meissner 3
Mercury Thiogallate 143
meta-Nitroaniline 143
Metal
  data of 108

optics of 98
vapor
  inert gas mixture 209, 210
  mixture of different 209, 210
Methane 226
Methane, Q 227
Methanol 224, 226, 229, 230, 238
Methanol-d_4 224
Methyl N-(2,4-Dinitrophenyl)-L-Alaninate 142
Methyliodide 229
Mg 210
Mg^+ 210
Mg:O-doped Lithium Niobate 143
MHBA 142, 157
Mid IR region
  optical parametric oscillation in 186
Miller
  coefficient 147
  delta formulation 147
Mirror
  concave
    spherical 114
  phase-conjugate 235
  self-adapted 238
  spherical 113
    concave 114
Misalignment 116
Mixed moment 56
Mixing
  frequency 208
  two-wave 242
Mixture
  metal-vapor inert gas 209, 210
  of different metal vapors 209, 210
  of gaseous media 209, 211
mNA 143, 157
Mode
  competition 26
  hopping 26
  matching 128
Model
  Gauss–Schell 53
Moment
  angular 56
  centered 55
  dipole 11, 15
  mixed 56
  second-order 55
    determination of 66
    radiation field, propagation of 111
  spatial 56
Momentum
  conservation of 208
Monoclinic 105
MOPA (Master Oscillator Power Amplifier) 240
Mueller calculus 77

Multimode oscillation 26
Mutual power spectrum 53

$n\pi$-pulse 35
N_2 211, 238
N_2O 226
N-(4-Nitrophenyl)-(L)-Propinol 142
N-[2-(Dimethylamino)-5-Nitrophenyl]-Acetamide 142
n-Hexane 229
n-Nitrotoluene 229
Na 210
$NaClO_3$ 225
Nanosecond optical parametric oscillation 176
Naphthalene 225, 231
Nd:glass high-power laser radiation
  generation of harmonics 162
  harmonics 162
Nd:YAG laser radiation
  fifth harmonic generation of 161
  fourth harmonic generation of 160
  harmonics of 161
  intracavity second harmonic generation of 158
  second harmonic generation of 156, 157, 159
  second harmonic generation, intracavity, of 158
  sixth harmonic generation of 161
  third harmonic generation of 159
Ne 211
Near field 56
Near IR
  radiation
    up-conversion of 170
  region
    cw optical parametric oscillation in 176
    femtosecond optical parametric oscillation in 184
    nanosecond optical parametric oscillation in 176
    picosecond optical parametric oscillation in 180
Negative
  refractive index material, data of 108
  uniaxial crystal 105, 146
$(NH_2)_2CO$ 143
NH_3 laser 166
$NH_4H_2AsO_4$ 143
$NH_4H_2PO_4$ 143
Nitroacetophenone 231
Nitrobenzaldehyde 231
Nitrobenzene 225, 226, 229, 231
Nitrogen 225, 226
NO 226
$NO_2C_6H_4NH_2$ 143
Nodal point 110
Noise input 220
Non-symmetrical optical system 112
Noncollinear phase matching 146

Nonlinear
  coefficient
    effective   150
  interaction length   152
  optics
    fundamental equations of   207
  regime
    second harmonic generation in   154
  susceptibility   205
    third-order   206, 221
Nonlinearity
  effective   147
Normalization
  of line shapes   27
Normalized
  line shape   27
  shape function   27
Notation
  complex   6
NPP   142, 179, 185
Number
  Abbe's   97
  Fresnel   88

o-Nitroaniline   231
o-Nitrophenol   231
o-Nitrotoluene   231
o-Xylene   225
Octanol   229
Optical
  crystal   105
  glass   97
    data of   108
  indicatrix   145
  material   95
  parametric oscillation   144, 153, 176
    continuous wave   176
    femtosecond   184
    in the mid IR region   186
    nanosecond   176
    picosecond   180
  radiometry   45
  self-focusing   220
  system
    non-symmetrical   112
    symplectic   116
Optics
  crystal   104
  diffractive   94
  Fourier   94
  geometrical   108
  linear   73
  nonlinear, fundamental equations of   207
  of metals   98
  of semiconductors   98
Ordinary beam   145
Orthorhombic   105

Oscillating dipole   8
Oscillation
  multimode   26
  optical parametric   144, 153, 176
    continuous wave   176
    femtosecond   184
    in the mid IR region   186
    nanosecond   176
    picosecond   180
Oscillator
  self-sustained   3
Oxide
  ferroelectric   242
Oxygen   226

p-Dichlorobenzene   229
p-Nitroanisol   231
p-Nitrotoluene   231
p-Xylene   225, 229
Parabolic system, general   112
Paraboloid
  phase   59
Paraxial range   108
Partially
  coherent beam   54
  polarized light   77
PCM (Phase-Conjugate Mirror)   235
Penetration depth   102
Permeability   6
Permittivity   6
Perturbation approach   206
Phase
  -conjugate mirror   235
  conjugation   235
    self-pumped   237
  conjugator   235
    fiber   240
    photorefractive   242
    self-pumped   235, 243
    vector   236
  curvature, generalized   59
  matching   144
    condition   208
    noncollinear   146
  paraboloid   59
Phonons
  acoustic   217
Photoacoustic calorimetry   50
Photoconductor   49
Photodiode   49
Photometric quantities   45
Photonic crystals, data of   108
Photorefraction   242
Photorefractive
  crystal   237
  material   242
  phase conjugator   242

Picosecond
    continuum generation   186
        in crystals   186
    optical parametric oscillation   180
Pipe
    heat   209
Piperidine   225
Planar plate   113
Planck's law   21
Planckian radiation   47
Plane   113
    cardinal   110
    wave   7, 79
        approximation   152
        representation   87
Plate
    Brewster-angle-tilted   76
    half-wave   76
    planar   113
    quarter-wave   76
Point
    cardinal   110
    group   150
    nodal   110
Polarization   12, 17, 75
    degree of   77
    electric   6
    magnetic   6
Polarized
    partially
        light   77
Polydiacetylene   225
Polymeric materials, data of   108
Polynomial
    Laguerre   83
POM   142, 157
Position
    waist   60
Positional stability
    beam   67
Positive uniaxial crystal   105, 146
Potassium Dideuterium Arsenate   143
Potassium Dideuterium Phosphate   143
Potassium Dihydrogen Phosphate   143
Potassium Niobate   143
Potassium Pentaborate Tetradeuterate   143
Potassium Pentaborate Tetrahydrate   143
Potassium Titanyl Arsenate   143
Potassium Titanyl Phosphate   143
Power
    density distribution   55
    radiant   46
    spectrum   53
        mutual   53
Poynting vector   7

Pressure
    broadening   28
    light   50
Primary standards   47
Principal
    axis   58
        of the beam   57
    value   145
Principle
    Huygens'   8, 85
Process
    pumping   16
Prokhorov   3
Propagation   60
    beam   120
        Gaussian   111
    free-space   61
    in doped media   10
    law
        hyperbolic   60
    of the second-order moments of the radiation field   111
    short-pulse   97
    three-dimensional   127
    two-dimensional   126
Proustite   142
Pseudo-symmetric beam   64
Pulse
    2π-   35
    amplification, adiabatic   25
    $n\pi$-   35
    π-   36
    resonant coherent   34
    short, propagation of   97
Pumping
    process   16
    schemes   22
Pyrargyrite   142
Pyridine   225, 229
Pyroelectric detector   49

q-parameter, complex   120
Quantities
    photometric   45
    radiometric   45
Quantum
    conversion efficiency   153
    detector   49
Quarter-wave plate   76
Quartz   175, 227, 229, 238
Quasistatic interaction length   151

Rabi frequency   33
Radiance   46
    geometric-optical   54
    spectral   45

Index

Radiant
  energy   45, 46
  excitance   46
  intensity   46
  power   46
Radiation
  field
    propagation of the second-order moments   111
  Planckian   47
  synchrotron   47
Radiator
  black-body   47
  lambertian   47
Radiometer
  cryogenic   47, 50
Radiometric
  quantities   45
  standards   47
Radiometry   45
  optical   45
Raman
  line
    vibration-rotation   218
    vibrational   218, 223
  parameters
    gases   227
    liquids   226
    solids   227
  scattering
    cross section   223
    gases   225
    liquids   224
    solids   225
    stimulated   217, 223, 226
Range
  paraxial   108
  Seidel's   109
Rate equations   20
Ray
  surface   105
  tracing   55
  transfer matrix   111
    general   117
Rayleigh
  distance   82
  length   61
  line scattering   217
    stimulated   228
  scattering
    line   217
    line, stimulated   228
    stimulated   217, 228, 230
    stimulated, thermal   220, 228
    wing   217
    wing, stimulated   228, 230, 231

–Sommerfeld
  approximation   86
  –Debye diffraction theory   89
  wing scattering   217
  wing scattering, stimulated   228, 230, 231
Rb   210
RbH_2AsO_4 143
RbH_2PO_4 143
$RbTiOAsO_4$ 143
RDA   143, 156, 159, 163
RDP   143, 156, 159, 163, 164
Real Bessel beam   80
Rectangular aperture   89
Reflection
  external   101
  internal   101
  total   101
Refraction   101
  at a sphere   114
Refractive index
  change, light-induced   242
  complex   8, 103
Region
  X-ray   47
Relation
  Kramers–Kronig   97
Relaxation
  energy   15
  entropy   16
Replica
  time-reversed   236
Representation
  angular-spectrum   87
  plane-wave   87
Resonance condition   5
Resonant coherent pulses   34
Responsivity   48
Rotating wave approximation   14
Rotator
  Faraday   76
Rowland concave grating   114
RTA   143, 179, 185
Rubidium Dihydrogen Arsenate   143
Rubidium Dihydrogen Phosphate   143
Rubidium Titanyl Arsenate   143
Ruby laser radiation
  second harmonic generation of   163

Saturable absorber   237
Saturation
  broadening   29
  intensity   18
  of the two-level transition   17
SBS (stimulated Brillouin scattering)   227, 236
  threshold   238, 239
Scalar diffraction theory   85

Scattering
  anti-Stokes 222
  anti-Stokes Raman, coherent 222
  Brillouin
    stimulated 217, 227, 229, 236
    stimulated, thermal 220, 227
  coefficient 226
  coherent
    anti-Stokes Raman 222
  cross section 217
    differential 227
    Raman 223
  Raman
    cross section 223
    gases 225
    liquids 224
    solids 225
    stimulated 217, 223, 226
  Rayleigh
    line 217
    line, stimulated 228
    stimulated 217, 228, 230
    stimulated, thermal 220, 228
    wing 217
    wing, stimulated 228, 230, 231
  spontaneous
    Stokes 219
  stimulated 205, 217
    Brillouin 217, 227, 229
    Brillouin, thermal 220, 227
    Raman 217, 223, 226
    Rayleigh 217, 228, 230
    Rayleigh, line 228
    Rayleigh, thermal 220, 228
    Rayleigh, wing 228, 230, 231
    Stokes 219
    transient 222
  Stokes
    spontaneous 219
    stimulated 219
Schawlow 3
Schrödinger equation 13
  of optics 9
Second harmonic generation 144, 152, 156
  in "nonlinear regime" 154
  of dye laser radiation 164
  of gas laser radiation 166
  of Nd:YAG laser radiation 156, 157, 159
  of Nd:YAG laser radiation, intracavity 158
  of ruby laser radiation 163
  of semiconductor laser radiation 164
  of Ti:sapphire laser radiation 163
Second-order moment 55
  determination of 66
  radiation field
    propagation of 111

Secondary standards 48
Seebeck coefficient 48
Seidel's range 109
Self-adapted mirror 238
Self-focusing
  optical 220
Self-pumped
  phase conjugation 237
  phase conjugator 235, 243
Self-sustained oscillator 3
Sellmeier's formula 97
Semiclassical laser theory 4
Semiconductor
  data of 108
  laser radiation
    second harmonic generation of 164
  optics of 98
SF_6 226, 238
Shape
  factor of a lens 117
  function, normalized 27
  line 26
Shift
  Goos–Hänchen 103
  Stokes 224–227
Short-pulse propagation 97
Siliciumtetrachloride 225
Sillenites 242
Silver Gallium Selenide 142
Silver Thiogallate 142
Simple astigmatic beam 63, 120, 121
Sixth harmonic generation 156
  of Nd:YAG laser radiation 161
Slit factor 92
Slowly varying envelope (SVE) 9
  approximation 9
  for diffraction 9
  equation 74, 80
Small signal
  gain factor 11
  solutions 19
Solid
  Raman
    parameters 227
    scattering 225
  scattering
    Raman 225
  -state laser material, data of 108
Solutions
  small-signal 19
  steady-state 17
  strong-signal 19
Spatial
  frequency 88
  hole-burning 237
  moment 56

Index

Spectral
  line shape 17
  radiance 45
Spectrum
  angular, representation 87
Sphere
  refraction at 114
Spherical
  aberration 117
    third-order 119
  concave mirror 114
  mirror 113
  surface 113
  vector wave 78
  wave 8, 79
Spontaneous
  emission 4, 15, 28
    Einstein coefficient 21
  scattering
    Stokes 219
  Stokes scattering 219
Sr 210
SRLS (stimulated Rayleigh line scattering) 228
SRS (stimulated Raman scattering) 223
SRWS (stimulated Rayleigh wing scattering) 228, 230, 231
Stability
  positional
    beam 67
Standards
  primary 47
  radiometric 47
  secondary 48
STBS (stimulated thermal Brillouin scattering) 227
Steady state
  condition 5
  solutions 17
Steradian 47
Stigmatic beam 62, 120, 121
Stimulated
  amplification 221
  Brillouin scattering 217, 227, 229, 236
    thermal 220, 227
  Raman scattering 217, 223, 226
  Rayleigh scattering 217, 228, 230
    line 228
    thermal 220, 228
    wing 228, 230, 231
  scattering 205, 217
    Brillouin 217, 227, 229
    Brillouin, thermal 220, 227
    Raman 217, 223, 226
    Rayleigh 217, 228, 230
    Rayleigh, line 228
    Rayleigh, thermal 220, 228
    Rayleigh, wing 228, 230, 231
    Stokes 219
  transient 222
  Stokes scattering 219
  thermal
    Brillouin scattering 220, 227
    Rayleigh scattering 220, 228
Stokes
  –anti-Stokes coupling 222
  emission
    higher-order 222
  line 218
  scattering
    spontaneous 219
    stimulated 219
  shift 224–227
  spontaneous scattering 219
  stimulated scattering 219
  vector 77
Stop 127
Strong-signal solutions 19
STRS (stimulated thermal Rayleigh scattering) 228
Styrene 225, 231
Sulfur 225
Sulfurhexafluoride 229
Sum frequency generation 144, 153, 167
  of UV radiation 167–169
Superradiance 37
Surface
  index 105, 146
  ray 105
  spherical 113
Susceptibility 19
  complex 10
  electric 6
  linear 205
  magnetic 6
  nonlinear 205
    third-order 206, 221
  third-order
    nonlinear 221
SVE (Slowly varying envelope) 9
  approximation 9
  for diffraction 9
SVE (slowly varying envelope)
  equation 74, 80
Symmetry conditions
  Kleinman 150
Symplectic optical system 116
Synchrotron radiation 47
System
  four-level 24
  general
    astigmatic 116
    parabolic 112

homogeneous 16
inhomogeneous 16
matrix 55
optical
  non-symmetrical 112
  symplectic optical 116
  three-level 23
  two-level 11

Te 143, 166
Tellurium 143
Tetrabromomethane 229
Tetrachloroethane 225
Tetrachloroethylene 225, 226
Tetrachloromethane 230
Tetragonal 105
Tetrahydrofuran 225
Thallium Arsenic Selenide 143
Theory
  scalar
    of diffraction 85
  vector
    of diffraction 85
Thermal
  detector 48
  grating 237
  lens 113
  scattering
    Brillouin, stimulated 220, 227
    Rayleigh, stimulated 220, 228
Thermodynamic considerations 21
Thermopile 48
Thick lens 110
  in air 113
Thin lens 113
Third harmonic generation 152, 156, 208
  of Nd:YAG laser radiation 159
  of Ti:sapphire laser radiation 163
Third-order
  nonlinear susceptibility 206, 221
  spherical aberration 119
Three-dimensional propagation 127
Three-level system 23
Three-wave interaction 144
Threshold
  condition 4
  SBS- 238, 239
Ti:sapphire laser radiation
  fourth harmonic generation of 164
  second harmonic generation of 163
  third harmonic generation of 163
Time
  constants 31
  decay, T_1 15
  decay, T_2 16
  -dependent diffraction theory 89
  -reversed replica 236

Tintetrabromide 225
Tintetrachloride 225
Tl 210
Tl_3AsSe_3 143, 167
Toluene 225, 226, 229, 231
Total reflection 101
Townes 3
Tracing
  ray 55
Transform
  Fourier 88
Transformation
  ABCD 120
  waist 124
Transient stimulated scattering 222
Triclinic 105
Trigonal 105
Twist 59
Two-dimensional propagation 126
Two-level
  system 11
  transition
    saturation intensity 17
Two-wave mixing 242

Uniaxial
  crystal 145
    negative 146
    positive 146
  negative
    crystal 105
  positive
    crystal 105
Up-conversion
  of CO_2 laser radiation 171
  of near IR radiation 170
Upper-level lifetime T_1 18
Urea 143, 161, 165, 179
UV
  radiation
    sum frequency generation of 167–169
  region
    cw optical parametric oscillation in 176
    femtosecond optical parametric oscillation in 184
    nanosecond optical parametric oscillation in 176
    picosecond optical parametric oscillation in 180

Value
  principal 145
Vanillin (4-Hydroxy-3-Methoxy-Benzaldehyde) 142
Vapor
  barium 225
  cesium 226

Variance
  ellipse 57
  matrix 56
Vector
  Bloch 33
  -ial Bessel beam 80
  Jones 75
  phase conjugator 236
  Stokes 77
  theory of diffraction 85
  wave
    cylindrical 78
    spherical 78
Vibration
  -al Raman line 218, 223
  -rotation Raman line 218
Visible region
  cw optical parametric oscillation in 176
  femtosecond optical parametric oscillation in 184
  nanosecond optical parametric oscillation in 176
  picosecond optical parametric oscillation in 180

Waist
  beam 82
  position 60
  transformation 124
Walk-off angle 145

Water 225, 226, 229, 230
Wave
  cylindrical 79
  equation 73, 78
  plane 7, 79
    approximation 152
    representation 87
  spherical 8, 79
  vector
    cylindrical 78
    spherical 78
Waveguide coupling 127, 128
Width
  beam 57
  long-term 69
Wigner distribution 53
Wing scattering
  Rayleigh, stimulated 228, 230, 231

X-ray region 47
Xe 211, 238

Yb 210

Zinc Germanium Phosphide 143
Zn 210
ZnGeP$_2$ 143, 166, 167, 171, 175, 186
ZnSe 175
ZnTe 175

Landolt-Börnstein License Agreement for

LB CD-ROM

Springer-Verlag Berlin Heidelberg New York Tokyo and the end-user of LB CD-ROM approve and enter into the following agreement for the use and registration of LB CD-ROM

§ I. User's Rights and Copyrights

1. The data on the CD-ROM, the retrieval system (the software) and the handbook are protected by copyright; all rights to them or their use accrue to the user exclusively through Springer-Verlag. Independently of this, both sides agree to apply the rules of copyright to LB CD-ROM.
2. The user has the non-exclusive authorization to use LB CD-ROM as described in the handbook. Any alternation, reworking, decompiling or reverse engineering of the software is not permitted.
3. The CD-ROM and the software may be used only on one computer and at one workstation at any one time. Any person using LB CD-ROM must be a member of the user's institution, for example an employee of the user's university or a library user.
4. The user may save or print items from the LB CD-ROM databank only for temporary purposes and only as described in the handbook. The CD-ROM, the software and the handbook may not be reproduced in any other respects.

§ II. Transfer

1. Any transfer (e.g., sale) of LB CD-ROM, or any o.. and the use thereof requires the written approval of Springer-Verlag.
2. Springer-Verlag will give this permission if the previous user makes a written application and the subsequent User makes a declaration that he or she will be bound by the terms of this agreement. Receipt of this permission ends any rights of the previous user to LB CD-ROM and the transfer may take place.

§ III. Registration

By returning the accompanying card, the user may register with Springer-Verlag. He or she will then receive regular information about the latest up-dates of LB CD-ROM and may make use of the Help Service, as described in § IV.

§ IV. Help Service

1. Springer-Verlag has created the possibility for registered users, as described in § III, to pose questions about the installation and use of LB CD-ROM. There is, however, no legal claim to this service.
2. Questions may be posed through either regular mail, electronic mail, FAX or telephone. Springer-Verlag will respond appropriately.

§ V. Warranty

1. Springer-Verlag is not the author of the data or of the software but is making them available for use. A warranty of the completeness of the data cannot be guaranteed. The user is aware that databanks and software of this type cannot be made completely error-free; the user will check the accuracy of any results of his or her searches in a suitable manner.

2. In cases of faulty materials, production defects or of damage in delivery, Springer-Verlag will replace the product. The user has rights to further claims only when he or she has purchased LB CD-ROM directly from Springer-Verlag. The warranty assumes that the user gives an exact description of the fault immediately and in writing.

§ VI. Liability of Springer-Verlag

Springer-Verlag is only liable for wilful intent, gross negligence and wilful misrepresentation. Representations are only valid when made in writing. There is no warranty for any information obtained according to § IV. Liability according to product liability law is not affected. Springer-Verlag reserves the right to assert that the user was acting at his or her own risk and the user must assume full or partial responsibility.

§ VII. Responsibilities of the user

1. The user is obliged to abide by the terms and conditions (§§ I and II) governing use and transfer of the product. Any violation is punishable and may be pursued under civil and criminal law. The user may then be made to pay damages to Springer-Verlag or to one of the grantors of its licenses.
2. In cases of severe violation, Springer-Verlag reserves the right to immediately withdraw the user's authorization to use LB CD-ROM and to demand the return of the product.

§ VIII. Data Protection

The user agrees that his or her name and address may be stored electronically.

§ IX. Conclusion of the Agreement

The user waives any claim to a written acknowledgement from Springer-Verlag of its agreement to the terms of this agreement.

§ X. Conclusion

1. This agreement is valid for the product that has already been delivered and for any versions of it delivered in the future.
2. Should any provision of this agreement be or become ineffective or should the agreement be incomplete, then the remainder of the agreement shall remain in effect. The ineffective provisions will be assumed to have been replaced by an effective provision that in meaning and purpose comes closest to the ineffective provision in financial terms. The same holds for any omission in the agreement.
3. The place of jurisdiction is Heidelberg, if the user is a registered trader or equivalent, is a legal entity according to public law, is a public separate property, or has no legal residence or place of business in Germany.
4. This agreement is governed exclusively by the laws of the Federal Republic of Germany, with the exception of the UNCITRAL laws of trade and commerce.

Please ask your librarian for the LB CD-ROM

For installation and further use, please see readme.txt on the CD.